AF750837

# Springer Series in Solid-State Sciences

Volume 212

**Series Editors**

Thierry Giamarchi, Department of Quantum Matter Physics, University of Geneva, Geneva, Switzerland

Armen Gulian, Institute for Quantum Studies, Chapman University, Ashton, USA

Bernhard Keimer, Max Planck Institute for Solid State Research, Stuttgart, Germany

Yoshio Kuramoto, Sendai, Japan

Hans-Joachim Queisser, MPI für Festkörperforschung, Stuttgart, Germany

Sven Rogge, Physics, UNSW, Sydney, Australia

Klaus von Klitzing, Max Planck Institute for Solid State Research, Stuttgart, Germany

Wu-Ming Liu, Institute of Physics, Beijing, China

The Springer Series in Solid-State Sciences features fundamental scientific books prepared by leading and up-and-coming researchers in the field. They strive to communicate, in a systematic and comprehensive way, the basic principles as well as new developments in theoretical and experimental solid-state science.

We welcome submissions for monographs or edited volumes from scholars across this broad domain. Topics of current interest include, but are not limited to:

- Semiconductors and superconductors
- Quantum phenomena
- Spin physics
- Topological insulators
- Multiferroics
- Nano-optics and nanophotonics
- Correlated electron systems and strongly correlated materials
- Vibrational and electronic properties of solids
- Spectroscopy and magnetic resonance

Seiji Miyashita

# Physics of Entropy

## Impact of Frustration to Thermodynamics of Matter

Springer

Seiji Miyashita
Department of Physics
University of Tokyo
Tokyo, Japan

ISSN 0171-1873 ISSN 2197-4179 (electronic)
Springer Series in Solid-State Sciences
ISBN 978-981-95-1757-2 ISBN 978-981-95-1758-9 (eBook)
https://doi.org/10.1007/978-981-95-1758-9

© The Editor(s) (if applicable) and The Author(s), under exclusive license to Springer Nature Singapore Pte Ltd. 2025

This work is subject to copyright. All rights are solely and exclusively licensed by the Publisher, whether the whole or part of the material is concerned, specifically the rights of translation, reprinting, reuse of illustrations, recitation, broadcasting, reproduction on microfilms or in any other physical way, and transmission or information storage and retrieval, electronic adaptation, computer software, or by similar or dissimilar methodology now known or hereafter developed.
The use of general descriptive names, registered names, trademarks, service marks, etc. in this publication does not imply, even in the absence of a specific statement, that such names are exempt from the relevant protective laws and regulations and therefore free for general use.
The publisher, the authors and the editors are safe to assume that the advice and information in this book are believed to be true and accurate at the date of publication. Neither the publisher nor the authors or the editors give a warranty, expressed or implied, with respect to the material contained herein or for any errors or omissions that may have been made. The publisher remains neutral with regard to jurisdictional claims in published maps and institutional affiliations.

This Springer imprint is published by the registered company Springer Nature Singapore Pte Ltd.
The registered company address is: 152 Beach Road, #21-01/04 Gateway East, Singapore 189721, Singapore

If disposing of this product, please recycle the paper.

# Preface

This book provides an overview of the role of entropy in physics. Entropy was introduced to treat phenomena related to heat and temperature. The concept of entropy originated from the work of Nicolas Sadi Carnot and was later named "entropy" by Rudolf Clausius. We begin by defining entropy in thermodynamics, where it is associated with heat transfer. Subsequently, it is formulated as Boltzmann's principle in statistical mechanics. The derivations of this quantity are rather sophisticated. Consequently, although entropy is a fundamental quantity in physics, it is not intuitively easy to understand. Indeed, unlike common physical quantities such as weight or volume, entropy cannot be directly observed.

Entropy plays a crucial role in the ordering processes of many-body systems. Phase transitions are generally understood as a competition between the ordering effect of energy and the disordering effect of entropy. However, when a system exhibits competing interactions, i.e., frustration among interactions, the ordering process becomes more complex. In such cases, entropy plays a significant role in determining which state is the most thermodynamically stable. Such phenomena are often referred to as "order by disorder."

Entropy provides a measure of randomness, i.e., the number of macroscopically indistinguishable states. The concept of entropy has been extended to non-equilibrium states expressed by probability distributions. Boltzmann introduced the $H$ function to study the relaxation process to equilibrium, and entropy is expressed in the form of the Boltzmann-Gibbs formula. The same form was introduced by Shannon in information theory. Furthermore, entropy has played important roles in various fields of physics.

After an introduction of the book (Chap. 1), in Part I, we review how "entropy" is introduced in both thermodynamics and statistical mechanics (Chap. 2). In thermodynamics, entropy is defined in relation to heat flows during quasistatic thermodynamic cycles, such as the Carnot cycle. In statistical mechanics, entropy is introduced via Boltzmann's principle, which relies on the assumption of equal a priori probabilities stating that, in equilibrium, all microstates with the same total energy are equally probable. This principle relates entropy to the number of states in an isolated system (i.e., in the microcanonical ensemble). These two definitions of entropy are equivalent.

Chapter 3 discusses the role of entropy in various systems, including simple ones such as the two-level system and the harmonic oscillator, as well as more complex cases where entropy plays a significant role, such as adiabatic demagnetization, spin-crossover phenomena, and mixing entropy.

In Part II, properties of frustrated systems are reviewed in detail where entropy plays important roles. The effect of competition among interactions on the ordering process has different aspects between systems with discrete variables and those with continuous variables.

First, frustrated Ising systems are reviewed (Chap. 4). In the case of discrete variables, frustration leads to a degeneracy of the ground state among macroscopically many competing configurations, resulting in the so-called residual entropy as known in the arrangement of hydrogens in ice.

The properties of the ground state are discussed for various models, such as antiferromagnets on the triangular lattice, the so-called Villain model, and antiferromagnets on the kagome lattice. The nature of degeneracy depends on the value of $S$, and it has been found that, even with macroscopic degeneracy, a long-range order can emerge in the ground state for large $S$ values at $T = 0$, while no long-range order exists for $S = 1/2$. This is a typical example of order-by-disorder.

Frustration also gives rise to peculiar properties at finite temperatures. The spin correlation function in frustrated systems may show non-monotonic temperature dependence; a temperature-dependent sign change, transitioning from ferromagnetic to antiferromagnetic behavior. These properties can result in phenomena such as reentrant phase transitions and temperature-chaos effects (rejuvenation) in spin glasses. Furthermore, peculiar ordering patterns may arise due to the frustration. The ordering process of the antiferromagnetic Ising model with next-nearest-neighbor interaction on the triangular lattice has been studied as the Mekata model, where frustration induces a new type of ordered state, referred to as the partially disordered phase, whose mechanism is studied in detail.

Some thermodynamic properties of two-dimensional Ising models can be solved exactly, providing deeper insights into subtle ordering behaviors in frustrated systems. The methods used for these exact solutions are briefly reviewed.

In Chap. 5, we study models with continuous spins, where the ground states exhibit non-collinear structures due to the compromise imposed by competing interactions, such as the 120° structure in triangular antiferromagnets. The ordering processes in these systems depend strongly on the anisotropy of the interaction, e.g., planar anisotropy (XY), isotropic (Heisenberg), or easy-axis anisotropy (Ising-like Heisenberg).

Entropy plays a crucial role in selecting the thermodynamically stable state among energetically degenerate (or nearly degenerate) ordering patterns. This entropy effect leads to a non-trivial sequence of phase transitions in the magnetization process. The entropy in a harmonic analysis of small deviation from given configurations at low temperatures is explained in detail.

In Chap. 6, the effects of quantum fluctuations in the ground state are reviewed. When the order parameter does not commute with the system Hamiltonian, quantum fluctuations lead to an energy gain, although they reduce the magnitude of classical order parameter. In frustrated systems, the interplay between frustration and quantum fluctuations results in a significant reduction of the order parameter of the corresponding classical system. It is well known that quantum fluctuations play an important role in determining the most stable state, similarly to the effects of entropy at finite temperatures.

In Part III (Chap. 7), we study the dynamics of entropy as expressed by the probabilities of states. This quantity is known as the Boltzmann-Gibbs entropy and is also referred to as Shannon entropy in information theory. In systems coupled to a thermal bath (i.e., the canonical ensemble), entropy does not necessarily increase during relaxation toward equilibrium, whereas free energy decreases. In such cases, instead of entropy, the relative entropy (also known as the Kullback-Leibler divergence) is used. We examine how these quantities converge to their thermal equilibrium values under the dynamics described by the master equation. Various extensions of the concept of entropy are also briefly introduced.

I sincerely thank Prof. Masuo Suzuki for his kind guidance in statistical physics, and also all the collaborators of the works related to the present book. I appreciate Eric Vincent, Tadashi Takayanagi, Hikaru Kawamura, Eiichi Nakamura, Hosho Katsura, Hiroaki Onishi, Shu Tanaka, and Yutaka Fujii for their valuable comments and kind suggestions, and also Mr. Akiyuki Tokuno for his invitation and kind support to write the book. Finally, I thank my wife, Ami, for her continuous support and also my first grandson, Saiki, for his cheerful smiles.

Tokyo, Japan — Seiji Miyashita
July 2025

# Contents

# Part II Frustration and Entropy

# Chapter 1
# Introduction

The thermal equilibrium state favors a low-energy configuration at low temperatures, while it favors a random state at high temperatures. The system may exhibit a phase transition between the low-temperature ordered phase and the high-temperature disordered phase, such as the paramagnetic-ferromagnetic transition in magnets and the gas-liquid transition in water. For these transitions, thermodynamics provides a precise quantitative description, even though the systems themselves undergo microscopic fluctuations. In equilibrium, it is not the energy $E$ itself, but the free energy $F(= E - TS)$, that determines the most stable state, where $T$ is the absolute temperature. Entropy $S$ is an essential quantity in thermodynamics, but it is a rather abstract concept, as is the temperature.

In the present book, in Chap. 2 we give brief reviews on introductions of entropy in thermodynamics based on the heat flow and also in statistical mechanics based on the principle of equal a priori probabilities, which leads to Boltzmann's principle [1]. Several fundamental properties related to entropy are overviewed.

Chapter 3 reviews several representative entropy-related phenomena. These include two-level systems, which show a Schottky peak in specific heat, and three-level systems, which may display double peaks. Harmonic oscillator systems are also discussed. Moreover, the chapter examines several phenomena in which entropy plays a key role, such as adiabatic demagnetization, spin-crossover transitions, and mixing entropy.

When a system has incompatible interactions, it becomes energetically frustrated. In such cases, entropy plays a crucial role in determining the most thermodynamically stable state, e.g., phenomena referred to as order-by-disorder. Phenomena in frustrated systems will be reviewed in detail. The characteristics of the ordering processes differ significantly between systems with discrete variables and those with continuous variables.

In Chap. 4, we review frustrated systems with discrete variables, such as Ising spins, where frustration leads to macroscopic degeneracy of ground states. The residual entropies of several frustrated Ising models are studied, for which exact methods for the two-dimensional Ising model play important roles. It is also found that, despite

© The Author(s), under exclusive license to Springer Nature Singapore Pte Ltd. 2025

S. Miyashita, *Physics of Entropy*, Springer Series in Solid-State Sciences 212,
https://doi.org/10.1007/978-981-95-1758-9_1

macroscopic degeneracy, the Ising model with large spin quantum number $S$ exhibits long-range order induced by entropy (i.e., the order-by-disorder phenomenon). At finite temperatures, frustration gives rise to intriguing ordered states. As a representative example, the antiferromagnetic Ising model on the triangular lattice with next-nearest-neighbor interactions (the Mekata model) is discussed in detail. Frustration can also produce non-monotonic temperature dependence in correlation functions, leading to reentrant ordering behavior. Finally, the effects of randomness on ordering processes in such systems are briefly reviewed.

In Chap. 5, we review frustrated systems with continuous variables, such as XY or Heisenberg spins, where the order parameters consist of several spins, e.g., the 120° structure resulting from compromises in frustrated interactions. The structures of order parameters of antiferromagnetic models on the triangular lattice are studied in detail. There, several non-trivial degeneracies emerge. Among them, entropy plays an important role in selecting the most thermodynamically stable state. Entropy due to small fluctuations in given structures is studied in detail. The composite order parameters lead to interesting ordering processes that depend on the symmetry of the order parameters. Even in continuous spin systems, a system may have a macroscopically degenerate ground state in corner-sharing lattices, such as the kagome lattice. The order-by-disorder mechanism in the system is reviewed. Although all the Ising, XY, and Heisenberg models on the kagome lattice do not have a phase transition at finite temperature, the Ising-like Heisenberg model exhibits an exotic ferromagnetic phase transition due to the structure of the order parameter, which is also reviewed.

Moreover, in Chap. 6, quantum effects on frustrated systems are reviewed, including quantum fluctuation-induced phase transitions. The zero-temperature fluctuation in a quantum system causes an energy gain that resolves degeneracy in the ground state of classical systems. It is found that the energy raising due to frustration can be largely resumed by quantum effect. In classical systems, zero-temperature entropy resolves the degeneracy. The similarities and differences between them are compared. The magnetic field dependence of ordered states of the anisotropic Heisenberg model on the triangular lattice is reviewed and shows similarity to that of temperature dependence in corresponding classical models. The combination of quantum fluctuations and frustration may lead to a structured energy level. The specific heat of the Heisenberg antiferromagnet ($S = 1/2$) on the kagome lattice shows double peaks depending on the system sizes, which is also reviewed.

In Chap. 7, we examine the dynamics of entropy. The second law of thermodynamics, which describes the irreversible process toward thermodynamic equilibrium, is expressed through the increase in entropy, as demonstrated by William Thomson (Lord Kelvin) and Rudolf Clausius. Since then, the concept of entropy has been extended to quantify the randomness of systems, particularly in terms of the distribution functions of states, where entropy is expressed as a function of probability distribution.

Boltzmann's famous $H$-theorem demonstrated the monotonic decrease of the $H$-function (proportional to entropy with an opposite sign) during the relaxation process. A generalized form of entropy, as a function of the distribution function, was later introduced by Gibbs and is now referred to as Boltzmann-Gibbs entropy.

This formulation shares the same mathematical structure as Shannon entropy in information theory. The relationship between information and entropy is discussed with reference to Maxwell's demon.

When a system is coupled to a thermal bath, entropy does not necessarily increase during the relaxation process toward equilibrium, whereas the free energy consistently decreases. The Kullback-Leibler divergence (relative entropy) provides a measure of how a system approaches equilibrium. The dynamics of the probability distribution is described by the master equation, which governs transitions based on the transition probabilities between states. During relaxation, the decrease in free energy is reflected in the reduction of the Kullback-Leibler divergence (relative entropy) as the system approaches its equilibrium value.

Various extensions of entropy, such as Tsallis entropy, Rényi entropy, and the entropy of black holes, are also briefly reviewed.

## References

1. For detailed explanation, look at text books: For example, A. Sommerfeld, *Thermodynamics and Statistical Mechanics* (Academic Press 1954); L.D. Landau, E.M. Lifshitz, *Statistical Physics* (1969) internet Arcive; H.B. Callen, *Thermodynamics and an Introduction to Thermostatistics*, 2nd edn., internet Arcive, L.E. Reichl, *A Modern Course in Statistical Physics* (Wiley, 1997); S. Miyashita, *NETSURIKIGAKU*, in Japanse (Tokyo TOSHO, 2019), and *TOUKEIRIKIGAKU*, in Japanse (Tokyo TOSHO, 2020)

# Part I
# What is Entropy?

Entropy is a fundamental concept in physics. Originally, it was introduced in the thermodynamics. Thermodynamics is one of the major branches of physics, alongside mechanics, electromagnetism, and quantum mechanics. However, it deals with quantities like "heat" and "temperature", which are not as simple to measure as length or weight. To handle such quantities in a physically well-defined manner, a more abstract approach is required, which can feel unfamiliar compared to other fields of physics. In this Part, we review how entropy has been introduced in thermodynamics and statistical mechanics, and also present examples of phenomena caused by entropy.

# Chapter 2
# Definition of Entropy

**Abstract** In this chapter, we review how entropy $S$ is introduced in thermodynamics and statistical mechanics. The definition of entropy is accompanied by the definition of temperature $T$. In thermodynamics, entropy and temperature are introduced through arguments based on heat flow in quasi-adiabatic processes (Carnot cycle): $dS = \delta Q/T$, where $\delta Q$ is the amount of heat exchanged during the process. In statistical mechanics, entropy is derived from the principle of equal a priori probabilities, which leads to Boltzmann's principle: $S = k_{\rm B} \ln W$, where $W$ is the number of microstates corresponding to a given energy.

## 2.1 Introduction

In mechanics, electromagnetism, and quantum mechanics, the fundamental principles are clearly defined at the microscopic level: Newton's laws in mechanics, Maxwell's equations in electromagnetism, and the commutation relations of canonical variables in quantum mechanics. In contrast, thermodynamics deals with macroscopic phenomena, and its principles are described in a less direct way. For instance, the first law of thermodynamics states that heat is an "incomprehensible form of energy transfer", while the second law states that heat spontaneously and irreversibly flows from a high-temperature source to a low-temperature source. Constructing a physically rigorous framework from these descriptive principles is not straightforward.

It is a kind of miracle to find the concept of "entropy" for thermodynamical states in such a situation. As a key feature of heat, Nicolas Sadi Carnot provided essential insights through his study of the Carnot cycle, and Rudolf Clausius formalized the idea of entropy, providing its definition. With the concept of entropy thus introduced, thermodynamics became a robust and universal field of physics. Because of its abstract nature, the concept of entropy has proven to have wide-ranging applicability. Definition of entropy and temperature in thermodynamics will be briefly reviewed in the next section (Sect. 2.2).

In statistical physics, entropy is defined based on the number of states with the same energy based on the principle of equal a priori probabilities, known as Boltz-

© The Author(s), under exclusive license to Springer Nature Singapore Pte Ltd. 2025 
S. Miyashita, *Physics of Entropy*, Springer Series in Solid-State Sciences 212,
https://doi.org/10.1007/978-981-95-1758-9_2

mann's principle. This definition is equivalent to that in thermodynamics when the same reference of temperature is used. The relationship between these two definitions is discussed in Sect. 2.3.

## 2.2 Definition of Entropy and Temperature in Thermodynamics

### 2.2.1 *Principles of Thermodynamics*

The principles of thermodynamics are as follows. The zeroth law of thermodynamics establishes the existence of a thermodynamic state that is macroscopically stable, meaning it does not change spontaneously unless influenced by external controls.

The first law of thermodynamics is expressed as

$$\Delta E = Q + W, \tag{2.1}$$

where $\Delta E$ represents the total energy transferred to a system during a process, $W$ is the work done on the system, and $Q$ is the heat transferred to the system. This law demonstrates the principle of energy conservation. However, the key point of this law is the distinction between distinguishable energy transfer, referred to as "work", and indistinguishable energy transfer, referred to as "heat".

On a microscopic scale, all energy transfers are entirely due to work, and no $Q$ exists. In macroscopic systems, however, some parts of energy transfer cannot be explicitly associated with work. This unaccounted-for energy transfer is defined as $Q$. Thus, the concept of heat is inherently tied to the level of recognition of the system. This is an essential assumption of thermodynamics: we cannot recognize some parts of processes, and we cannot control those processes. Indeed, this concept of the existence of unrecognized energy transfer, as introduced in the first law of thermodynamics, was formulated after the second law of thermodynamics for which the specific nature of heat was not really a matter of concern.

The second law of thermodynamics states that heat, regardless of its precise property, naturally flows from the "high-temperature" side to the "low-temperature" side without performing work, and the reverse process does not occur spontaneously. At this stage, "high-temperature" and "low-temperature" are not yet quantitatively defined. However, based on the direction of energy transfer, it is possible to identify which side is at a higher temperature.

Unlike Newton's laws of mechanics, the second law cannot be expressed by a simple equation. Carnot introduced a groundbreaking idea by considering thermal efficiency to provide a quantitative description of the fundamental nature of heat transfer. He demonstrated that the maximum thermal efficiency achievable between two thermal baths is uniquely determined, from which temperature can be quantitatively defined.

Thermal efficiency is defined as the ratio of the work $W$ gained in the process to the input heat from the high-temperature bath $Q_{\mathrm{in}}$ as

$$\eta = \frac{W}{Q_{\mathrm{in}}}. \tag{2.2}$$

According to the first law of thermodynamics, $W = Q_{\mathrm{in}} - Q_{\mathrm{out}}$, where $Q_{\mathrm{out}}$ represents the heat discarded during the process. Using this relationship, entropy and temperature are defined as complementary quantities, which will be explained in detail in the next section.

### 2.2.2 *Cycle: Process with Thermal Transfer Among Thermodynamic States*

As mentioned earlier, heat spontaneously transfers from the "high-temperature" side to the "low-temperature" side. When systems with different temperatures are simply connected, heat flows without performing any work. This transfer can be regarded as a leakage of heat. In the old days, heat was mainly used for warming or heating other materials. However, humanity invented the "heat engine", a device capable of extracting work $W$ from heat flow. The invention of the heat engine marked a revolutionary step in the practical use of heat.

The idea of the heat engine is depicted in Fig. 2.1, where the cycle conceptually represents the heat engine where input heat $Q_{\mathrm{H}}$ is used to produce work $W$. For example, the cycle could represent a steam engine, where the high-temperature system corresponds to the combustion chamber. If no heat engine is used, the heat simply leaks into the low-temperature system, resulting in $W = 0$.

When using the cycle, what is the maximum possible work $W$ that can be taken from the incoming heat $Q_{\mathrm{H}}$? For this problem, Carnot introduced an essential idea of the ratio $W/Q_{\mathrm{H}}$ which is referred to as the "thermal efficiency" $\eta$ (2.2):

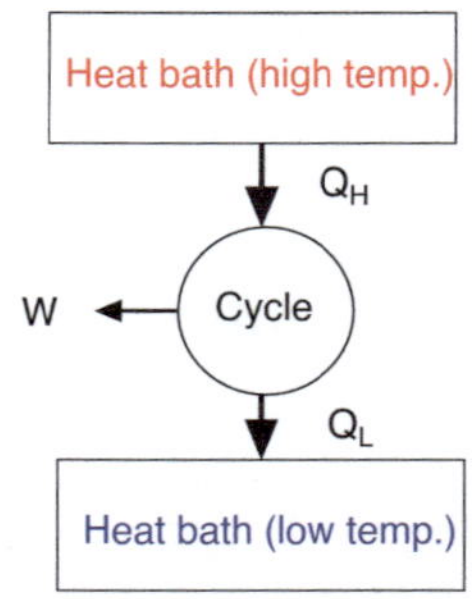


**Fig. 2.1** Heat engine: extraction of work from heat transfer

$$\eta_C = \eta(T_H, T_L), \quad 1 - \eta(T_H, T_L) = \frac{Q_L}{Q_H}, \tag{2.3}$$

where $T_H$ and $T_L$ are temperatures of both baths which is not yet quantitatively defined, although $T_H > T_L$. Carnot found that the maximum efficiency is uniquely determined for given high and low temperature thermal baths.

For the proof of this property, the second law plays an essential role. Assume that we can use the so-called reversible cycle in which we can move heat from the low temperature system to the high temperature system using the work extracted from the process in which all the arrows in Fig. 2.1 are inversed. If we combine this inverse cycle with the cycle considered above, the heat transfers cancel each other out, and the total system does not change at all.

Now, let us assume that there are two different reversible cycles, A and B, with different thermal efficiencies, $\eta^A$ and $\eta^B$:

$$Q_L^A = (1 - \eta^A) Q_H^A, \quad Q_L^B = (1 - \eta^B) Q_H^B. \tag{2.4}$$

Assuming that $\eta^A > \eta^B$, we use cycle A to extract work $W^A (= \eta^A Q_H^A)$ and we pump up the same amound of heat $Q_L^B (= Q_L^A)$ by cycle B. There, $Q_H^B$ is given by

$$Q_H^B = \frac{Q_L^B}{1 - \eta^B} = \frac{Q_L^A}{1 - \eta^B} = Q_H^A \frac{1 - \eta^A}{1 - \eta^B}. \tag{2.5}$$

Thus,

$$Q_H^A - Q_H^B = Q_H^B \left( \frac{1 - \eta^B}{1 - \eta^A} - 1 \right) = Q_H^B \frac{\eta^A - \eta^B}{1 - \eta^A} > 0. \tag{2.6}$$

In this process, the heat $Q_H^A - Q_H^B$ of system A is transferred to the work $W^A - W^B$ without any change in the total system, which contradicts the second law of thermodynamics. Thus, all reversible cycles must have the same thermal efficiency. Therefore, the thermal efficiency of a reversible cycle $\eta(T_H, T_L)$ is a uniquely determined quantity for the given two systems. We call this maximum efficiency "Carnot efficiency" and denote it $\eta_C(T_H, T_L)$. Thus, making use of this fact, we can begin a quantitative argument for relations of systems with different "temperatures", which will be explained in the following.

### 2.2.3 Introduction of Quantitative Temperature

We aim to define "temperature" quantitatively using the thermal efficiency of a reversible cycle $\eta(T_H, T_L)$. In the above argument, $T_H$ and $T_L$ are merely labels of the states and were not quantitatively defined. However, the function $\eta_C(T_H, T_L)$ is a physically well-defined quantity as a function of $T_H$ and $T_L$. For multiple systems with different temperatures, $\eta_C$ must be defined consistently. To find conditions for

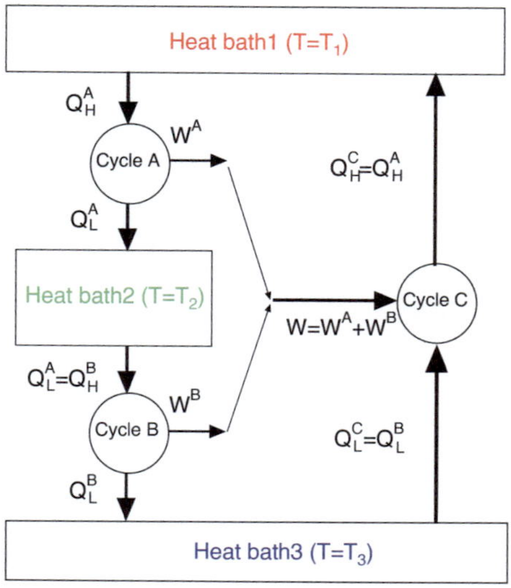


**Fig. 2.2** Three systems connected by reversible cycles

consistency, we consider a situation where three systems with different temperatures ($T_1 > T_2 > T_3$) are connected by reversible cycles (Fig. 2.2).

The heat transfer occurs from system 1 to system 2 via the reversible cycle A. There, $Q_{\mathrm{H}}^{\mathrm{A}}(>0)$ flows in the cycle, and the work $W^{\mathrm{A}}(>0)$ is extracted. Then, $Q_{\mathrm{L}}^{\mathrm{A}} = Q_{\mathrm{H}}^{\mathrm{A}} - W^{\mathrm{A}}(>0)$ flows out to system 2. The thermal efficiency of cycle A is given by

$$1 - \eta^{\mathrm{A}} = 1 - \eta_{\mathrm{C}}(T_1, T_2) = \frac{Q_L^{\mathrm{A}}}{Q_{\mathrm{H}}^{\mathrm{A}}}. \tag{2.7}$$

where $\eta_{\mathrm{C}}(T_1, T_2)$ is the Carnot efficiency between systems 1 and 2.

Next, for the heat transfer between systems 2 and 3, we consider cycle B, where the heat $Q_H^{\mathrm{B}}$ $(= Q_L^{\mathrm{A}})$ flows in and $Q_L^{\mathrm{B}}$ flows out to system 3, extracting the work $W^{\mathrm{B}}(>0)$ . Here, we obtain the following relation:

$$1 - \eta^{\mathrm{B}} = 1 - \eta_{\mathrm{C}}(T_2, T_3) = \frac{Q_L^{\mathrm{B}}}{Q_{\mathrm{H}}^{\mathrm{B}}} = \frac{Q_L^{\mathrm{B}}}{Q_{\mathrm{L}}^{\mathrm{A}}}. \tag{2.8}$$

The total thermal efficiency between systems 1 and 3 is given by

$$\eta^{\mathrm{AB}} = \frac{W^{\mathrm{AB}}}{Q_{\mathrm{H}}^{\mathrm{A}}} = 1 - \frac{Q_{\mathrm{L}}^{\mathrm{B}}}{Q_{\mathrm{H}}^{\mathrm{A}}}, \tag{2.9}$$

where $W^{\mathrm{AB}}$ is the total work extracted:

$$W^{\mathrm{AB}} = W^{\mathrm{A}} + W^{\mathrm{B}} = Q_{\mathrm{H}}^{\mathrm{A}} - Q_{\mathrm{L}}^{\mathrm{B}} > 0. \tag{2.10}$$

Now, we consider direct heat transfer from system 3 to system 1, using the work $W^{\mathrm{A}} + W^{\mathrm{B}}$, extracting $Q_L^{\mathrm{B}}$ from system 3. Here, the heat is transferred to system 1 together with $W^{\mathrm{A}} + W^{\mathrm{B}}$. The amount of heat to system 1 must be the same as $Q_{\mathrm{H}}^{\mathrm{A}}$ because all the processes are reversible and the whole system must come back to the original state. Therefore, the following relation must hold:

$$Q_{\mathrm{L}}^{\mathrm{B}} + W^{\mathrm{A}} + W^{\mathrm{B}} = Q_{\mathrm{H}}^{\mathrm{A}}. \tag{2.11}$$

The thermal efficiency is

$$1 - \eta^{\mathrm{C}} = 1 - \eta_{\mathrm{C}}(T_1, T_3) = \frac{Q_L^{\mathrm{B}}}{Q_{\mathrm{H}}^{\mathrm{A}}}. \tag{2.12}$$

From the relations (2.7), (2.8), (2.9) and (2.12), we obtain the following relation:

$$1 - \eta^{\mathrm{AB}} = (1 - \eta^{\mathrm{A}})(1 - \eta^{\mathrm{B}}). \tag{2.13}$$

Using the temperatures, this relation is expressed as

$$1 - \eta(T_1, T_3) = (1 - \eta(T_1, T_2))(1 - \eta(T_2, T_3)). \tag{2.14}$$

To satisfy this relation for any set of systems where $T_1 > T_2 > T_3$, the function of $\eta(T_{\mathrm{H}}, T_{\mathrm{L}})$ must take the form:

$$1 - \eta(T_{\mathrm{H}}, T_{\mathrm{L}}) = \frac{f(T_{\mathrm{L}})}{f(T_{\mathrm{H}})}, \tag{2.15}$$

where $f(T)$ is an arbitrary monotonically increasing function. Using this relation, we can define the ratio $f(T)$ based on the thermal efficiency. Since $f(T)$ is any monotonic function, we may choose its form arbitrarily as long as it monotonically increases with $T$.

Here, we use the properties of an ideal gas, specifically the ideal gas law (Boyle-Charles's law):

$$PV = nRT^{\text{gas-thermometer}}, \tag{2.16}$$

where $R = 8.31$J/K is the gas constant, defined as the product of Avogadro's number ($N_{\mathrm{a}} = 6.02 \times 10^{23}$) and Boltzmann constant ($k_{\mathrm{B}} = 1.38 \times 10^{-23}$J/K), i.e., $R = N_{\mathrm{a}} \times k_{\mathrm{B}}$, $P$ is the pressure, $V$ is the volume, $n$ is the mole number of the particles, and $T^{\text{gas-thermometer}}$ is the absolute temperature which we normally use.

We calculate the thermal efficiency of the Carnot cycle, which is a reversible cycle and will be defined in the following section. Using the relation (2.16), we find the following relation (see Sect. 2.2.5):

$$1 - \eta(T_{\mathrm{H}}^{\mathrm{gas-thermometer}}, T_{\mathrm{L}}^{\mathrm{gas-thermometer}}) = \frac{T_{\mathrm{L}}^{\mathrm{gas-thermometer}}}{T_{\mathrm{H}}^{\mathrm{gas-thermometer}}}. \tag{2.17}$$

It is convenient to define the "temperature" in thermodynamics to align with the conventional absolute temperature $T^{\mathrm{gas-thermometer}}$. Thus, we choose the form of $f(t)$ as

$$f(T) = T, \tag{2.18}$$

and then,

$$1 - \eta(T_{\mathrm{H}}, T_{\mathrm{L}}) = \frac{T_{\mathrm{L}}}{T_{\mathrm{H}}}. \tag{2.19}$$

As the benchmark of the absolute temperature $T^{\mathrm{gas-thermometer}}$, the triple point of water (defined at $P$=611 Pa, $T$=273.16 K) is used for the thermodynamic temperature. Hereafter we simply use $T^{\mathrm{gas\ thermometer}}$ as the temperature $T$. Another form of $f(T)$ would also work defining another temperature, which would not be convenient.

### 2.2.4 Introduction of the Entropy

In thermodynamics, the concept of the so-called "state function", which is uniquely determined for a given equilibrium state, such as the internal energy, volume, pressure, and temperature, plays an important role. The state function does not change after a cyclic operation, since the system returns to its original state.

For instance, the volume $V$ satisfies

$$\oint_{\mathrm{cycle}} dV = 0. \tag{2.20}$$

Now, let us consider whether heat can be treated as a state function.

First, we examine the work $W$ produced by a cycle. Using the pressure $P$ and the volume $V$ as representative variables for the force and its conjugate quantity, the work is expressed as

$$\delta W = -P(V)dV \implies W = -\oint_{\mathrm{cycle}} P(V)dV \neq 0. \tag{2.21}$$

Thus, $W$ is not a state function since the integral over the cycle does not vanish. Next, consider internal energy $U$, which is a state function, and the integral is zero for any cyclic process. The change of internal energy is given by $W$ and the amount of heat transferred during the process $Q$:

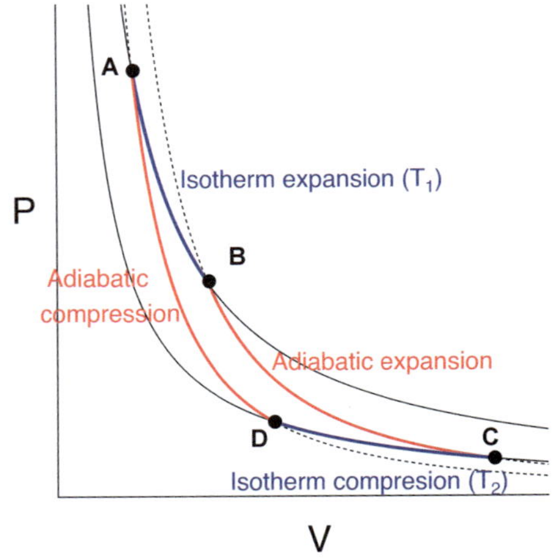


**Fig. 2.3** Carnot cycle consisting of isothermal volume changes and adiabatic volume changes between two temperatures

$$0 = \oint_{\text{cycle}} dU = W + Q. \tag{2.22}$$

From this relationship, it follows that neither $W$ nor $Q$ are state functions because of (2.21).

Now, let us seek a state function associated with the heat $Q$. To make the discussion concrete, we examine the Carnot cycle of the ideal gas, depicted in Fig. 2.3.

The Carnot cycle consists of four processes: AB (isothermal expansion at a constant temperature $T_1$), BC (adiabatic expansion during which the temperature decreases from $T_1$ to $T_2$), CD (isothermal compression at a constant temperature $T_2$), and DA (adiabatic compression during which the temperature increases from $T_2$ back to $T_1$). Then, the system comes back to the original state A. The quantitative analysis for changes of variables will be given in the next subsection (Sect. 2.2.5).

During the adiabatic processes (BC and DA), there is no heat transfer by definition. Therefore, the heat flow occurs in the isothermal processes (AB and CD). In AB, the heat $Q_1$ flows into the system from the high-temperature bath ($T_1$), and in CD, the heat $Q_2$ flows out of the system to the low-temperature bath ($T_2$). During the isothermal process, the heat transfer occurs entirely at the same temperature. In such a process, there is no "leaking" of heat from a high-temperature system to a low-temperature system. Thus, the isothermal process is reversible. Hence, all processes are reversible, and the Carnot cycle is reversible.

As will be shown in the next subsection (Sect. 2.2.5), the relation:

$$1 - \eta = \frac{Q_2}{Q_1} = \frac{T_2}{T_1} \tag{2.23}$$

holds. This relation can be written as

$$\frac{Q_1}{T_1} = \frac{Q_2}{T_2}. \tag{2.24}$$

So far, $Q_1$ and $Q_2$ represent the absolute value of heat transfer. Now, we consider their signs where heat flow into the system is positive. In this case, relation (2.24) becomes

$$\frac{Q_1}{T_1} = -\frac{Q_2}{T_2}. \tag{2.25}$$

From this, we deduce

$$\frac{Q_1}{T_1} + \frac{Q_2}{T_2} = 0. \tag{2.26}$$

This relation indicates that the quantity $Q/T$ corresponds to a kind of state function. In this analysis, we consider the specific example of the Carnot cycle (Fig. 2.3), but any thermodynamic cycle can be decomposed into a sum of small Carnot cycles. Thus, the above analysis is valid for arbitrary cycles.

Denoting the heat transfer during any infinitesimal process as $\delta Q$ and the temperature at that point as $T$, the expression (2.26) is generalized as

$$\oint_{\text{cycle}} \frac{\delta Q}{T} = 0, \tag{2.27}$$

which allows us to introduce the entropy, defined as

$$dS = \frac{\delta Q}{T}, \quad S(T) = \int_{T_0}^{T} \frac{\delta Q}{T'} + S(T_0), \tag{2.28}$$

where $S(T_0)$ is the value of entropy at a reference temperature $T_0$[1]. By the definition in (2.28), the heat transfer can be expressed in terms of the temperature and entropy:

$$\delta Q = TdS, \tag{2.30}$$

and the fundamental equation of thermodynamics is given in the following form:

$$dU = TdS - PdV + \cdots. \tag{2.31}$$

---

[1] The entropy at absolute zero temperature ($S(T = 0)$) is called residual entropy. According to the third law of thermodynamics, at absolute zero temperature ($T_0 = 0$), entropy approaches zero:

$$S(0) = 0. \tag{2.29}$$

See Sect. 2.2.6.

Generally, energy change is given by a sum of force×(change of the conjugate quantity). The relation (2.30) has this form, where $T$ is a force conjugate to entropy $S$. We would state that the essence of thermodynamics is the expression $TdS$ representing the heat change, which has been derived from the second law of thermodynamics.

### 2.2.5 Carnot Cycle: Isothermal Process and Adiabatic Process

Here, we study the change of the pressure $P$ as a function of the volume $V$ in the isothermal and adiabatic processes of the ideal gas. The equation of state of the ideal gas at a given temperature $T$ is given by the Boyle-Charles law (2.16):

$$PV = nRT. \tag{2.32}$$

In the adiabatic state, there is no heat exchange, and the entropy does not change. When the volume changes adiabatically, the temperature changes following the relation:

$$\left(\frac{\partial T}{\partial V}\right)_S = -\frac{1}{\left(\frac{\partial V}{\partial S}\right)_T \left(\frac{\partial S}{\partial T}\right)_V} = -\frac{\left(\frac{\partial S}{\partial V}\right)_T}{\left(\frac{\partial S}{\partial T}\right)_V}. \tag{2.33}$$

Using the Maxwell relation and the definition of the specific heat, i.e.,

$$\left(\frac{\partial S}{\partial V}\right)_T = \left(\frac{\partial P}{\partial T}\right)_V, \quad C_V = T\left(\frac{\partial S}{\partial T}\right)_V, \tag{2.34}$$

we obtain

$$\left(\frac{\partial T}{\partial V}\right)_S = -\frac{T}{C_V}\left(\frac{\partial P}{\partial T}\right)_V. \tag{2.35}$$

Substituting (2.32) into this expression gives

$$\left(\frac{\partial T}{\partial V}\right)_S = -\frac{nRT}{C_V V}. \tag{2.36}$$

This relation indicates that

$$T \propto V^{-nR/C_V}. \tag{2.37}$$

Using Mayer's relation $R = C_P - C_V$, we rewrite this as

$$T \propto V^{1-\gamma}, \quad \gamma = \frac{C_P}{C_V}. \tag{2.38}$$

This relation is known as the "Poisson relation". For monoatomic gases, $\gamma = 5/3$, and for diatomic gases $\gamma = 7/5$. Using (2.32), this relation (2.38) can be expressed as

$$P = \frac{nRT}{V} \propto V^{-\gamma}. \tag{2.39}$$

Because $\gamma > 1$, the pressure change in the adiabatic process is greater than in the isothermal process, as depicted in Fig. 2.3.

Now we prove the relation (2.17) using the relations (2.32) and (2.38). From these,

$$T_{\mathrm{B}} V_{\mathrm{B}}^{\gamma-1} = T_{\mathrm{C}} V_{\mathrm{C}}^{\gamma-1}, \quad T_{\mathrm{D}} V_{\mathrm{D}}^{\gamma-1} = T_{\mathrm{A}} V_{\mathrm{A}}^{\gamma-1}, \tag{2.40}$$

and the amounts of heat transfer are

$$Q_{\mathrm{H}} = -\int_{V_{\mathrm{A}}}^{V_{\mathrm{B}}} \frac{nRT_{\mathrm{H}}}{V} dV = nRT_{\mathrm{H}} \ln \frac{V_{\mathrm{A}}}{V_{\mathrm{B}}}, \tag{2.41}$$

and

$$Q_{\mathrm{L}} = -\int_{V_{\mathrm{D}}}^{V_{\mathrm{C}}} \frac{nRT_{\mathrm{L}}}{V} dV = nRT_{\mathrm{L}} \ln \frac{V_{\mathrm{D}}}{V_{\mathrm{C}}}. \tag{2.42}$$

From (2.40), and noting $T_{\mathrm{A}} = T_{\mathrm{B}} = T_1$ and $T_{\mathrm{C}} = T_{\mathrm{D}} = T_2$, we have the relation:

$$\frac{V_{\mathrm{D}}}{V_{\mathrm{C}}} = \frac{V_{\mathrm{A}}(T_1/T_2)^{\frac{1}{\gamma-1}}}{V_{\mathrm{B}}(T_1/T_2)^{\frac{1}{\gamma-1}}} = \frac{V_{\mathrm{A}}}{V_{\mathrm{B}}}. \tag{2.43}$$

Therefore,

$$\frac{Q_{\mathrm{H}}}{Q_{\mathrm{L}}} = \frac{T_1}{T_2} \Rightarrow \eta = 1 - \frac{Q_{\mathrm{L}}}{Q_{\mathrm{H}}} = 1 - \frac{T_2}{T_1}. \tag{2.44}$$

### 2.2.6 The Third Law of Thermodynamics: Nernst-Planck Theorem

The difference in internal energy between a temperature $T$ and absolute zero is given by

$$U(T) = U(0) + \int_0^T \left(\frac{\partial U}{\partial T}\right)_x dT, \tag{2.45}$$

where $U(0)$ is an arbitrary reference value, which can be chosen freely without affecting the thermodynamic analysis.

The entropy is also a state function given by

$$S(T) = S(0) + \int_0^T \left(\frac{\partial S}{\partial T}\right)_x dT. \tag{2.46}$$

Here, only the difference $S(T) - S(0)$ has physical significance. However, as an important property of entropy, it has been established that entropy is zero at absolute zero temperature:

$$S(0) = 0. \tag{2.47}$$

This conclusion arises from considerations of the equilibrium constant in the law of mass action for chemical reactions.

This property is known as the "Nernst-Planck theorem", and it is regarded as one of the fundamental principles of thermodynamics. This theorem results from a fundamental aspect of nature described by quantum mechanics.

Although the value of $S(0)$ does not directly alter the theoretical structure of thermodynamics, it imposes constraints on the temperature dependence of thermodynamic quantities. For instance, the specific heat is related to entropy as

$$C_x = T\left(\frac{\partial S}{\partial T}\right)_x, \tag{2.48}$$

and the entropy can be expressed as

$$S(T) = S(0) + \int_0^T \frac{C}{T} dT. \tag{2.49}$$

To avoid divergence of entropy at zero temperature, the integral must converge, requiring the condition

$$\lim_{T \to 0} C = 0. \tag{2.50}$$

In classical systems with continuous variables, this relation is violated. For example, the specific heat of a monoatomic ideal gas is given by

$$C_V = \frac{3}{2} nR, \tag{2.51}$$

and the corresponding entropy is

$$S(0) = S(T) - \frac{3}{2} nR \int_0^T \frac{1}{T} dT = S(T) - \frac{3}{2} nR \ln T + \frac{3}{2} nR \ln 0 = -\infty. \tag{2.52}$$

In real gases, liquefaction and solidification occur at low temperatures. Thus, historically, this issue was not considered seriously. However, the same problem of non-zero specific heat arises in solids. Classically, it is a constant as described by the Dulong-Petit law. However, deviations from this law have been explicitly observed, which provided a breakthrough to quantum mechanics. These deviations are explained by the Debye model, which incorporates quantum mechanical treatments of harmonic oscillations.

Interestingly, certain materials exhibit a so-called residual entropy, where $S(0) > 0$. For example, the degree of freedom in the configuration of hydrogens in ice results in residual degeneracy (known as the "ice problem", discussed in Chap. 4). Similarly, materials in a glassy state do not undergo solidification within observable time scales. In such cases, the residual entropy can be attributed to non-equilibrium effects.

### 2.2.7 Entropy as a Label of Adiabatically Isolated States

The concept of entropy can be introduced through the notion of adiabatic accessibility, which examines whether one state of a system can be transformed into another without the exchange of heat. This idea was first proposed by Carathéodory [1], who demonstrated the necessity of a quantity that classifies thermal equilibrium states based on their adiabatic accessibility. Later, Lieb and Yngvason provided a more general and rigorous mathematical formulation of this concept [2].

### 2.2.8 Thermodynamic Potential

The quantity which takes the extremum value in the equilibrium state is referred to as the "thermodynamic potential". Such a quantity depends on the specific constraints imposed. For example, entropy $S$ is maximized when the energy is fixed in an isolated system, while the Helmholtz free energy ($F = E - TS$) is minimized when the temperature is fixed by contact with a thermal bath. Moreover, when the pressure is fixed instead of the volume, the enthalpy ($H = E + PV$) is the thermodynamic potential, and when both $T$ and $P$ are fixed, the Gibbs free energy ($G = E - TS + PV$) serves as the thermodynamic potential.

### 2.2.9 Clausius-Clapeyron Equation for the Latent Heat at the First-Order Phase Transition

At a first-order phase transition, the phase changes discontinuously. For example, at the boiling point of water, the volume $V$ undergoes a discontinuous change from the value of liquid water to that of vapor. Similarly, the internal energy $U$ also changes discontinuously at this point, requiring external heat, commonly referred to as boiling heat.

The heat required for a first-order phase transition is referred to as the latent heat $\Delta Q$, which is directly related to the entropy difference $\Delta S$ between the two phases:

$$\Delta Q = T_C \Delta S. \tag{2.53}$$

There is an interesting relation among the discontinuous changes at first-order phase transitions. Consider a phase diagram in the $(P, T)$ space, where phases A and B undergo a first-order phase transition, as depicted in Fig. 2.4. The phase boundary is described by the function $T = T_C(P)$.

At a point on the boundary $(T_C, P_C)$, both phases coexist, and their chemical potentials are equal:

$$\mu_A(T_C, P_C) = \mu_B(T_C, P_C). \tag{2.54}$$

At a nearby point $(T_C + dT_C, P_C + dP_C)$ on the boundary, the chemical potentials of the phases are also equal:

$$\mu_A(T_C + dT_C, P_C + dP_C) = \mu_B(T_C + dT_C, P_C + dP_C). \tag{2.55}$$

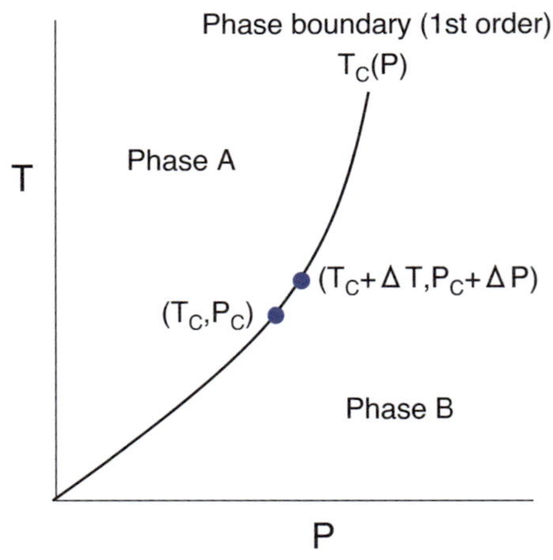


**Fig. 2.4** Phase diagram of phases A and B. The boundary at which a first-order phase transition occurs is given by $T = T_C(P)$

Expanding $\mu_\mathrm{A}(T_\mathrm{C} + dT_\mathrm{C}, P_\mathrm{C} + dP_\mathrm{C})$ to the first order in $dT_\mathrm{C}$ and $dP_\mathrm{C}$, we obtain

$$\mu_\mathrm{A}(T_\mathrm{C} + dT_\mathrm{C}, P_\mathrm{C} + dP_\mathrm{C}) = \mu_\mathrm{A}(T_\mathrm{C}, P_\mathrm{C}) + \left(\frac{\partial \mu_\mathrm{A}}{\partial T}\right)_P dT_\mathrm{C} + \left(\frac{\partial \mu_\mathrm{A}}{\partial P}\right)_T dP_\mathrm{C}. \tag{2.56}$$

and similarly for $\mu_\mathrm{B}(T_\mathrm{C} + dT_\mathrm{C}, P_\mathrm{C} + dP_\mathrm{C})$:

$$\mu_\mathrm{B}(T_\mathrm{C}, P_\mathrm{C}) + \left(\frac{\partial \mu_\mathrm{B}}{\partial T}\right)_P dT_\mathrm{C} + \left(\frac{\partial \mu_\mathrm{B}}{\partial P}\right)_T dP_\mathrm{C}. \tag{2.57}$$

Using the relations (2.54) and (2.55), the following holds:

$$\left(\frac{\partial \mu_\mathrm{A}}{\partial T}\right)_P dT_\mathrm{C} + \left(\frac{\partial \mu_\mathrm{A}}{\partial P}\right)_T dP_\mathrm{C} = \left(\frac{\partial \mu_\mathrm{B}}{\partial T}\right)_P dT_\mathrm{C} + \left(\frac{\partial \mu_\mathrm{B}}{\partial P}\right)_T dP_\mathrm{C}. \tag{2.58}$$

Rearranging terms, we find

$$\frac{dP_\mathrm{C}}{dT_\mathrm{C}} = \frac{\left(\frac{\partial \mu_\mathrm{A}}{\partial T}\right)_P - \left(\frac{\partial \mu_\mathrm{B}}{\partial T}\right)_P}{\left(\frac{\partial \mu_\mathrm{B}}{\partial P}\right)_T - \left(\frac{\partial \mu_\mathrm{A}}{\partial P}\right)_T}. \tag{2.59}$$

Using the Gibbs-Duhem relation, $-SdT + VdP - Nd\mu = 0$, the derivatives of the chemical potential can be expressed as:

$$\left(\frac{\partial \mu_\mathrm{A}}{\partial T}\right)_P = -\frac{S_\mathrm{A}}{N_\mathrm{A}}, \quad \left(\frac{\partial \mu_\mathrm{A}}{\partial P}\right)_T = \frac{V_\mathrm{A}}{N_\mathrm{A}}, \tag{2.60}$$

and similarly for phase B. Substituting these relations into (2.59), we obtain

$$\frac{dP_\mathrm{C}}{dT_\mathrm{C}} = \frac{-\frac{S_\mathrm{A}}{N_\mathrm{A}} + \frac{S_\mathrm{B}}{N_\mathrm{B}}}{\frac{V_\mathrm{B}}{N_\mathrm{B}} - \frac{V_\mathrm{A}}{N_\mathrm{A}}}. \tag{2.61}$$

The jump in the volume per particle is

$$\Delta V = \frac{V_\mathrm{A}}{N_\mathrm{A}} - \frac{V_\mathrm{B}}{N_\mathrm{B}}, \tag{2.62}$$

and the latent heat per particle is

$$\Delta Q = T_\mathrm{C}\left(\frac{S_\mathrm{A}}{N_\mathrm{A}} - \frac{S_\mathrm{B}}{N_\mathrm{B}}\right). \tag{2.63}$$

Thus, the slope of the phase boundary is given by

$$\frac{dP_C}{dT_C} = \frac{\Delta Q}{T_C \Delta V}, \tag{2.64}$$

which is known as the Clausius-Clapeyron equation.

For second-order phase transitions, the quantities change continuously. In such cases, limiting forms of (2.59) apply, leading to what are known as Ehrenfest's relations.

## 2.3 Definition of Entropy in Statistical Mechanics

Statistical mechanics establishes the relationship between the microscopic details of a system, described by its Hamiltonian, and the macroscopic quantities of thermodynamics. Macroscopic quantities are obtained as averages over an ensemble of microscopic states.

Let us consider the average quantity of $A$. To simplify the discussion, we assume the system has $M$ microscopic states. If necessary, we take the limit $M \to \infty$. If the probability of a microscopic state $i$ is given by $p(i)$, the average is defined as

$$\langle A \rangle = \sum_{\text{all states } i} p(i) A(i). \tag{2.65}$$

The fundamental principle of statistical mechanics is the so-called the "principle of equal a priori probabilities (equal weight principle)". For an isolated system, the total energy is conserved. Such a system is referred to as the "microcanonical ensemble". In this ensemble, all the states have the same energy, and we cannot distinguish them macroscopically. Consequently, we assume that the probability of each microscopic state is identical.

Here, we need to be careful to define what "a microscopic state" is. In systems of discrete variables, the state is well-defined. For example, consider a system consisting of 10 coins. There are $2^{10}$ possible states ($+ + \cdots +$ to $- - \cdots -$), where $+$ (head) and $-$ (tail) represent the two faces of a coin. Suppose the head state has energy $\varepsilon$, and the tail state has energy $-\varepsilon$. The number of microscopic states $W(E)$ having the energy $E$ can be calculated using combinatorics. For example, for $E = 10\varepsilon$, all 10 coins are $+$ (head). There is only one such configuration, and thus

$$W(10\varepsilon) = \frac{10!}{1!10!} = 1, \tag{2.66}$$

For $E = 0$, five coins are $+$ (head) and the other five coins are $(-)$ (tail). The number of such configurations is given by

$$W(0) = \frac{10!}{5!5!} = 252. \tag{2.67}$$

On the other hand, in systems with continuous variables, the definition of states is not straightforward. Consider a particle confined in a box. The state is described by the particle position and momentum $(\boldsymbol{x}, \boldsymbol{p})$. However, since both $\boldsymbol{x}$ and $\boldsymbol{p}$ vary continuously, counting the number of states becomes difficult. In such a case, we consider probability density $P(\boldsymbol{x}, \boldsymbol{p})$. The probability of finding the particle in the region $\boldsymbol{x} \sim \boldsymbol{x} + d\boldsymbol{x}$ and $\boldsymbol{p} \sim \boldsymbol{p} + d\boldsymbol{p}$ is given by $P(\boldsymbol{x}, \boldsymbol{p}) d\boldsymbol{x} d\boldsymbol{p}$. According to Liouville's theorem in analytical mechanics, the hyper-volume in the phase space does not change during the time evolution. This invariance allows us to assign "a state" to each unit hyper-volume. Let

$$C^d = \Delta\boldsymbol{x}\Delta\boldsymbol{p} \tag{2.68}$$

be a unit of state, where $d$ is the spatial dimension. Concrete values of $\Delta\boldsymbol{x}$ and $\Delta\boldsymbol{p}$ are arbitrary although they should be small enough that the state within the unit can be regarded as uniform.

Hereafter, we consider three-dimensional systems ($d = 3$). Thus, the number of states of a system with $N$ particles, $W(E)$, is defined as

$$W(E) = \frac{\mathcal{V}}{C^{3N}}, \tag{2.69}$$

where $\mathcal{V}$ is the hyper-volume of the phase space corresponding to the energy $E$. The phase space of $N$ particles in three dimensions is $6N$-dimensional space. There, the region with energy exactly equal to $E$ corresponds to a $(6N - 1)$-dimensional surface. The volume of the surface is zero. To address this inconvenience, we consider the region where the energy lies in the range $E \sim E + \Delta E$, and define the hyper-volume associated with $E$ as

$$\mathcal{V}(E \sim E + \Delta E) = A(E)\Delta E, \tag{2.70}$$

where $A(E)$ is the area of the $(6N - 1)$-dimensional surface. The number of states within the energy range $E \sim E + \Delta E$ is then given by

$$W(E \sim E + \Delta E) = \frac{\mathcal{V}(E \sim E + \Delta E)}{C^{3N}} \simeq D(E)\Delta E, \tag{2.71}$$

where $D(E)$ is the density of state as a function of the energy $E$.

Here, we encounter ambiguity in defining $\Delta E$. For usual physical quantities such as internal energy and pressure etc., the ratio of the number of states is what matters, and the value of $\Delta E$ is irrelevant as long as $\Delta E$ is small. However, this ambiguity does affect the entropy, as we will discuss later.

The idea of the equal weight principle originates from the concept of ergodicity. Ergodicity assumes that, over a sufficiently long time, the system explores all accessible states, making the long-time average equivalent to the phase-space average. Although the trajectory of a system with constant energy is a one-dimensional orbit and cannot fully cover the energy surface, it is assumed that the trajectory comes sufficiently close to all points on the surface. This assumption, while reasonable, raises an issue: even if ergodicity holds, traversing the entire trajectory can take an impractically long time. This conflicts with the fact that macroscopic variables in equilibrium are observed almost instantaneously.

Therefore, it seems that a stronger property is required to account for the rapid realization of equilibrium averages. In this context, the concept of ETH (Eigenstate Thermalization Hypothesis) has been proposed and studied, particularly for isolated systems [3]. To explain the instantaneous establishment of equilibrium, the extensivity of the system seems to play a crucial role. Extensivity implies that when a macroscopic system is divided into smaller, but still macroscopic subsystems, each subsystem exhibits the same macroscopic properties. This characteristic helps in understanding the rapid convergence to equilibrium. While further exploration of such approaches is expected, we will not delve into this topic in detail in the present book.

### 2.3.1 Microcanonical Ensemble

As mentioned earlier, in an isolated system, the energy of the system is conserved. According to the equal weight principle, all states are equally probable. Therefore, the probability of a state $i$ is given by

$$p_{\text{microcanonical}}(i) = \frac{1}{W(E)}, \tag{2.72}$$

where $W(E)$ is the number of states with energy $E$ (or within the range $E \sim E + \Delta E$). This ensemble of states with the same energy is called microcanonical ensemble. In this ensemble, the average of physical quantities is calculated using (2.72), which makes the computation straightforward. At this stage, however, the concepts of heat and temperature have not yet been introduced.

### 2.3.2 Introduction of Temperature in Statistical Physics

Heat transfer occurs when two systems with different temperatures come into contact. To analyze this phenomenon, let us consider two isolated systems, denoted as system-1 and system-2. Initially, their energies are $E_1^0$ and $E_2^0$, respectively, and the number of microscopic states for this energy is given by $W_1(E_1^0)$ for system-1 and $W_2(E_2^0)$ for system-2.

When the two systems are brought into contact, the combined system can be treated as an isolated system with total energy $E = E_1^0 + E_2^0$. After contact, some amount of energy $\Delta E$ is transferred between the two systems. As a result, the energy of system-1 becomes $E_1 = E_1^0 + \Delta E$ and that of system-2 becomes $E_2 = E_2^0 - \Delta E$

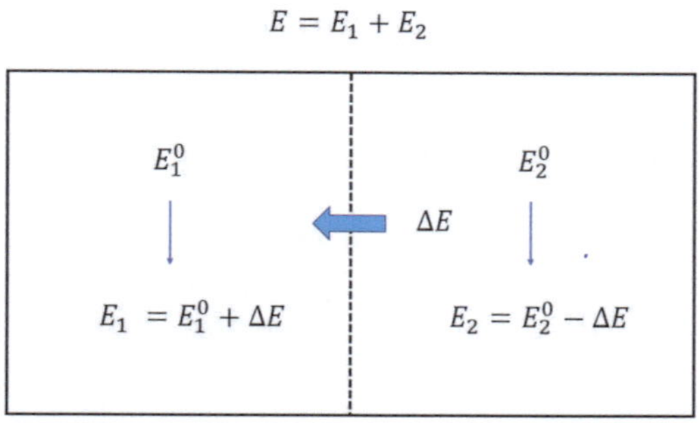


**Fig. 2.5** Energy (heat) transfer between system-1 and system-2

as depicted in Fig. 2.5. Importantly, the total energy of the combined system remains conserved:

$$E = E_1^0 + E_2^0 = E_1 + E_2. \tag{2.73}$$

Thus, the entire system is treated as an isolated system.

Here, we assume that both system-1 and system-2 are very large, and the energy associated with the contact between them can be ignored.[2] Under these assumptions, the total number of states for the combined system is given by

$$W(E) = W_1(E_1^0 + \Delta E)W_2(E_2^0 - \Delta E). \tag{2.74}$$

The thermal equilibrium state is regarded as the state with maximum probability. Hence, we expect that heat flow ($\Delta E$) will occur to achieve this state. The condition of the maximum probability is given by

$$\frac{dW}{d\Delta E} = 0, \tag{2.75}$$

which leads to:

$$\begin{aligned}&\frac{dW(E)}{d\Delta E}\\ &= \left.\frac{dW_1(E')}{dE'}\right|_{E'=E_1} W_2(E_2^0 - \Delta E) - W_1(E_1^0 + \Delta E)\left.\frac{dW_2(E')}{dE'}\right|_{E'=E_2} = 0.\end{aligned} \tag{2.76}$$

Dividing both sides $W_1(E_1)W_2(E_2)$, we arrive at the relation:

$$\frac{\left.\frac{dW_1(E)}{dE}\right|_{E=E_1}}{W_1(E_1)} = \frac{\left.\frac{dW_2(E)}{dE}\right|_{E=E_2}}{W_2(E_2)}. \tag{2.77}$$

---

[2] The functions $W_1(E_1)$ and $W_2(E_2)$ do not change significantly after the contact. Generally, $\ln W(E)$ is proportional to $N$ for large $N$, due to the property of extensivity, that is $\ln W(E) \propto O(N)$. The contribution from the contact energy is of order $o(N)$ in systems with short-range interactions. Therefore, the leading $O(N)$ term remains unaffected as long as the extensivity holds.

Equivalently, this can be written as

$$\left.\frac{d\ln W_1(E)}{dE}\right|_{E=E_1} = \left.\frac{d\ln W_2(E)}{dE}\right|_{E=E_2}. \tag{2.78}$$

In this relation, right and left sides represent quantities determined independently in system-1 and in system-2, respectively. Thus, we found that the quantity $d\ln W_i/dE$ must be the same for both systems in equilibrium under energy balance. Now, let us recall the condition for thermal equilibrium; when two systems are in thermal contact and can exchange heat, equilibrium is achieved when their temperatures are equal.

Therefore, it follows that $d\ln W_i/dE$ is a function of the temperature:

$$\left.\frac{d\ln W_i(E)}{dE}\right|_{E=E_i} = f(T_i). \quad i = 1, 2 \tag{2.79}$$

Here, $f(T)$ is an arbitrary monotonic function and the temperature is defined as

$$T = f^{-1}\left(\frac{d\ln W(E)}{dE}\right). \tag{2.80}$$

In thermodynamics, we adopted the definition of temperature to be the same as the temperature used in the ideal gas. Thus, here, we also determine the form of $f(T)$ in the same way. In the ideal gas, the internal energy is given by the relation:

$$U = E = \frac{3}{2}RT, \tag{2.81}$$

where $R$ is the gas constant.

As discussed below (Sect. 2.3.3), the number of states for the ideal gas, $W_{\text{ideal gas}}(E)$, is proportional to $E^{3N/2}$. This leads to the relation:

$$\frac{d\ln W_{\text{ideal gas}}(E)}{dE} = \frac{3N}{2E} + o(N). \tag{2.82}$$

Comparing (2.81) and (2.82), we define $f(T)$ as

$$f(T) = \frac{d\ln W(E)}{dE} = \frac{3N}{2E} = \frac{N}{RT} = \frac{1}{k_B T}. \tag{2.83}$$

That is, the temperature in statistical mechanics is defined through $W(E)$ as

$$\frac{d\ln W(E)}{dE} = \frac{1}{k_B T^{\text{statistical mechanical}}}. \tag{2.84}$$

This definition coincides with the thermodynamic temperature, $T^{\text{thermodynamics}}$. Hence, we will hereafter denote both simply as $T$.

Now, we compare (2.84) with the thermodynamic relation between temperature and entropy:

$$k_{\mathrm{B}}\frac{d\ln W(E)}{dE}=\frac{1}{T}\Longleftrightarrow\left(\frac{\partial S}{\partial E}\right)_{V,N}=\frac{1}{T}. \tag{2.85}$$

From this, we arrive at the key relation:

$$S(E)=k_{\mathrm{B}}\ln W(E). \tag{2.86}$$

This relation is known as "Boltzmann's principle", which connects the microscopic and macroscopic descriptions of a system. Sometimes, this relation is considered to be the fundamental principle of statistical mechanics, replacing the equal weight principle.

### 2.3.3 *W*(*E*) *of the Ideal Gas*

Let us consider a system of $N$ free particles (ideal gas) with the mass $m$ in a box of volume $V$. The positions and momenta of particles are denoted by $(x_1, y_1, z_1), \cdots (x_N, y_N, z_N)$ and $(p_1^x, p_1^y, p_1^z), \cdots (p_N^x, p_N^y, p_N^z)$, respectively. The energy of the system is given by

$$E=\sum_{i=1}^{N}\sum_{\alpha=x,y,z}\frac{1}{2m}\left(p_i^{\alpha}\right)^2, \tag{2.87}$$

which depends only on the momenta but not on the positions.

Generally, the hyper-volume of an $n$-dimensional sphere of radius $r$ is given by

$$\Omega_n(r)=\frac{r^n\pi^{n/2}}{\Gamma(n/2+1)}, \tag{2.88}$$

where $\Gamma(x)$ is the Gamma function defined as

$$\Gamma(z)=\int_0^{\infty}t^{z-1}e^{-t}dt,\ (Re(z)>0),\quad \Gamma(x+1)=x\Gamma(x),$$
$$\text{for integer } n,\quad \Gamma(n+1)=n! \tag{2.89}$$

For large $x$, it is approximated (Starling's relation) as

$$\Gamma(x)\simeq\sqrt{2\pi}x^{x+1/2}e^{-x},\quad \ln\Gamma(x)\simeq x\ln x-x,\quad x\gg 1. \tag{2.90}$$

In the present case, $n = 3N, r = \sqrt{2mE}$, and the area of surface of constant energy $E$ is given by

$$A(E) = \frac{d\Omega_{3N}(\sqrt{2mE})}{dE} = \frac{3N}{2}2m\frac{(2mE)^{(3N/2-1)}\pi^{3N/2}}{\Gamma(3N/2+1)}.$$

The volume of the energy shell with width $\Delta E$ is $A(E) \times \Delta E$. The contribution from the positions is simply $V^N$, so the hyper-volume of the energy shell is

$$\mathcal{V} = V^N A(E)\Delta E. \tag{2.91}$$

Thus, the number of states is given by

$$W(E) = \frac{V^N A(E)\Delta E}{C^{3N}} = V^N \frac{E^{3N/2-1}}{\Gamma(3N/2+1)} \times \frac{3N}{2C^{3N}}(2\pi m)^{3N/2}\Delta E. \tag{2.92}$$

Using the Starling's relation $\ln\Gamma(x) \simeq x\ln x - x$ for large $N$ (where approximately $3N/2 + 1 \simeq 3N/2$), we obtain

$$\begin{aligned} W(E) &\propto V^N \frac{E^{3N/2}}{(3N/2)^{3N/2}} e^{3N/2}\frac{3N}{2}\left(\frac{2\pi m}{C^2}\right)^{3N/2}\Delta E \\ &= V^N\left(\frac{E}{N}\right)^{3N/2} e^{3N/2}\frac{3N}{2}\left(\frac{4\pi m}{3C^2}\right)^{3N/2}\Delta E. \end{aligned} \tag{2.93}$$

Thus, the number of states for the ideal gas is proportional to $E^{3N/2}$:

$$W(E) \propto E^{3N/2}. \tag{2.94}$$

It should be noted that we assumed that all the particles are distinguishable in the above calculation. However, in the case where the particles are indistinguishable, we overcount a configuration in which particles are at positions $(x_1, y_1, z_1), \cdots (x_N, y_N, z_N)$ by a factor of $N!$. Therefore, in this case, we have to divide $W(E)$ by $N!$ as follows:

$$\begin{aligned} W(E) &= \frac{1}{N!}V^N\left(\frac{E}{N}\right)^{3N/2} e^{3N/2}\frac{3N}{2}\left(\frac{4\pi m}{3C^2}\right)^{3N/2}\Delta E \\ &\simeq \left(\frac{V}{N}\right)^N e^N\left(\frac{E}{N}\right)^{3N/2} e^{3N/2}\frac{3N}{2}\left(\frac{4\pi m}{3C^2}\right)^{3N/2}\Delta E \end{aligned}$$

$$= \left(\left(\frac{V}{N}\right)\left(\frac{E}{N}\right)^{3/2} e^{5/2}\right)^{N} \frac{3N}{2}\left(\frac{4\pi m}{3C^2}\right)^{3N/2} \Delta E. \tag{2.95}$$

Without the factor of $1/N!$, the entropy is given by

$$\frac{S(E)}{k_{\mathrm{B}}} = N \ln V + \frac{3N}{2} + \frac{3N}{2} \ln\left(\frac{E}{N}\right) + \frac{3N}{2} \ln\left(\frac{4\pi m}{3C^2}\right) + o(N).$$

In the first term, $V$ is proportional to $N$, so the entropy has a term proportional to $N \ln N$. This implies that the entropy of this case is not extensive.

On the other hand, with the factor of $1/N!$, using $\ln N! \simeq N \ln N - N$, we obtain

$$\frac{S(E)}{k_{\mathrm{B}}} = N \ln V - N \ln V + N + \frac{3N}{2} + \frac{3N}{2} \ln\left(\frac{E}{N}\right) + \frac{3N}{2} \ln\left(\frac{4\pi m}{3C^2}\right) + o(N), \tag{2.96}$$

which is extensive, and

$$\frac{S(E)}{N k_{\mathrm{B}}} = \ln \frac{V}{N} + \frac{5}{2} + \frac{3}{2} \ln\left(\frac{E}{N}\right) + \frac{3}{2} \ln\left(\frac{4\pi m}{3C^2}\right). \tag{2.97}$$

This is the entropy per particle for the ideal gas in the microcanonical ensemble.

### 2.3.4 Canonical Ensemble

In statistical physics, when a system is in contact with a thermal bath at a fixed temperature $T$, the distribution of microscopic states is described by the canonical ensemble. In this framework, the independent variables specifying the system are the temperature $T$, the volume $V$, and the number of particles $N$, in contrast to the case of microcanonical ensemble, where the energy $E$, $V$, and $N$ are the independent variables.

To understand the canonical ensemble, consider a large isolated system with a total energy $E_{\mathrm{tot}}$. We define a subsystem as "the system" and treat the remainder as the "thermal bath". Both the system and the bath are assumed to be macroscopic and in thermal equilibrium.

Let $W_{\mathrm{sys}}(E_{\mathrm{sys}})$ represent the number of states of the system, and $W_{\mathrm{bath}}(E_{\mathrm{bath}})$ the number of states of the bath with energy $E_{\mathrm{bath}} = E_{\mathrm{tot}} - E_{\mathrm{sys}}$. The total number of states of the combined system is given by

$$W(E, E_{\mathrm{tot}} - E_{\mathrm{sys}}) = W_{\mathrm{sys}}(E_{\mathrm{sys}}) W_{\mathrm{bath}}(E_{\mathrm{tot}} - E_{\mathrm{sys}}), \tag{2.98}$$

neglecting the effect of interaction between the system and the bath. Using Boltzmann's principle, the number of states of the bath is expressed as

$$W_{\text{bath}}(E_{\text{tot}} - E_{\text{sys}}) = e^{S_{\text{bath}}(E_{\text{tot}} - E_{\text{sys}})/k_{\text{B}}}, \tag{2.99}$$

where $S_{\text{bath}}$ is the entropy of the bath. Assuming the bath is much larger than the system, we expand the entropy of the bath as

$$S_{\text{bath}}(E_{\text{tot}} - E_{\text{sys}}) \simeq S_{\text{bath}}(E_{\text{tot}}) - \left(\frac{\partial S_{\text{bath}}}{\partial E}\right)_{V,N} E_{\text{sys}} = S_{\text{bath}}(E_{\text{tot}}) - \frac{E_{\text{sys}}}{T_{\text{bath}}}, \tag{2.100}$$

where $T_{\text{bath}}$ is the temperature of the bath defined by

$$\left(\frac{\partial S_{\text{bath}}}{\partial E}\right)_{V,N} = \frac{1}{T_{\text{bath}}}. \tag{2.101}$$

Consequently, the number of states of the bath becomes

$$W_{\text{bath}}(E_{\text{tot}} - E_{\text{sys}}) = e^{S_{\text{bath}}(E_{\text{tot}})/k_{\text{B}}} e^{\frac{-E_{\text{sys}}}{k_{\text{B}} T_{\text{bath}}}}. \tag{2.102}$$

The probability $P(E_{\text{sys}})$, that the system has the energy $E_{\text{sys}}$, is equal to probability that the total energy is separated into $E_{\text{sys}}$ and $E_{\text{tot}} - E_{\text{sys}}$. Thus,

$$P(E_{\text{sys}}) \propto W_{\text{sys}}(E_{\text{sys}}) e^{\frac{-E_{\text{sys}}}{k_{\text{B}} T_{\text{bath}}}}. \tag{2.103}$$

Since all $W_{\text{sys}}(E_{\text{sys}})$ states of the system has the same energy $E_{\text{sys}}$, all the states are equally probable. Thus, the probability of any specific microstate $i$ of the system with the energy $E_{\text{sys}}$ is given by

$$p_i = \frac{1}{Z} e^{-\beta E_i}. \quad \beta = \frac{1}{k_{\text{B}} T}, \tag{2.104}$$

which is called "Boltzmann factor"' and $Z$ is the normalization factor. This $Z$ is known as "the partition function":

$$Z = \sum_{\text{all states i}} e^{-\beta E_i}. \tag{2.105}$$

The partition function plays a central role in the canonical ensemble, allowing us to calculate thermodynamic properties. For example, the free energy is given by

$$F = -k_{\text{B}} T \ln Z, \tag{2.106}$$

and the internal energy is given by

$$E = -\frac{\partial}{\partial \beta} \ln Z. \tag{2.107}$$

These expressions bridge microscopic information of the systems with thermodynamic quantities, providing a powerful framework for analyzing systems in contact with a thermal bath.

The second law of thermodynamics is often expressed as the law of increasing entropy, which is called entropy principle. The heat flows from a system at a higher temperature system ($T_1$) to one at a low temperature ($T_2 < T_1$). Then, the entropy change in this process is given by

$$\Delta S = \frac{Q}{T_2} - \frac{Q}{T_1} = Q\frac{T_1 - T_2}{T_1 T_2} > 0, \tag{2.108}$$

where $Q$ is the heat transferred. Thus, inhomogeneous temperature distribution in an isolated system becomes uniform and the entropy of the whole system increases.

### 2.3.5 Entropy of Ideal Gas

Let us derive the entropy of the ideal gas by the method of the canonical ensemble. The partition function is given by

$$Z(\beta) = \frac{1}{C^{3N}}\frac{1}{N!}\prod_{j=1}^{N}\int dx_j \int dy_j \int dz_j \int_{-\infty}^{\infty} dp_{jx} \int_{-\infty}^{\infty} dp_{jy} \int_{-\infty}^{\infty} dp_{jz} \\ \times e^{-\beta \sum_{j=1}^{N}(p_{jx}^2+p_{jy}^2+p_{jz}^2)/2m}, \tag{2.109}$$

where $C$ is the factor to convert the volume of the phase space to the number of states (2.69). Using the following relations:

$$\int dx_j \int dy_j \int dz_j = V, \quad \int_{-\infty}^{\infty} dp e^{-\beta p^2/2m} = \sqrt{2\pi m k_B T}, \tag{2.110}$$

the partition function is given by

$$Z(\beta) = \frac{1}{C^{3N}}\frac{1}{N!}(2\pi m k_B T)^{3N/2} V^N, \tag{2.111}$$

from which the internal energy is given as

$$U = \langle E \rangle = -\frac{\partial}{\partial \beta}\ln Z(\beta) = \frac{3}{2}N k_B T, \tag{2.112}$$

and the pressure is given as

$$P = \left(\frac{\partial F}{\partial V}\right)_{T,N} = \frac{\partial}{\partial V}k_B T \ln Z(\beta) = \frac{N k_B T}{V}. \tag{2.113}$$

This is the Boyle-Charles equation of the ideal gas. The free energy given by the relation $F = -k_B T \ln Z$ (2.111) is

$$F = -k_B T \left( N \ln V - \ln N! + \frac{3N}{2} \ln \left( \frac{2\pi m k_B T}{C^2} \right) \right). \tag{2.114}$$

With the Stirling's relation $\ln N! \simeq N \ln N - N, \quad N \gg 1$, we have

$$F = -k_B T \left( N \ln V - N \ln N + N + \frac{3N}{2} \ln \left( \frac{2\pi m k_B T}{C^2} \right) \right)$$

$$= -N k_B T \left( \ln \frac{V}{N} + 1 + \frac{3}{2} \ln \left( \frac{2\pi m k_B T}{C^2} \right) \right). \tag{2.115}$$

The entropy is given by

$$S = -\left( \frac{\partial F}{\partial T} \right)_{V,N} = N k_B \left( \left( \ln \frac{V}{N} + 1 + \frac{3}{2} \ln \left( \frac{2\pi m k_B T}{C^2} \right) + \frac{3}{2} ) \right) \right), \tag{2.116}$$

which is the same as (2.97) as it must be.

### 2.3.6 *Free Expansion of the Ideal Gas*

Let us consider the free expansion of the gas into the vacuum chamber after removal of the partition wall as depicted in Fig. 2.6.

In this process, there is no heat exchange. However, the process is not quasi-static, meaning it cannot be described by using thermodynamic quantities. Nevertheless,

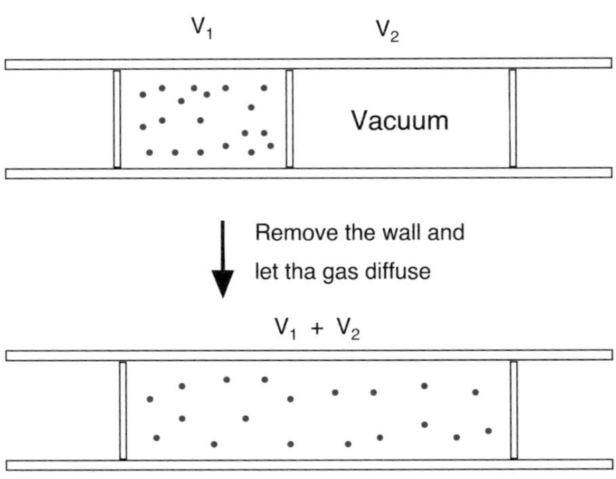


**Fig. 2.6** Free expansion of the ideal gas

the initial and final states are equilibrium states, allowing us to calculate changes in thermodynamic quantities using an equivalent quasi-static process that connects the same initial and final states. For example, consider slowly moving the partition wall to the right. Since the right chamber is in vacuum, there is no pressure, and no work is required for the expansion. In this process, the internal energy is conserved. The change in temperature is determined by the following relation:

$$\left(\frac{\partial T}{\partial V}\right)_U = \frac{1}{C_V}\left[P - T\left(\frac{\partial P}{\partial T}\right)_V\right]. \tag{2.117}$$

Substituting the ideal gas relation $P = nRT/V$ (2.32), we find

$$\left(\frac{\partial T}{\partial V}\right)_U = 0. \tag{2.118}$$

This result shows that the temperature of an ideal gas, at constant internal energy, is independent of its volume. This property is known as Joule's law of ideal gases.

Since

$$\left(\frac{\partial S}{\partial V}\right)_U = \frac{P}{T}, \tag{2.119}$$

the change of entropy is given by

$$\Delta S = \int_{V_1}^{V_2} \left(\frac{\partial S}{\partial V}\right)_U dV = \int_{V_1}^{V_2} \frac{P}{T} dV. \tag{2.120}$$

For an ideal gas, using the relation, $PV = nRT$, the change of entropy due to the volume change is given by

$$\Delta S = \int_{V_1}^{V_2} \frac{nR}{V} dV = nR \ln\left(\frac{V_2}{V_1}\right). \tag{2.121}$$

## References

1. C. Carathéodory, Mathematische Annalen (in German). **67** (3), 355-386 (1909)
2. E.H. Lieb, J. Yngvason, Jakob. Phys. Rep. **310**(1), 1–96 (1999)
3. J.M. Deutsch: Phys. Rev. A **43**, 2046 (1991); M. Srednicki, Phys. Rev E **50**, 888 (1994). M. Rigol, V. Dunjko, M. Olshanii, Nature (London) **452**, 854 (2008); M. Rigol, M. Srednicki, Phys. Rev. Lett. **108**, 110601, (2012); T. Mori, N. Shiraishi, Phys. Rev. E **96**, 022153 (2017)

# Chapter 3
# Examples of Phenomena Caused by Entropy

**Abstract** In this chapter, we explain how entropy contributes to thermodynamic behavior in various systems. Entropy gives rise to several non-trivial effects, such as entropic forces in rubber, mixing phenomena in multicomponent systems, and degeneracy-induced phenomena (e.g., spin-crossover systems) and so on. Phase transitions can also be driven by entropy effects. Entropy plays a crucial role in systems with competing interactions (frustration), which will be discussed in the next part.

## 3.1 Entropy of Two-Level System and Adiabatic Demagnetization

When a system has two microscopic states, A and B, it is referred to as a two-level system. Let the energies of states A and B be denoted as $E_A$ and $E_B$, respectively as depicted in Fig. 3.1. We consider a macroscopic system composed of $N$ two-level systems in thermal equilibrium at a temperature $T$. By representing the states A and B by $\sigma = +1$ and $-1$, respectively, the Hamiltonian of the system is given by

$$\mathcal{H} = NE_0 - \varepsilon \sum_{i=1}^{N} \sigma_i, \quad E_0 = \frac{E_A + E_B}{2}, \quad \varepsilon = \frac{E_B - E_A}{2}. \tag{3.1}$$

For simplicity, we set $E_0 = 0$ unless specified otherwise.

This model is applicable to a variety of physical systems, such as uniaxial magnetic systems with spin $S = 1/2$, where $\sigma = \pm 1$ represents the spin direction ($s_i^z = \pm 1/2$), binary alloys (distinguishing atom types A and B), adsorption systems and lattice gas models (presence or absence of a particle), and even social phenomena (e.g., binary choices such as "aye" or "no").

© The Author(s), under exclusive license to Springer Nature Singapore Pte Ltd. 2025

S. Miyashita, *Physics of Entropy*, Springer Series in Solid-State Sciences 212,
https://doi.org/10.1007/978-981-95-1758-9_3

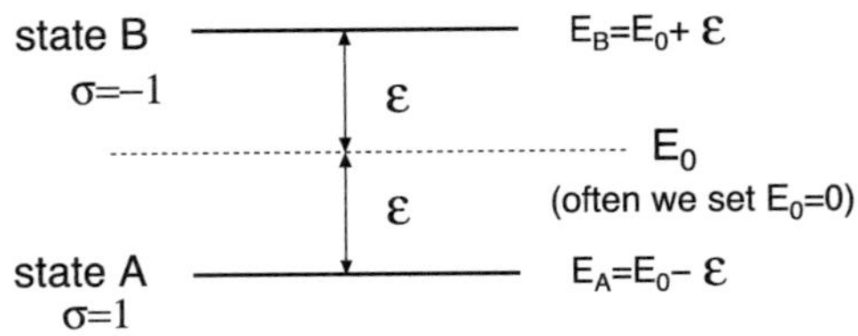


**Fig. 3.1** Two-level system

### *3.1.1 Entropy in the Microcanonical Ensemble*

When $N_+$ elements take $\sigma_i = 1$ (and $N_- = N - N_+$ elements take $\sigma_i = -1$), the energy of the whole system is

$$E = \varepsilon(N_+ - N_-). \tag{3.2}$$

The number of states corresponding to this energy is

$$W(E) =_N C_{N_+} = \frac{N!}{N_+!N_-!}. \tag{3.3}$$

Since $N = N_+ + N_-$ and $E = \varepsilon N_+ - \varepsilon N_-$, $N_+$ and $N_-$ are expressed in terms of $E$ as

$$N_+ = \frac{N + E/\varepsilon}{2}, \quad N_- = \frac{N - E/\varepsilon}{2}. \tag{3.4}$$

Thus, the number of states $W(E)$ is expressed as a function of $E$:

$$W(E) = \frac{N!}{\left(\frac{N+E/\varepsilon}{2}\right)!\left(\frac{N-E/\varepsilon}{2}\right)!}. \tag{3.5}$$

For a large $N$, using the Stirling's relation, $\ln N! \simeq N \ln N - N + o(N)$, the entropy given by $S(E) = k_B \ln W(E)$ is obtained as

$$\frac{S(E)}{k_B} \simeq N \log N - N - \left(\frac{N + E/\varepsilon}{2}\right) \ln\left(\frac{N + E/\varepsilon}{2}\right) + \frac{N + E/\varepsilon}{2}$$

$$- \left(\frac{N - E/\varepsilon}{2}\right) \ln\left(\frac{N - E/\varepsilon}{2}\right) + \frac{N - E/\varepsilon}{2}$$

$$= N \log N - \left(\frac{N + E/\varepsilon}{2}\right) \ln\left(\frac{N + E/\varepsilon}{2}\right) - \left(\frac{N - E/\varepsilon}{2}\right) \ln\left(\frac{N - E/\varepsilon}{2}\right)$$

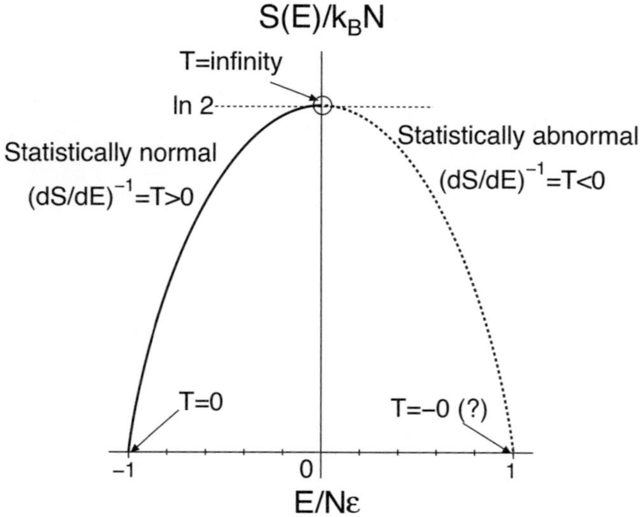


**Fig. 3.2** $S(E)/k_BN$ (3.6) of the two-level system as a function of $E/(N\varepsilon)$

$$= -N\left(\left(\frac{1+E/N\varepsilon}{2}\right)\ln\left(\frac{1+E/N\varepsilon}{2}\right)+\left(\frac{1-E/N\varepsilon}{2}\right)\ln\left(\frac{1-N/\varepsilon}{2}\right)\right). \tag{3.6}$$

In Fig. 3.2, $S(E)/N$ is plotted as a function $E/N$.

Here, it should be noted that the states with $E > 0$ ($N_- > N_+$) exist. In normal thermodynamic states, the number of excited states is smaller than that of ground states, i.e., $N_- > N_+$. Indeed, $N_- = N_+$ at $T = \infty$. Thus, the part of $E > 0$ corresponds to an abnormal state in thermodynamics. Indeed, using the relation:

$$\frac{1}{T} = \frac{\partial S}{\partial E} = \frac{k_B}{2\varepsilon}\ln\left(\frac{N - E/\varepsilon}{N + E/\varepsilon}\right), \tag{3.7}$$

the temperature in this region is negative. Such system is unstable when it contacts with other systems. That is, if the system absorbs heat in this region, its temperature decreases, causing the system to absorb even more heat. Therefore, the system cannot relax to an equilibrium state. This kind of state is called "inversion state" and has interesting properties. For example, it serves as the underlying mechanism for lasers. But we do not treat them in this book.

Hereafter, We focus on the region where $E < 0$. From (3.7),

$$\frac{2\varepsilon}{k_BT} = \ln\left(\frac{N - E/\varepsilon}{N + E/\varepsilon}\right) \to e^{2\beta\varepsilon} = \frac{N - E/\varepsilon}{N + E/\varepsilon}, \tag{3.8}$$

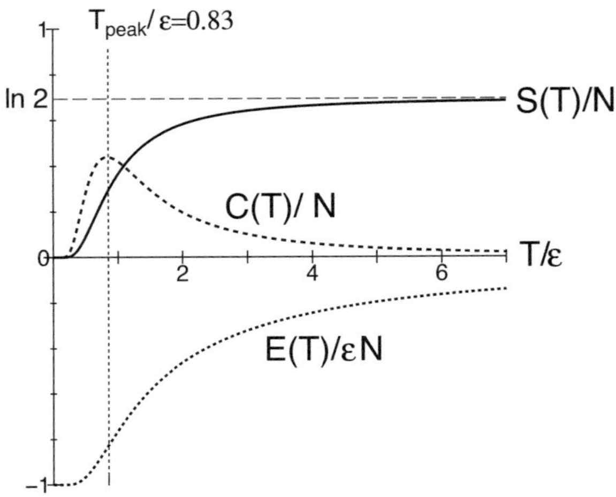


**Fig. 3.3** Temperature dependences of $S(E)$(solid curve), $E(T)$(dotted curve), and $C(T)$(short-dashed curve: Schottky type). The dashed line denotes ln 2. The vertical dotted line denotes the temperature of the peak of $C(T)$: $T_{\text{peak}}/\varepsilon \simeq 0.83$

which is rewritten as

$$E(T) = -\varepsilon N \tanh(\beta\varepsilon). \tag{3.9}$$

The specific heat is given by

$$C = \frac{\partial \langle E \rangle}{\partial T} = \frac{\partial \beta}{\partial T}\frac{\partial \langle E \rangle}{\partial \beta} = N\frac{1}{k_B T^2}\frac{\varepsilon^2}{\cosh^2(\beta\varepsilon)} = k_B N \frac{\varepsilon^2 \beta^2}{\cosh^2(\beta\varepsilon)}. \tag{3.10}$$

Temperature dependence of $S(E)$, $E(T)$, and $C(T)$ are plotted in Fig. 3.3. The specific heat exhibits a peak at $T/\varepsilon \simeq 0.83$, where the thermal energy becomes comparable to the energy difference $2\varepsilon$. Around this temperature, the system begins to be significantly excited, leading to a large change in entropy. This characteristic peak in the specific heat is known as the "Schottky anomaly".

### 3.1.2 Entropy in the Canonical Ensemble

In the canonical ensemble, the partition function is given by

$$Z = \sum_{\sigma_1=\pm 1} \cdots \sum_{\sigma_N=\pm 1} e^{\beta\varepsilon \sum_{i=1}^N \sigma_i}. \tag{3.11}$$

Here, the summations are taken over all possible configurations of the $N$ elements, i.e., $(++\cdots+) \sim (--\cdots-)$. Since the $N$ elements are independent, the partition function can be simplified as

$$Z = \left(e^{\beta\varepsilon} + e^{-\beta\varepsilon}\right)^N = (2\cosh(\beta\varepsilon))^N, \tag{3.12}$$

from which the free energy is derived as

$$F = -k_B T N \log(2\cosh(\beta\varepsilon)). \tag{3.13}$$

The internal energy is obtained as

$$\langle E \rangle = -\frac{\partial \log Z}{\partial \beta} = -N\varepsilon \tanh(\beta\varepsilon), \tag{3.14}$$

which agrees with the result obtained in (3.9).

The entropy is expressed as

$$S = \frac{E - F}{T} = \frac{-N\varepsilon \tanh(\beta\varepsilon) + k_B T N \log(2\cosh(\beta\varepsilon))}{T}. \tag{3.15}$$

These expressions of the entropy (3.6) and (3.15) must be the same.[1]

### 3.1.3 Paramagnetic State of Ising Model

The variable introduced in the previous section $\sigma_i = \pm 1$ is referred to as "Ising spin", which was introduced as a simplified model of ferromagnetic system [1]:

$$\mathcal{H} = \sum_{i,j} J_{ij}\sigma_i\sigma_j - H\sum_i^N \sigma_i, \quad M = \sum_i^N \sigma_i, \quad \sigma_i = \pm 1, \tag{3.18}$$

where $J_{ij}$ is exchange coupling constant between spins at the sites $i$ and $j$, $H$ represents the magnetic field and $N$ is the total number of spins. In magnetic systems, spins interact with an external magnetic field $H$ via the Zeeman interaction, which gives the second term of the Hamiltonian (3.18). In this context, $\sigma_i = \pm 1$ represents the upward and downward spin states at the $i$-th lattice point of a magnetic material, as illustrated in Fig. 3.4.

At high temperatures, the exchange energy becomes irrelevant, and the spins behave independently, which is called paramagnetic state. Representing such a case, setting $J_{ij} = 0$, the system is described by

---

[1] From the relation $E = -N\tanh\beta\varepsilon$, setting $x = \beta\varepsilon$,

$$\frac{E}{\varepsilon N} = -\tanh x, \quad x = \beta\varepsilon. \tag{3.16}$$

and then

$$\frac{1 \pm \frac{E}{\varepsilon N}}{2} = \frac{1 \mp \tanh x}{2} = \frac{e^{\pm x}}{e^x + e^{-x}}. \tag{3.17}$$

Substituting this relationship to (3.6), we recover the form (3.15).

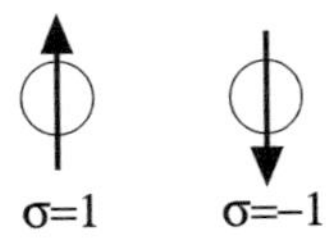


**Fig. 3.4** Ising spin $\sigma = \pm 1$

$$\mathcal{H} = -H \sum_{i=1}^{N} \sigma_i, \quad \sigma_i = \pm 1, \tag{3.19}$$

which has the same form as that of the two-level model setting $\varepsilon = H$.

As explained in the previous section, (3.12), the partition function of this case is given by

$$Z = (2\cosh(\beta H))^N. \tag{3.20}$$

The internal energy is given by

$$E = -\frac{\partial \ln Z}{\partial(\beta)} = -H \tanh(\beta H). \tag{3.21}$$

The magnetization $M$ is obtained by differentiating the logarithm of the partition function with respect to the scaled field, $\beta H$:

$$M = \frac{\mathrm{Tr} \sum_{i=1}^{N} \sigma_i e^{-\beta \mathcal{H}}}{\mathrm{Tr} e^{-\beta \mathcal{H}}} = \frac{\partial \ln Z}{\partial(\beta H)} = \tanh(\beta H). \tag{3.22}$$

The entropy $S$ is given by

$$S = -\frac{\partial F}{\partial T} = N k_{\mathrm{B}} \left( \ln(2\cosh(\beta H)) - \beta H \tanh(\beta H) \right), \tag{3.23}$$

where $F = -k_{\mathrm{B}} T \ln Z$ is the free energy.

### *3.1.4 Entropy of the Ising Model as a Function of the Magnetization at Infinite Temperature*

Distribution of the magnetization $M$ is related to the entropy for the specific value of the magnetization. For the Ising model (3.18), at infinite temperature ($\beta = 0$), the interactions $J_{ij}$ do not affect the state. In this case, the entropy of a state with magnetization $M = N, N-2, \cdots, -N$ is given by

$$S(M, \beta = 0) = k_{\mathrm{B}} \ln \frac{N!}{\left(\frac{N-M}{2}\right)! \left(\frac{N+M}{2}\right)!}. \tag{3.24}$$

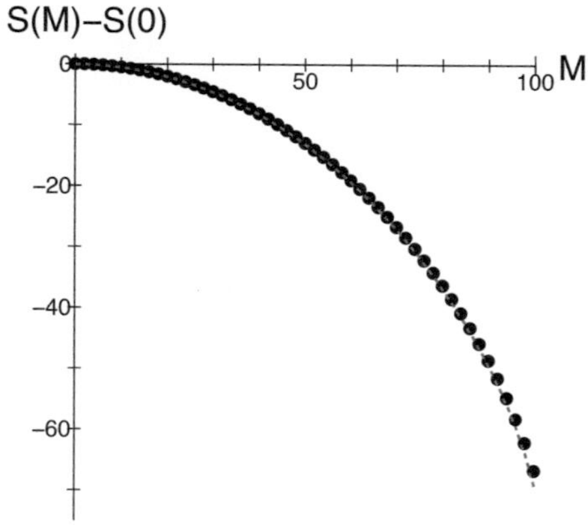


**Fig. 3.5** $S(M, \beta = 0) - S(0, \beta = 0)$ of the Ising model. Dots denote exact values for $N = 100$ and dotted line shows the value obtained using Stirling's relation (3.27)

Using the following relation:

$$\ln \frac{N!}{\left(\frac{N-M}{2}\right)!\left(\frac{N+M}{2}\right)!} = \ln \frac{N(N-1)\cdots((N+M)/2+1)}{\frac{N-M}{2}\left(\frac{N-M}{2}-1\right)\cdots 1} = \sum_{k=1}^{(N-M)/2} \ln\left(\frac{k+\frac{N+M}{2}}{k}\right), \tag{3.25}$$

the entropy is given by

$$S(M, \beta = 0) = k_{\rm B} \sum_{k=1}^{(N-M)/2} \ln\left(1 + \frac{N+M}{2k}\right). \tag{3.26}$$

This expression can be explicitly obtained numerically for finite $N$.

As discussed for (3.6), the entropy for large $N$ is estimated using Stirling's relation, $N! \simeq N^N e^{-N}$, yielding

$$S(M, \beta = 0) \simeq -k_{\rm B}\left[\frac{N-M}{2}\ln\frac{N-M}{2N} + \frac{N+M}{2}\ln\frac{N+M}{2N}\right]. \tag{3.27}$$

We compare numerically evaluated results of (3.26) and (3.27) for $N = 100$ in Fig. 3.5, where we find good agreement [2].

#### 3.1.4.1 Entropy of the Paramagnetic Three-Dimensional Vector Model (Heisenberg Spin)

We now derive $S(M, \beta = 0)$ for the tree-dimensional vector model, i.e., the Heisenberg spin system, where $\boldsymbol{S}_i = (S_i^x, S_i^y, S_i^z)$, $|\boldsymbol{S}_i| = 1$. The Hamiltonian of a magnetic system of Heisenberg spins is described as

$$\mathcal{H} = \sum_{i,j} J_{ij} \boldsymbol{S}_i \cdot \boldsymbol{S}_j - D \sum_i^N (S_i^z)^2 - H \sum_i^N S_i^z, \quad M = \sum_i^N S_i^z. \tag{3.28}$$

The partition function is given by

$$Z(M, \beta) = \int_0^\pi \int_0^{2\pi} \sin\theta_i d\theta_i d\phi_i e^{-\beta\mathcal{H}} \delta(M = \sum_i^N S_i^z). \tag{3.29}$$

Using the relation:

$$\ln Z = \frac{S}{k_\mathrm{B}} - \frac{E}{k_\mathrm{B} T} \to \frac{S}{k_\mathrm{B}}, \quad (T \to \infty), \tag{3.30}$$

the entropy is given by

$$S(M, \beta = 0) = k_\mathrm{B} \ln \int_0^\pi \int_0^{2\pi} \sin\theta_i d\theta_i d\phi_i \delta(M = \sum_i^N S_i^z). \tag{3.31}$$

Using the integral representation of tha delta-function:

$$\delta(x) = \frac{1}{2\pi} \int_{-\infty}^{\infty} e^{i\alpha x} d\alpha, \tag{3.32}$$

the partition function becomes

$$Z(M, \beta = 0) = (2\pi)^N \int_0^\pi \sin\theta_i d\theta_i \frac{1}{2\pi} \int_{-\infty}^{\infty} e^{i\alpha(M - \sum_i^N S_i^z)} d\alpha. \tag{3.33}$$

Defining $m = M/N$, this can be rewritten as

$$\begin{aligned} Z(M, \beta = 0) &= (2\pi)^N \int_{-1}^{1} dx_i \frac{1}{2\pi} \int_{-\infty}^{\infty} e^{i\alpha \sum_i (m - \sum_i^N x_i)} d\alpha \\ &= (2\pi)^N \frac{1}{2\pi} \int_{-\infty}^{\infty} d\alpha \left[ \frac{2\sin\alpha}{\alpha} e^{i\alpha m} \right]^N . \end{aligned} \tag{3.34}$$

Using the saddle-point method to evaluate the integral, the condition for the saddle point is

$$\cot\alpha - \frac{1}{\alpha} + im = 0. \tag{3.35}$$

Substituting $\alpha = ix$, the above relation becomes

$$\coth x - \frac{1}{x} - m = 0, \rightarrow m = L(x), \tag{3.36}$$

where $L(x)$ is the Langevin function. This relation describes magnetization as a function field $H$ for a single classical spin, setting $x = \beta H$. For a given value of $m$, we have a real solution $x$. Using this solution, the entropy is given by

$$S(M, \beta = 0) = N\left(\ln 2\pi + \ln\left[\frac{2\sinh x}{x} e^{-xm}\right]\right). \tag{3.37}$$

The $m$-dependence of the above formula Eq.(3.37) is found to agree with the numerical estimation of the entropy at the infinite temperature [2] obtained by the Wang-Landau method [3].

The solution of Eq.(3.37) gives the value of the entropy for $M = 0$:

$$S(M = 0, \beta = 0) = N\ln 4\pi. \tag{3.38}$$

It should be noted that this value is the same as the entropy for the case without constraint on $M$ at the infinite temperature $S(\beta = 0)$. For finite systems, $S(M = 0, \beta = 0)$ must be smaller than $S(\beta = 0)$, but they become the same at the thermodynamic limit, where the canonical ensemble and the grand-canonical ensemble yield identical thermodynamic properties.

### 3.1.5 Three-Level System

The specific heat of a two-level system exhibits a peak at a temperature approximately half the energy gap (Schottky anomaly detected in Fig. 3.3). Here, we extend this analysis to a three-level system ($E_1 < E_2 < E_3$), focusing specifically on the case where

$$\Delta_1 = E_2 - E_1 \ll \Delta_2 = E_3 - E_1, \tag{3.39}$$

as depicted in Fig. 3.6.

The partition function of this system is given by

$$Z = e^{-\beta E_1} + e^{-\beta E_2} + e^{-\beta E_1} = e^{-\beta E_1}\left(e^{-\beta\Delta_1} + e^{-\beta\Delta_2} + 1\right). \tag{3.40}$$

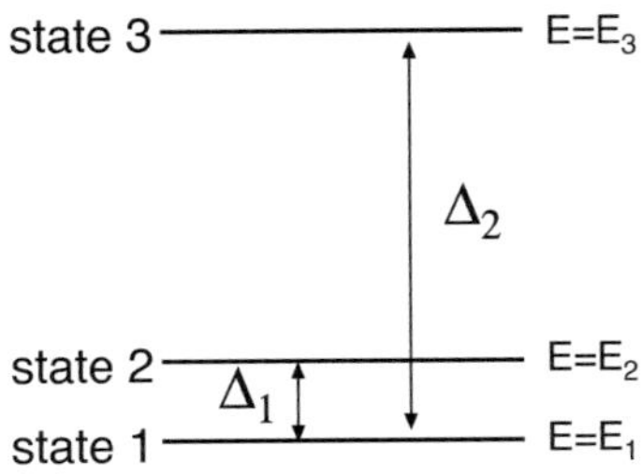


**Fig. 3.6** Three-level system with $\Delta_1 \ll \Delta_2$

The internal energy and the specific heat are

$$E = -\frac{\partial \ln Z}{\partial \beta} = E_1 - \frac{\Delta_1 e^{-\beta\Delta_1} + \Delta_2 e^{-\beta\Delta_2}}{1 + e^{-\beta\Delta_1} + e^{-\beta\Delta_2}}, \tag{3.41}$$

and

$$C = \frac{\partial E}{\partial T} = -\frac{1}{k_B T^2}\frac{\partial E}{\partial \beta}$$

$$= \frac{1}{k_B T^2}\left(\frac{\Delta_1^2 e^{-\beta\Delta_1} + \Delta_2^2 e^{-\beta\Delta_2}}{1 + e^{-\beta\Delta_1} + e^{-\beta\Delta_2}} - \left(\frac{\Delta_1 e^{-\beta\Delta_1} + \Delta_2 e^{-\beta\Delta_2}}{1 + e^{-\beta\Delta_1} + e^{-\beta\Delta_2}}\right)^2\right), \tag{3.42}$$

respectively.

The temperature dependences of the entropy $S$, internal energy $E$ and the specific heat $C$ are shown in Fig. 3.7. At the temperature of the order $\Delta_1$, the third level of $E_3$ is much higher than the thermal energy ($k_B T$). Thus, the model can be regarded as a two-level model, and thus the specific heat exhibits a peak of the two-level system around $T = 0.83 \times (\Delta_1/2)$. At temperatures on the order $\Delta_2$, the lowest two states are almost degenerate, and the model is essentially a two-level model. Despite the twofold degeneracy of the lowest energy level, the system still shows a peak in the specific heat around $k_B T \sim \Delta_2$.

In this manner, depending on the structure of the energy levels, the specific heat of the system can exhibit multiple peaks at characteristic temperatures.

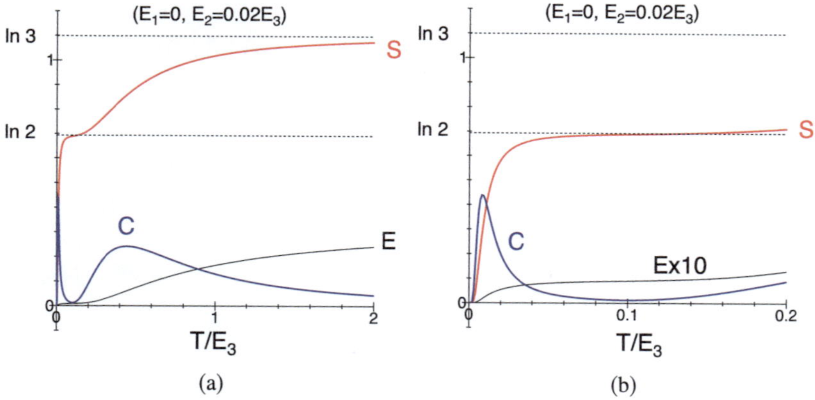


**Fig. 3.7** **a** Temperature dependences of the entropy $S$, internal energy $E$ and the specific heat $C$ of the three-level system. **b** Magnification at low temperatures

## 3.2 Adiabatic Demagnetization Cooling (Magnetic Refrigeration)

In thermodynamics, "adiabatic process" refers to a quasistatic process in which no heat exchange occurs, i.e., $\Delta Q = 0$. In such a process, the entropy of the system does not change, because

$$\Delta S = \frac{\Delta Q}{T} = 0. \tag{3.43}$$

Although there is no heat exchange, the temperature of the system may change during the adiabatic process. For example, temperature of a magnetic material changes in response to the application or removal of an external magnetic field under adiabatic conditions. This type of operation is called magnetocaloric effect (MCE). Making use of this fact, it is possible to control the temperature of the system by varying external magnetic field while keeping the entropy constant. This cooling technique is known as adiabatic demagnetization cooling, or magnetic refrigeration.

The entropy of paramagnets which composed of $N$ non-interacting independent spins ($S = 1/2$) is given by (3.23). Figure 3.8 shows, temperature dependence of the entropy for several magnetic field strengths.

Entropy is a monotonic function of the temperature for a given magnetic field ($H$). At high field, the spins are more aligned, reducing the entropy at the same temperature. In Fig. 3.8, the system is equilibrated at the temperature $T_0$ in the field $H = 2$. When we change the field adiabatically from $H = 2$ to $H = 0.5$, the temperature changes from $T_0$ to $T_1$. In this process, the system loses some energy, which is extracted by the change of the external field. Namely, the apparatus outside which changes the applied field gives negative work on the system. In this way, the temperature of the system decreases, and if some material contacts with the present

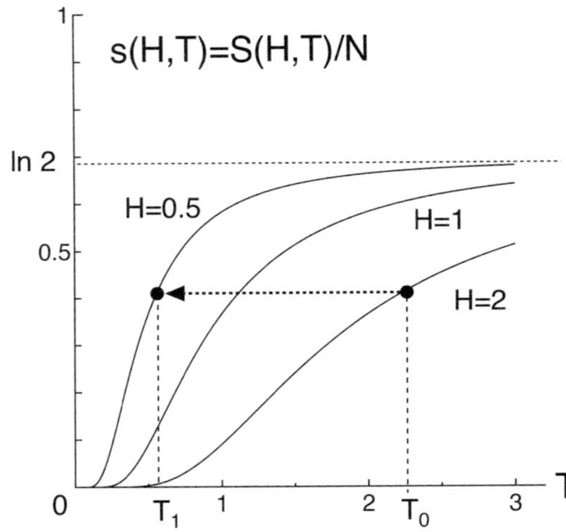


**Fig. 3.8** Temperature dependence of the entropy of the paramagnet system described by (3.19) under magnetic field $H = 0.5, 1$ and 2. The arrow indicates the temperature change when the field decreases from $H = 2$ to $H = 0.5$

paramagnetic system, the material is cooled down. This is the mechanism of adiabatic demagnetization cooling.

To make this method work properly, the magnetic system must remain paramagnetic even at low temperatures. If the magnetic system becomes frozen in an ordered state, the adiabatic change following the magnetic field becomes difficult, and the adiabatic demagnetization becomes ineffective. At lower temperatures, many magnets that are paramagnetic at room temperature are ordered into ferromagnetic, antiferromagnetic, or randomly frozen states. Therefore, we need to find materials that retain their paramagnetic properties at the target low temperatures.

Potassium chrome alum and cerium magnesium nitrate (CMN) are well-known examples of materials that remain paramagnetic at millikelvin (mK) temperatures. The former is effective around 20 mK, while the latter works at temperatures as low as 3 mK. For temperatures below these ranges, nuclear spins of paramagnetic metals such as Cu and Ag are utilized. The magnetic moments of nuclear spins are much smaller than those of electron spins and are almost free from interactions, making them suitable for adiabatic demagnetization at extremely low temperatures ($10^{-8}$ K for Cu, $10^{-10}$ K for Ag). Samples in thermal contact with these metals can be cooled to approximately $10^{-5}$ K.

In Fig. 3.9, an experimental unit for the nuclear demagnetization is shown. The lower half of the apparatus contains Cu, which provides nuclear spins. This section is enclosed in a cryostat and surrounded by superconducting magnets.

Another commonly used refrigeration technique is dilution refrigeration, where $^3$He atoms diffuse into $^4$He liquid, absorbing heat through the heat of vaporization. However, $^3$He has become very expensive in recent years, making this method less commonly used.

**Fig. 3.9** Nuclear-adiabatic demagnetization unit (Ishimoto Labo. ISSP, 2002, Adapted with permission from [4]©Seiji Miyashita, 2004)

## 3.3 Entropy of Systems of Harmonic Oscillators

A particle confined to a harmonic potential can be a model for small-amplitude oscillations at the bottom of trapping potentials. Here, we study the simplest one-dimensional, single-component model. (The extension to systems with more degrees of freedom is straightforward.)

$$\mathcal{H} = \frac{1}{2m}p^2 + \frac{k}{2}x^2 = \frac{1}{2m}p^2 + \frac{m\omega^2}{2}x^2, \quad \omega = \sqrt{\frac{k}{m}}, \tag{3.44}$$

where $m$ is the mass of the particle, $k$ is the stiffness constant, and $\omega$ is the angular frequency of the oscillation. The partition function of the model is given by

$$Z = \int_{-\infty}^{\infty} e^{-\frac{\beta}{2m}p^2} dp \int_{-\infty}^{\infty} e^{-\beta k x^2} dx = \sqrt{\frac{2\pi m}{\beta}}\sqrt{\frac{2\pi}{k\beta}} = 2\pi\sqrt{\frac{m}{k}}\frac{1}{\beta}, \tag{3.45}$$

and the energy and the free energy are obtained as

$$E = -\frac{\partial \ln Z}{\partial \beta} = \frac{1}{\beta} = k_{\mathrm{B}}T, \tag{3.46}$$

and

$$F = -k_{\mathrm{B}}T \ln Z = -k_{\mathrm{B}}T \left( \ln(2\pi) + \frac{1}{2}\ln\frac{m}{k} - \ln\beta \right). \tag{3.47}$$

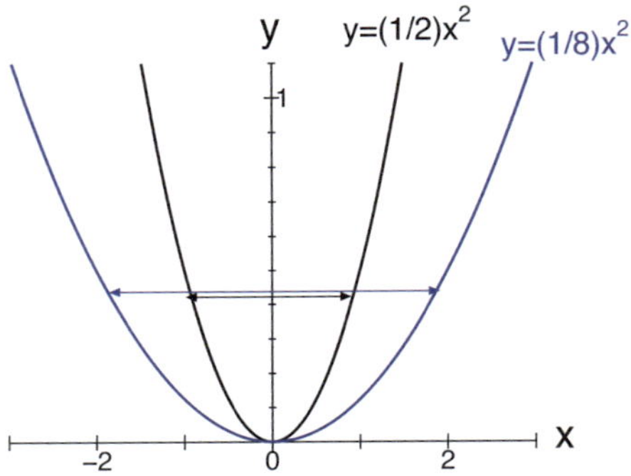


**Fig. 3.10** $x$ dependence of $y = (k/2)x^2$ for $k = 1$ and $1/4$. The arrows denote ranges of $x$ for fixed energy. The length of arrow of $k = 1/4$ is twice long

Using these, the entropy is given by

$$S = \frac{E - F}{T} = k_B \left( 1 + \ln(2\pi) + \frac{1}{2} \ln \frac{m}{k} - \ln \beta \right). \tag{3.48}$$

If we compare the entropies of systems with different stiffness constants $k$ and $k'$, the difference is

$$S(k, \beta) - S(k', \beta) = \frac{k_B}{2} \ln \frac{k'}{k}, \tag{3.49}$$

which indicates that the system with a smaller stiffness constant has a larger entropy. This dependence can be understood by considering the difference in the range of $x$ for a given energy, as depicted in Fig. 3.10. For a fixed energy, the range of $x$ is proportional to $1/\sqrt{k}$.

Thus, the entropy is proportional to $-(1/2) \ln k$. In other words, a particle in a loose binding potential has large entropy. This simple relation plays an important role in the entropic selection of energetically degenerate states, as we will explore in Part II.

### 3.3.1 Thermal Expansion

At finite temperatures, particles at lattice points vibrate. Here, we examine whether this vibration causes lattice expansion or not. Let the potential energy as a function of the distance between two particles be $V(|\boldsymbol{x}_1 - \boldsymbol{x}_2|)$. A typical form of the function is the Lennard-Jones potential:

$$V(r) = 4\epsilon \left[ \left( \frac{r}{\sigma} \right)^{12} - \left( \frac{r}{\sigma} \right)^{6} \right], \tag{3.50}$$

where $\epsilon$ is the strength of the force, and $\sigma$ represents a typical range of the interaction.

The distance $r_0$ at which the potential reaches its minimum energy is calculated from the derivative of $V(r)$:

$$\frac{dV(r)}{dr} = -24\epsilon\left[2\left(\frac{r}{\sigma}\right)^{13} - \left(\frac{r}{\sigma}\right)^{7}\right] = 0 \Longrightarrow r_0 = 2^{1/6}\sigma \simeq 1.096\sigma. \quad (3.51)$$

Around the minimum, the potential is approximated as

$$V(r) \simeq V(r_0) + \frac{1}{2}\left(\frac{d^2V(r)}{dr^2}\right)_{r=r_0}(r-r_0)^2 + \frac{1}{3!}\left(\frac{d^3V(r)}{dr^3}\right)_{r=r_0}(r-r_0)^3$$
$$+\frac{1}{4!}\left(\frac{d^4V(r)}{dr^4}\right)_{r=r_0}(r-r_0)^4 + \cdots. \quad (3.52)$$

Denoting $x = r - r_0$, and the potential is expressed as

$$V(r) \simeq V(r_0) + cx^2 - gx^3 - fx^4 + \cdots, \quad (3.53)$$

where $c$, $g$, and $f$ are constants related to the second, third, and fourth derivatives of $V(r)$ evaluated at $r_0$.

The partition function is given by

$$Z = \int_{-\infty}^{\infty} e^{-\beta(cx^2-gx^3-fx^4+\cdots)}dx \simeq \int_{-\infty}^{\infty} e^{-\beta cx^2}(1+\beta gx^3+\beta fx^4+\cdots)dx. \quad (3.54)$$

The average deviation of $x$ is

$$\langle x\rangle = \frac{\int_{-\infty}^{\infty} xe^{-\beta cx^2}(1+\beta gx^3+\beta fx^4+\cdots)dx}{\int_{-\infty}^{\infty} e^{-\beta cx^2}(1+\beta gx^3+\beta fx^4+\cdots)dx}$$
$$\simeq \frac{\beta g\int_{-\infty}^{\infty} x^4e^{-\beta cx^2}dx}{\int_{-\infty}^{\infty} e^{-\beta cx^2}dx} = \frac{3g}{4c^2}k_BT. \quad (3.55)$$

Up to second order (considering terms only up to $cx^2$), $\langle x\rangle = 0$, and the lattice spacing does not change. However, when the asymmetry around $r_0$ is taken into account, the lattice spacing increases proportionally to the temperature.

The linear thermal expansion coefficient is defined by the following form and is obtained as

$$\frac{1}{L}\left(\frac{\partial L}{\partial T}\right)_P = \frac{\langle x\rangle/T}{r_0} = \frac{3g}{4c^2r_0}k_B, \quad (3.56)$$

and the volume thermal expansion coefficient is

$$\frac{1}{V}\left(\frac{\partial V}{\partial T}\right)_P = 3\frac{\langle x\rangle/T}{r_0} = \frac{9g}{4c^2 r_0}k_B. \tag{3.57}$$

This expansion arises due to the asymmetry of the potential, where the particle is more loosely confined on the right side. In this sense, it is a kind of entropy effect.

## 3.4 Thermodynamic Force Due to Entropy

Rubber exhibits a restoring force when it is stretched, similar to a spring; however, the mechanism behind this force is fundamentally different. The elasticity of rubber originates from entropy, in contrast to the elasticity of a spring, which arises from forces in the atomic structure of the material [5].

The thermodynamic force (strain) $f$ is defined as the derivative of the free energy $F = U - TS$ with respect to the length $L$:

$$f = -\left(\frac{\partial F}{\partial L}\right)_T \tag{3.58}$$

For the force in the spring, the force comes from a mechanical origin and the effect of entropy is usually small, and thus $F \simeq U$, resulting in

$$f = -\left(\frac{\partial F}{\partial L}\right)_T = -\left(\frac{\partial U}{\partial L}\right)_T. \tag{3.59}$$

On the other hand, for the force in the rubber, the energy part is nearly zero, and thus $F \simeq -TS$, resulting in

$$f = \left(\frac{\partial F}{\partial L}\right)_T = -T\left(\frac{\partial S}{\partial L}\right)_T. \tag{3.60}$$

Let us examine the force due to entropy. Most simply, the elasticity of rubber can be modeled as a chain composed of small units that can freely fold at their joints without energy loss. Let us consider a system consisting of $N$ small bonds, each of length $a$, connected by free joints. The direction of each bond can point either to the left or to the right, which we denote as $\sigma_i = \pm 1$, respectively.

The total length $L$ between the two ends of the chain is given by

$$L = a\sum_{i=1}^{N}\sigma_i. \tag{3.61}$$

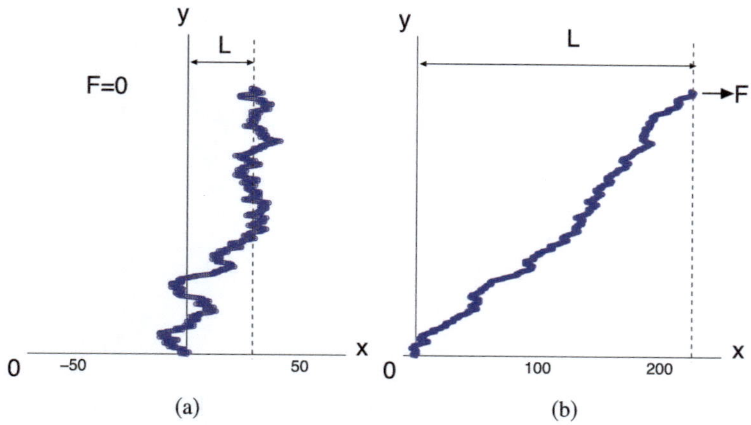


**Fig. 3.11** **a** A configuration of a one-dimensional polymer of free folding ($N = 1000$). **b** Configuration of a one-dimensional polymer under force with $p(1) = 0.6$ and $p(-1) = 0.4$ ($N = 1000$). To better observe the folding, the bonds are shifted in the $y$-direction

An example of a configuration is depicted in Fig. 3.11a. This is equivalent to a random walk where the average length is 0 and the variance is proportional to $N$. This property can be verified through the following calculation. Let the probability for $\sigma_i = \pm 1$ be $p_i(\pm 1)$, respectively. The average length is given by

$$\langle L \rangle = a \sum_{\sigma_1=\pm 1} \cdots \sum_{\sigma_N=\pm 1} p_1(\sigma_1) \cdots p_N(\sigma_N) \, (\sigma_1 + \cdots \sigma_N) \, . \tag{3.62}$$

Because all of $\sigma_i$ change independently, the sum is simplified as

$$\langle L \rangle = a \sum_{i=1}^{N} \sum_{\sigma_i=\pm 1} p_i(\sigma_i)\sigma_i = aN\langle \sigma_i \rangle, \tag{3.63}$$

Then, $\langle L \rangle = 0$, and the variance is given by

$$\langle L^2 \rangle - \langle L \rangle^2 = a^2 \sum_{\sigma_1=\pm 1} \cdots \sum_{\sigma_N=\pm 1} p_1(\sigma_1) \cdots p_N(\sigma_N) \, (\sigma_1 + \cdots \sigma_N)^2 - \langle L \rangle^2. \tag{3.64}$$

Expanding $(\sigma_1 + \cdots \sigma_N)^2$, we have

$$(\sigma_1 + \cdots \sigma_N)^2 = \sigma_1^2 + \cdots \sigma_N^2 + 2\sigma_1\sigma_2 + \cdots + 2\sigma_{N-1}\sigma_N \, . \tag{3.65}$$

Using the relations $\sigma_i^2 = 1$,

$$\langle L^2 \rangle = a^2 \left(N + 2\langle\sigma_1\rangle\langle\sigma_2\rangle + \cdots + 2\langle\sigma_{N-1}\rangle\langle\sigma_N\rangle\right) = a^2 \left(N - \sum_{i \neq j} \langle\sigma_i\rangle\langle\sigma_j\rangle\right)$$

$$= a^2 \left(N - \langle L\rangle^2 - \sum_{i=1}^{N} \langle\sigma_i\rangle^2\right), \tag{3.66}$$

leading to

$$\langle L^2 \rangle - \langle L \rangle^2 = a^2 \left(N - \sum_{i=1}^{N} \langle\sigma_i\rangle^2\right). \tag{3.67}$$

For the case of $p_i(1) = p_i(-1) = 1/2$,

$$\langle L \rangle = 0, \quad \langle L^2 \rangle - \langle L \rangle^2 = a^2 N. \tag{3.68}$$

When a force $f$ is applied to elongate the polymer, the probabilities $p(1)$ and $p(-1)$ become different, leading to an increase in length. In this case, the average and the variance are given by (3.63) and (3.67), respectively. An example for $p(1) = 0.6$ and $p(-1) = 0.4$ is depicted in Fig. 3.11b.

The force is balanced with the elasticity of the rubber, but there is no restoring mechanical force in the present model. The configuration shown in Fig. 3.11b is highly improbable in the case of zero force, even though the energy remains the same ($E = 0$). The balanced situation is given by minimizing the free energy.

The distribution of the polymer length $L$ follows a binary distribution, and the probability of the length $L$ is given by

$$P(L) = \frac{W(L)}{2^N} = \frac{1}{2^N} {}_N C_{\frac{N-L/a}{2}} = \frac{1}{2^N} \frac{N!}{\left(\frac{N-L/a}{2}\right)! \left(\frac{N+L/a}{2}\right)!}. \tag{3.69}$$

The entropy associated with this distribution is

$$S(L) = k_B \ln W(L), \tag{3.70}$$

which, for large $N$, can be approximated as

$$\simeq -k_B N \left( \left(\frac{1 - L/aN}{2}\right) \ln\left(\frac{1 - L/aN}{2}\right) + \left(\frac{1 + L/aN}{2}\right) \ln\left(\frac{1 + L/aN}{2}\right)\right). \tag{3.71}$$

The free energy is given by

$$F = -TS, \tag{3.72}$$

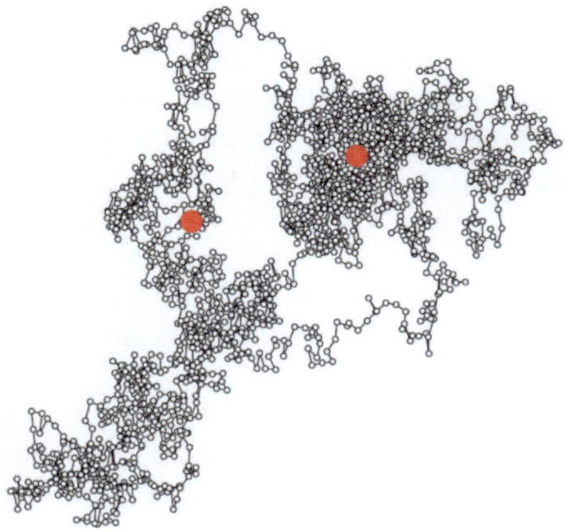

**Fig. 3.12** A configuration of a three-dimensional free polymer ($N = 4000$) projected onto the XY plane. The red marks indicate the starting and ending points

because $E = 0$, and the force is given by (3.60). Using the entropy (3.71), the force is given by

$$f = \frac{k_B T}{a} \ln \sqrt{\frac{aN + L}{aN - L}}. \tag{3.73}$$

From this relation, the expected length $\langle L \rangle$ under an applied force $f$ is estimated as

$$\langle L \rangle = aN \tanh(\beta f a). \tag{3.74}$$

For small values of $f$, the length $\langle L \rangle$ increases proportionally to $f$, resembling the behavior of a typical spring. However, as we see here, this is not a mechanical restoring force but an entropy-induced force. Consequently, the force is proportional to the temperature $T$ and vanishes as $T \to 0$.

It should be noted that the entropy decreases when the length of the rubber is elongated, which is opposite to the behavior of an ideal gas, where the entropy increases as the volume expands. Thus, if the rubber is suddenly stretched adiabatically, the temperature rises in contrast to an expanding gas, where the temperature decreases during expansion. The decrease in entropy here leads to a heat release, given by $\Delta Q = T \Delta S$. This explains why the rubber feels warm when stretched quickly. You can easily verify this phenomenon by conducting a simple experiment with a rubber band.

The present model can be straightforwardly extended to three dimensions. In this case, the total length vector $L$ is given by

$$\boldsymbol{L} = a \sum_i \left( \boldsymbol{e}_x \sigma_i^x + \boldsymbol{e}_y \sigma_i^y + \boldsymbol{e}_z \sigma_i^z \right), \quad \sigma_i^\alpha = \pm 1, \quad \alpha = x, y, z, \tag{3.75}$$

which represents the simplest model for a three-dimensional polymer. An example of a configuration is depicted in Fig. 3.12.

In the above argument, we assumed that there is no restriction on all the combinations $\{\sigma_i^\alpha = \pm 1\}$. However, in real polymers, the polymer cannot occupy the same position, a constraint known as the condition of self-avoiding walk [6]. Taking this constraint into account, the linear dimension of a polymer $L$ consisting of $N$ elements increases as

$$\langle L^2 \rangle \propto N^{2\nu}, \tag{3.76}$$

where the exponent $\nu = 3/5(> 1/2)$ is known as Flory's constant. This value is derived using a type of mean-field theory [7], as will be explained in the next subsection.

More precise values of the exponent $\nu$ have been obtained to be $0.588\cdots$ in three dimensions. In two dimensions, $\nu = 3/4$ while for $d > 4$, $\nu$ returns to the free case value, i.e., $\nu = 1/2$. At the critical dimension $d = 4$, a logarithmic correction appears on $\nu = 1/2$ [6].

### 3.4.1 Flory's Constant

The free energy of a polymer consisting of $N$ elements whose end-to-end distance is $L$, can be estimated as follows. Here, the length of each element is denoted by $a$. In the same way as in (3.6), the entropy is given by

$$\begin{aligned} S(L) &= k_B N \ln W(L) \\ &\simeq -N\left(\left(\frac{1+\frac{L}{aN}}{2}\right)\ln\left(\frac{1+\frac{L}{aN}}{2}\right)+\left(\frac{1-\frac{L}{aN}}{2}\right)\ln\left(\frac{1-\frac{L}{aN}}{2}\right)\right) \end{aligned} \tag{3.77}$$

For the case $x = \frac{L}{aN} \ll 1$,

$$\left(\left(\frac{1+\frac{L}{aN}}{2}\right)\ln\left(\frac{1+\frac{L}{aN}}{2}\right)+\left(\frac{1-\frac{L}{aN}}{2}\right)\ln\left(\frac{1-\frac{L}{aN}}{2}\right)\right) \simeq -\ln 2 + \left(\frac{L}{aN}\right)^2. \tag{3.78}$$

Next, we approximate the interaction energy among polymer segments as

$$F_{\text{interaction}} = \frac{v}{2}\int_{\text{sphere of radius}} (\text{density of segments})^2 d\boldsymbol{r} \propto vL^3\left(\frac{N}{L^3}\right)^2, \tag{3.79}$$

where $v$ represents the strength of the interaction. The total free energy is then given by

$$F = vL^3\left(\frac{N}{L^3}\right)^2 - k_B T N\left(\frac{L}{aN}\right)^2 + \text{constant}. \tag{3.80}$$

To find the value of $L$, we minimize the free energy:

$$\frac{dF}{dL} = 0 \to -\frac{v}{a^2}3\frac{N^2}{L^4} - \frac{k_B T}{a^2}2N\frac{L}{N^2} = 0 \to L^5 \propto N^3, \tag{3.81}$$

which gives the exponent 3/5:

$$L \propto N^{3/5}. \tag{3.82}$$

### 3.4.2 Canonical Ensemble Method with an Effective Force

The relation in Eq. (3.74) can be derived using the canonical ensemble method with the Hamiltonian:

$$\mathcal{H} = -fa\sum_{i=1}^{N} \sigma_i. \tag{3.83}$$

Here, $f$ represents the effective force applied to the system, and $a$ is the length of each element. When the system is in thermal equilibrium at a temperature $T$, the partition function is given by

$$Z = \left(\sum_{\sigma_i=\pm 1} e^{-\beta f a \sigma_i}\right)^N = (2\cosh(\beta f a))^N, \tag{3.84}$$

where $\beta = 1/k_B T$ is the inverse temperature.

The expectation value of the length $L$ is then obtained as

$$\langle L \rangle = \frac{\partial \ln Z}{\partial(\beta f)} = aN\tanh(\beta f a), \tag{3.85}$$

which agrees with (3.74).

In the case of three dimensions, each polymer element is represented by a three-dimensional vector $a\boldsymbol{n}$:

$$\boldsymbol{n} = (n_x, n_y, n_z), \quad |\boldsymbol{n}| = 1. \tag{3.86}$$

Assuming the force is applied along the $z$ direction, the Hamiltonian is expressed as

$$\mathcal{H} = -fa\sum_{i=1}^{N} n_i^z. \tag{3.87}$$

When the system is in equilibrium at temperature $T$, the partition function is given by

$$Z = \left(\int_0^{2\pi} d\phi_i \int_0^{\pi} \sin\theta_i d\theta_i\right) = \left(\frac{2\pi(e^{\beta f a} - e^{-\beta f a})}{\beta f a}\right)^N. \tag{3.88}$$

The expectation value of the projection of the polymer length along the $z$-direction, $L_z$, is calculated as

$$\langle L_z \rangle = \frac{\partial \ln Z}{\partial(\beta f)} = aN\left(\coth(\beta f a) - \frac{1}{\beta f a}\right) = aNL(\beta f a), \tag{3.89}$$

where $L(x)$ is the Langevin function, defined as

$$L(x) = \coth(x) - \frac{1}{x}. \tag{3.90}$$

The force required to achieve a given length $\langle L_z \rangle = L_0$ is given using the inverse function of the Langevin function:

$$f = \frac{k_B T}{a} L^{-1}\left(\frac{L_0}{aN}\right). \tag{3.91}$$

#### 3.4.2.1 Entropic Force and the Mechanical Force

As previously mentioned, there are two types of elastic forces (or stress), i.e., the mechanical stress $\sigma_E$ (3.59) and the entropic stress $\sigma_S$ (3.60). The entropic stress $\sigma_S$ is proportional to the temperature ($\sigma_S \propto T\sigma_{S0}$), while the mechanical stress exhibits little temperature dependence. Thus, the total strass $\sigma$ is decomposed as

$$\sigma = T\sigma_{S0} + \sigma_E. \tag{3.92}$$

In natural rubber, the entropic stress $\sigma_S$ dominates ($\sigma_S \gg \sigma_E$) and $\sigma_E$ is typically small and positive. However, in some gel, which is a material swelling with solvents, the mechanical stress estimated by $\sigma_E = \sigma - T\sigma_{S0}$ appears to be negative. This phenomenon is attributed to the interaction between the polymer network and the solvent molecules [8].

## 3.5 Entropy Change Due to Mixing of Gases (Entropy of Mixing)

Entropy increase is a fundamental characteristic of irreversible processes. Thus far, we have primarily examined entropy increase associated with heat flow. However, entropy also increases during any irreversible processes, regardless of heat transfer. One typical irreversible process without heat flow is the mixing of substances via diffusion (the free expansion studied in Sect. 2.3.6). Here, we study the entropy change during the mixing and related phenomena.

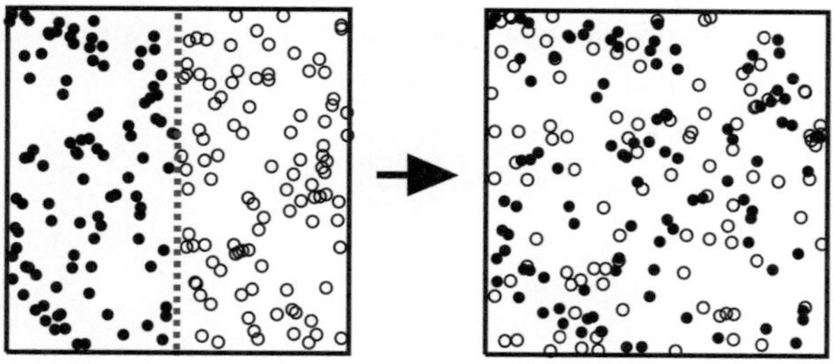

**Fig. 3.13** Mixing of two types of gases

For simplicity, we analyze the phenomenon using ideal gases. The diffusion process cannot be described using macroscopic quantities directly. Therefore, the entropy change is calculated as the difference between the entropies of the initial and final states, without considering the details of the intermediate process.[2]

Consider a scenario where two types of ideal gases, A and B, are initially separated by a wall, as shown in Fig. 3.13 (left). The quantities of gases are $N_A$ and $N_B$, respectively. Assume that temperature $T$ and pressure $P$ are uniform across the system. Under these conditions, the volumes of the gases are $V_A$ and $V_B$, satisfying the ratio $V_A : V_B = N_A : N_B$.

Then, we remove the wall (Fig. 3.13(right)), where the gases are mixed in the final state. Because both A and B are ideal gases, there is no interaction among the particles and the internal energy does not change. Thus, this process can be regarded as expansion of the gases (A and B) to the vacuum region, and the change of the entropy due to the mixing is given by sum of entropy changes of free expansion (diffusion process) of the gases. The change of entropy of the gas $A$ is given by

$$\Delta S_A = \int_{V_A}^{V_A+V_B} N_A \left(\frac{\partial S}{\partial V}\right)_T dV = \int_{V_A}^{V_A+V_B} N_A \left(\frac{\partial P}{\partial T}\right)_V dV$$

$$= \int_{V_A}^{V_A+V_B} N_A \frac{k_B}{V} dV = N_A k_B \ln \frac{V_A + V_B}{V_A}, \tag{3.93}$$

and similarly for B:

$$\Delta S_B = N_B k_B \ln \frac{V_A + V_B}{V_B}. \tag{3.94}$$

If we use the number of moles and the gas constant, we have

$$N_A k_B = n_A R, \quad N_B k_B = n_B R. \tag{3.95}$$

[2] A more comprehensive description of entropy changes requires the concept of entropy in non-equilibrium systems, which will be discussed in Chap. 7.

In this section, we will consider quantities per mole from now on.

The total entropy change due to the mixing is given by

$$\Delta S_{\text{mix}} = \Delta S_{\text{A}} + \Delta S_{\text{B}} = n_{\text{A}} R \ln \frac{V_{\text{A}} + V_{\text{B}}}{V_{\text{A}}} + n_{\text{B}} R \ln \frac{V_{\text{A}} + V_{\text{B}}}{V_{\text{B}}}. \tag{3.96}$$

Using the mole fractions of the gases:

$$x_{\text{A}} = \frac{N_{\text{A}}}{N_{\text{A}} + N_{\text{B}}}, \quad x_{\text{B}} = \frac{N_{\text{B}}}{N_{\text{A}} + N_{\text{B}}}, \tag{3.97}$$

the increase of entropy is given by

$$\Delta S = -R\left(n_{\text{A}} \ln x_{\text{A}} + n_{\text{B}} \ln x_{\text{B}}\right). \tag{3.98}$$

In general, when $M$ different gases are mixed, the change of the entropy is given by

$$\Delta S = -R \sum_{j=1}^{M} n_j \ln x_j, \tag{3.99}$$

which is called "mixing entropy". This change of entropy causes various interesting phenomena as explained in the following.

### 3.5.1 Gibbs Free Energy of Mixing

The Gibbs free energy of a gas in an isolated container can be expressed in terms of the chemical potential as

$$G^0(T, P, \mu^0) = n\mu^0(T, P), \tag{3.100}$$

where $n$ is the number of moles, $T$ is the temperature, and $P$ is the pressure.

For a system composed of $M$ separate gases, the total Gibbs free energy is given by

$$G^0(T, P, \{\mu_j^0\}) = \sum_{j=1}^{M} n_j \mu_j^0(T, P). \tag{3.101}$$

When these $M$ different gases are mixed, the energy terms remain unchanged. However, taking into account the entropy of mixing (3.99), the Gibbs free energy of the mixed system becomes:

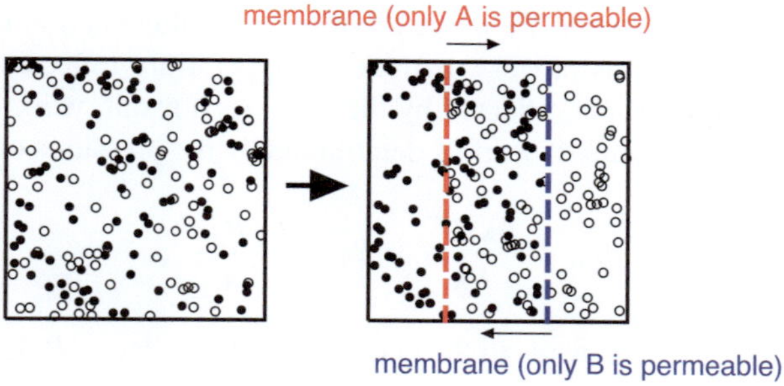


**Fig. 3.14** Separation of mixed gases using a semipermeable membrane

$$G(T, P) = \sum_{j=1}^{M} n_j \mu_j^0(T, P) - T\Delta S = \sum_{j=1}^{M} n_j \left(\mu_j^0(T, P) + RT \ln x_j\right), \quad (3.102)$$

where $x_j$ represents the mole fraction of the $j$-th gas.

### 3.5.2 *Gibbs Paradox*

The increase in entropy during the mixing process gives rise to a paradox in a thought experiment known as the Gibbs' paradox. Even in a one-component gas, we can hypothetically distinguish particles initially located in the left and right parts of the system at a given time $t$. If we track these particles, the gas in the left half of the container will diffuse into the entire container, as will the gas in the right half. This would imply an increase in the entropy of mixing. In this scenario, entropy would continue to increase indefinitely over time, suggesting that entropy is not well-defined.

How can we resolve this paradox? The thermodynamic resolution is that we observe only the macroscopic state of the system. That is, a gas occupying a volume $V$ at temperature $T$ and pressure $P$ is in a thermodynamic equilibrium state, which does not change with time. Therefore, it is meaningless to analyze microscopic details, such as tracing whether particles originate from the left or right side.

In the case of gases with different components, the entropy of mixing is given by (3.99). This formula assumes that the different species can be distinguished. For instance, we can separate the components using walls that are selectively permeable to specific gases, as shown in Fig. 3.14.

In the semipermeable membrane shown in Fig. 3.14, only particles of type A (represented by black circles) can pass through the wall shown as a dotted line, while only particles of type B (represented by white circles) can pass through the wall shown as a dashed line. When we move these walls as illustrated in the figure, the gases are separated. In this process, particles of type $A$ are reflected by the wall, while particles of type $B$ are reflected by the wall. As a result, work is required to move the walls. The amount of work is determined by the pressures of the gases:

$$P_{\mathrm{A}} = \frac{n_{\mathrm{A}} RT}{V_{\mathrm{A}}}, \quad P_{\mathrm{B}} = \frac{n_{\mathrm{B}} RT}{V_{\mathrm{B}}}. \tag{3.103}$$

The total work to move the walls to the ends, where $P_{\mathrm{A}} = P_{\mathrm{B}} = P$, is given by

$$\Delta W = -\int_{V}^{V/2} P_{\mathrm{A}} dV_{\mathrm{A}} - \int_{V}^{V/2} P_{\mathrm{B}} dV_{\mathrm{B}}. \tag{3.104}$$

This result is equal to the mixing entropy $T\Delta S$ (3.99). Thus, recovering the separated configuration macroscopically requires work.

In contrast, in the Gibbs paradox for single-component gas or indistinguishable particles, no such process for separation exists. As a result, there is no change in the system thermodynamically even if gases in the left and right parts of the container mix. In other words, if we lack a method to distinguish that the gas consists of two different types, we perceive the gas as a uniform substance, and the concept of mixing entropy does not arise. Namely, without the semipermeable membrane, the mixed gas is simply a uniform gas. This is somewhat philosophical but crucial: the inability to recognize certain aspects of a system gives the essence of the very concepts of "heat" and "entropy."

By controlling the level of recognition, the applicability of thermodynamics changes. The relationship between information manipulation and energy dissipation has been extensively studied, starting with the famous concept of Maxwell's demon, which has often been linked to the idea of perpetual motion. This topic will be discussed further in Chap. 7.

## 3.6 Dilute Solutions

As studied in the previous section, mixing leads to a change in entropy. The mixing of a solute in its solvent introduces interesting phenomena in dilute solutions (Fig. 3.15), similar to the mixing of gases.

A dilute solution is composed of a pure liquid (the solvent) and a small amount of solute. The free energy of this system can be described by the change in the Gibbs free energy (3.102), which is determined by the chemical potential $\mu(T, P)$. In this section, we introduce several typical examples to illustrate these phenomena.

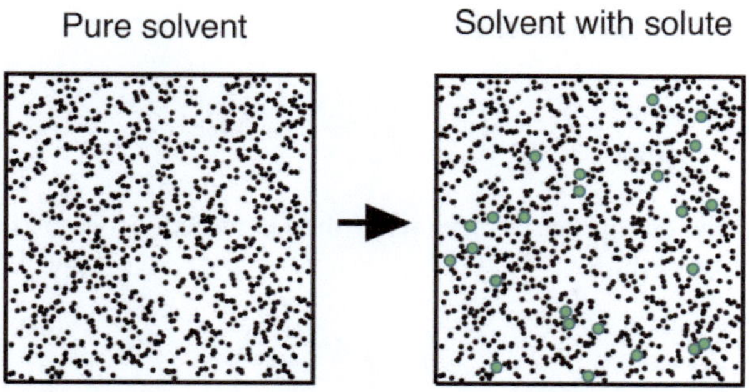


**Fig. 3.15** Dilute solution with solute. (left) Pure solvent; (right) Solvent with solute particles (green circles)

## *3.6.1 Elevation of Boiling Point*

The boiling temperature increases with the addition of a solute to the solvent. At the boiling temperature, the chemical potentials of the solvent in the gas and liquid phases are equal. For a pure solvent at a given pressure $P$, the boiling temperature $T_0$ is determined by the relation:

$$\mu^0_{\text{gas}}(T_0, P) = \mu^0_{\text{liquid}}(T_0, P). \tag{3.105}$$

When a solute is added at a concentration $x$, the condition for the boiling temperature $T_{\text{C}}(x)$ becomes

$$\mu^0_{\text{gas}}(T_{\text{C}}(x), P) = \mu^0_{\text{liquid}}(T_{\text{C}}(x), P) + RT_{\text{C}}(x)\ln(1-x), \tag{3.106}$$

where it is assumed that the solute does not evaporate into the gas phase.

For small values of $x$, expanding (3.106) in terms of $x$, the change in the boiling temperature, $T_{\text{C}}(x) - T_0$, is given by the relation:

$$\left(\frac{\partial \mu^0_{\text{gas}}}{\partial T}\right)_P (T_{\text{C}}(x) - T_0) = \left(\frac{\partial \mu^0_{\text{liquid}}}{\partial T}\right)_P (T_{\text{C}}(x) - T_0) - RT_0 x, \tag{3.107}$$

where $\left(\frac{\partial \mu^0}{\partial T}\right)_P$ is the temperature derivative of the chemical potential at constant pressure. Using this expression, the change in boiling temperature is obtained as

$$T_{\text{C}}(x) - T_0 = \frac{RT_0 x}{\left(\frac{\partial \mu^0_{\text{liquid}}}{\partial T}\right)_P - \left(\frac{\partial \mu^0_{\text{gas}}}{\partial T}\right)_P}. \tag{3.108}$$

Recalling the thermodynamic relation:

$$\left(\frac{\partial \mu^0}{\partial T}\right)_P = -S, \tag{3.109}$$

we can rewrite the denominator in terms of the latent heat of vaporization, $\Delta Q_{\text{boil}}$, at the boiling temperature:

$$\Delta Q_{\text{boil}} = T\left(S_{\text{gas}} - S_{\text{liquid}}\right) = T\left(-\left(\frac{\partial \mu^0_{\text{liquid}}}{\partial T}\right)_P + \left(\frac{\partial \mu^0_{\text{gas}}}{\partial T}\right)_P\right). \tag{3.110}$$

This leads to the expression for the elevation of the boiling point:

$$T_C(x) - T_0 = \frac{RT_0^2}{\Delta Q_{\text{boil}}} x, \tag{3.111}$$

which shows the elevation of boiling point that the boiling temperature increases proportionally to the solute concentration $x$, as long as $x$ is small.

### 3.6.2 Depression of Freezing Point

Using a similar argument as in the previous subsection on the elevation of boiling point, a formula for the depression of the freezing point can be derived. Given the latent heat of solidification of the solvent material, $\Delta Q_{\text{solidification}}$, the change in the freezing point temperature is expressed as

$$T_C(x) - T_0 = -\frac{RT_0^2}{\Delta Q_{\text{solidification}}} x, \quad \Delta Q_{\text{solidification}} = T_0(S_{\text{liquid}} - S_{\text{solid}}), \tag{3.112}$$

where $T_C(x)$ is the freezing point of the solution, and $T_0$ is the freezing point of the pure solvent. This formula shows the depression of freezing point that the freezing point decreases proportionally to the solute concentration $x$, as long as $x$ is small.

This effect has practical applications in daily life, e.g.,

Making ice cream: To achieve a lower temperature for freezing, salt is mixed with ice. A salted water solution with a concentration of $x = 25\%$ has a freezing point of approximately $-22$°C, compared to 0°C for pure water, and

De-icing roads: Materials such as calcium chloride ($CaCl_2$) are used to lower the freezing point of water to melt snow and ice.

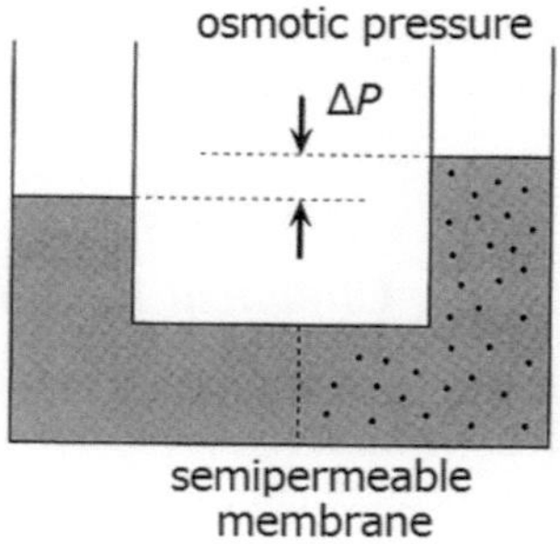


**Fig. 3.16** Osmotic pressure across a semipermeable membrane

## *3.6.3 Osmotic Pressure*

A film (or wall) that is permeable only to the solvent and not to the solute is called a semipermeable membrane. If a liquid containing a solute is separated from a pure solvent by such a membrane, the concentration difference ($x$) between the two sides causes a difference in the chemical potential of the solvent. This results in a pressure difference, which is referred to as osmotic pressure, as shown in Fig. 3.16.

In the absence of a solute ($x = 0$), the chemical potentials on both sides are equal:

$$\mu^0_{\text{left}}(T, P_0) = \mu^0_{\text{right}}(T, P_0), \tag{3.113}$$

where the temperatures of both sides are the same. If a solute is added to the right side, creating a concentration $x$, the equilibrium condition becomes

$$\mu^0_{\text{left}}(T, P_0) = \mu^0_{\text{right}}(T, P_{\text{C}}(x)) + RT\ln(1-x), \tag{3.114}$$

where $P_{\text{C}}(x)$ is the pressure on the right side.

For small $x$, expanding the logarithmic term in (3.114) and solving for $P_{\text{C}}(x) - P_0$, we obtain

$$\left(\frac{\partial \mu^0_{\text{right}}}{\partial P}\right)_T \left(P_{\text{C}}(x) - P_0\right) = V\left(P_{\text{C}}(x) - P_0\right) = RTx, \tag{3.115}$$

where $V$ is the molar volume of the solvent. This leads to the formula for osmotic pressure:

$$P_{\text{C}}(x) - P_0 = \frac{RT}{V}x, \tag{3.116}$$

valid for small $x$.

This effect is significant in everyday life, e.g.,

Water absorption in vegetables: When vegetables are placed in water, osmotic pressure causes water to be absorbed into their cells making them vivid, and conversely,

Pickling process: Placing vegetables in a dense saline solution causes water to leave the cells, leading to shrinkage as pickles.

### 3.6.4 Chemical Reaction and the Equilibrium Constant

We now consider the balance of concentration of components in chemical reactions. A general chemical reaction involving materials $A_1$, $A_2$, ..., $A_{M+1}$, $A_{M+2}$, ... $A_N$ is represented as

$$\nu_1 A_1 + \nu_2 A_2 + \cdots \nu_M A_M \leftrightarrow \nu_{M+1} A_{M+1} + \nu_{M+2} A_{M+2} + \cdots + \nu_N A_N \quad (3.117)$$

where $\nu_j$ denotes the stoichiometric coefficients of each component.

The molar fraction $x_j$ is defined in terms of the molar amount $n_j$ as

$$x_j = \frac{n_j}{\sum_{k=1}^{N} n_k}. \quad (3.118)$$

Including the entropy of mixing, the Gibbs free energy, as given in (3.102), can be expressed as

$$G = \sum_{j=1}^{N} n_j \left( \mu_j^0(T, P) + RT \ln x_j \right), \quad (3.119)$$

where $\mu_j^0(T, P)$ is the chemical potential of the pure $j$-th component.

In equilibrium, the Gibbs free energy reaches its minimum value. The condition for this extremum is

$$\delta G = \sum_{j=1}^{N} \left( \mu_j^0(T, P) + RT \ln x_j \right) \delta n_j + RT \sum_{j=1}^{N} n_j \delta(\ln x_j) = 0, \quad (3.120)$$

for any infinitesimal variations $\delta n_j$.

Using the definition in (3.118), the second term in (3.120) can be rewritten as

$$n_j \delta(\ln x_j) = n_j \sum_{\ell} \frac{\partial \ln x_j}{\partial n_\ell} \delta n_\ell = \frac{n_j}{x_j} \sum_{\ell} \frac{\partial x_j}{\partial n_\ell} \delta n_\ell = n \sum_{\ell} \frac{\partial x_j}{\partial n_\ell} \delta n_\ell. \quad (3.121)$$

Since the molar fractions satisfy $\sum_{j=1}^{N} x_j = 1$, we have

$$\sum_j n_j \delta(\ln x_j) = n \sum_{\ell} \frac{\partial}{\partial n_\ell} \left( \sum_j x_j \right) \delta n_\ell = 0.$$

Thus, the condition simplifies to

$$\delta G = \sum_j \left(\mu_j^0(T, P) + RT \ln x_j\right) \delta n_j = 0. \tag{3.122}$$

Since the molar changes $\{\delta n_j\}$ follow the stoichiometric coefficients of the reaction equation (3.117), they are not independent but satisfy the proportionality:

$$\delta n_1 : \delta n_2 : \cdots : \delta n_{M+1} : \delta n_{M+2} : \cdots = -\nu_1 : -\nu_2 : \cdots : \nu_{M+1} : \nu_{M+2} : \cdots, \tag{3.123}$$

where the negative signs are applied to the coefficients of the reactants in the left side of (3.117) because their amounts decrease during the reaction.

Using this relation, the condition (3.122) can be written as

$$\begin{aligned}
&-(\mu_1^0(T, P) + RT \ln x_1)\nu_1 - (\mu_2^0(T, P) + RT \ln x_2)\nu_2 - \cdots \\
&+(\mu_{M+1}^0(T, P) + RT \ln x_{M+1})\nu_{M+1} + (\mu_{M+2}^0(T, P) \\
&+RT \ln x_{M+2})\nu_{M+2} + \cdots = 0.
\end{aligned}$$

This equation simplifies to

$$-\sum_{j\in\text{left}} \nu_j \ln x_j + \sum_{k\in\text{right}} \nu_k \ln x_k = \frac{1}{RT}\left(\sum_{j\in\text{left}} \nu_j \mu_j^0(T, P) - \sum_{k\in\text{right}} \nu_k \mu_k^0(T, P)\right). \tag{3.124}$$

Here, the difference in free energy is defined as

$$-\frac{\Delta G^0(T, P)}{RT} = \frac{1}{RT}\left(\sum_{j\in\text{left}} \nu_j \mu_j^0(T, P) - \sum_{k\in\text{right}} \nu_k \mu_k^0(T, P)\right). \tag{3.125}$$

The exponentiation of (3.124) gives the following relation:

$$e^{-\sum_{j\in\text{left}} \nu_j \ln x_j + \sum_{k\in\text{right}} \nu_k \ln x_k} = \frac{\prod_{k\in\text{right}} x_k^{\nu_k}}{\prod_{j\in\text{left}} x_j^{\nu_j}} = e^{-\Delta G^0(T,P)/RT}, \tag{3.126}$$

which relates the equilibrium molar fractions $\{x_j\}$ and the Gibbs' free energies per mol (i.e., the chemical potential) $\mu_j^0(T, P)$ in the pure system. It is called the law of mass action for chemical reactions.

In (3.126), the chemical potentials $\mu_k^0(T, P)$ at the given pressure $P$ are used. However, it is convenient to express the relation in terms of the chemical potential $\mu_k^0(T, P_0)$ at the standard pressure $P_0$ (e.g., the standard atmospheric pressure: 1atm=$1.01325\times10^5$Pa).

Now we rewrite the relation using $\mu_k^0(T, P_0)$. The difference in Gibbs free energy is given by

$$G(T, P) - G(T, P_0) = \int_{P_0}^{P} \left(\frac{\partial G}{\partial P}\right)_T dP' = \int_{P_0}^{P} V(P')dP'. \tag{3.127}$$

For an ideal gas,

$$\int_{P_0}^{P} V(P')dP' = \int_{P_0}^{P} \frac{nRT}{P'}dP' = nRT\,(\ln P - \ln P_0)\,. \tag{3.128}$$

Therefore,

$$G(T, P) = G(T, P_0) + nRT \ln P - nRT \ln P_0. \tag{3.129}$$

The $P$ dependence of the chemical potential $\mu^0(T, P)$ is then given as

$$\mu^0(T, P) = \mu^0(T, P_0) + RT \ln\left(\frac{P}{P_0}\right). \tag{3.130}$$

The partial pressure of ideal gases is expressed as

$$P_j = \frac{n_j RT}{V}, \tag{3.131}$$

and thus, considering that $P = \sum_j P_j$ and $n = \sum_j n_j$, the following relation holds:

$$x_j = \frac{P_j}{P}. \tag{3.132}$$

Substituting this relation into (3.126), we obtain the following relation:

$$\frac{\prod_{k\in\text{right}}(P_k/P_0)^{\nu_k}}{\prod_{j\in\text{left}}(P_j/P_0)^{\nu_j}} = \exp\left(-\sum_{k\in\text{right}} \frac{\nu_k \mu_k^0(T, P_0)}{RT} + \sum_{j\in\text{left}} \frac{\nu_j \mu_j^0(T, P_0)}{RT}\right). \tag{3.133}$$

This equation represents the law of mass action for pressure. Here, the quantity:

$$K_P(T, P) \equiv \exp\left(-\sum_{k\in\text{right}} \frac{\nu_k \mu_k^0(T, P_0)}{RT} + \sum_{j\in\text{left}} \frac{\nu_j \mu_j^0(T, P_0)}{RT}\right) \tag{3.134}$$

is called the equilibrium constant. Using the molar fraction (3.118), the law of mass action is expressed as

$$K_P(T, P) = \frac{\prod_{k \in \text{right}} x_k^{\nu_k}}{\prod_{j \in \text{left}} x_j^{\nu_j}} \left(\frac{P}{P_0}\right)^{\sum_{k \in \text{right}} \nu_k - \sum_{j \in \text{left}} \nu_j}. \tag{3.135}$$

The temperature dependence of the equilibrium constant is given by

$$\begin{aligned}\frac{d \ln K_P}{dT} &= -\frac{d}{dT}\left(\frac{\Delta G^0}{RT}\right) = \frac{\Delta G^0}{RT^2} - \frac{1}{RT}\left(\frac{\partial \Delta G^0}{\partial T}\right)_{P,N} \\ &= \frac{\Delta G^0}{RT^2} - \frac{\Delta S}{RT} = \frac{\Delta G^0 - T\Delta S}{RT^2} = \frac{\Delta H}{RT^2}.\end{aligned} \tag{3.136}$$

This indicates that the temperature dependence is determined by the standard enthalpy change $\Delta H$.

The law of mass action for a given volume $V$ is expressed in terms of the concentration $c_j$ of the $j$-th component:

$$c_j = \frac{n_j}{V} \equiv [A_j]. \tag{3.137}$$

For a system at constant volume, the Helmholtz free energy must be minimized instead of the Gibbs free energy. Considering the relation:

$$\left(\frac{\partial F^0}{\partial n_j}\right)_{T,V,\mu} = \mu_j^0(T, V), \tag{3.138}$$

the condition for equilibrium is given by

$$\delta F = \sum_j \left(\mu_j^0(T, V) + RT \ln x_j\right) \delta n_j + RT \sum_j n_j \delta(\ln x_j) = 0. \tag{3.139}$$

For an ideal gas, the molar fraction $x_i$ is expressed as

$$x_i = \frac{n_i}{n} = \frac{n_i}{V}\frac{RT}{P}. \tag{3.140}$$

Substituting this relation into (3.126),

$$\begin{aligned}&e^{-\sum_{j \in \text{left}} \nu_j \ln x_j + \sum_{k \in \text{right}} \nu_k \ln x_k} \\ &= e^{-\sum_{j \in \text{left}} \nu_j (\ln(\frac{n_j}{V} RT) - \ln P) + \sum_{k \in \text{right}} \nu_k (\ln(\frac{n_k}{V} RT) - \ln P)} \\ &= e^{\sum_{i \in \text{left}} \nu_i (\mu_i^{(0)}(T, P_0) + RT \ln P)/k_B T - \sum_{j \in \text{right}} \nu_j (\mu_j^{(0)}(T, P_0) + RT \ln P)/k_B T}.\end{aligned}$$

Introducing the notation:

$$\Delta\nu = \sum_{j \in \text{left}} \nu_j - \sum_{k \in \text{right}} \nu_k, \tag{3.141}$$

we can rewrite the equation as

$$\frac{\prod_{k\in\text{right}} c_k^{\nu_k}}{\prod_{j\in\text{left}} c_j^{\nu_j}} \times (RT)^{\Delta\nu} = e^{-\Delta G^0/RT}. \tag{3.142}$$

Thus, the law of mass action is expressed as

$$\frac{\prod_{k\in\text{right}} c_k^{\nu_k}}{\prod_{j\in\text{left}} c_j^{\nu_j}} = e^{-\Delta G^0/RT}(RT)^{-\Delta\nu} \equiv K_\text{c}(T). \tag{3.143}$$

The factor $K_\text{c}(T)$ is also referred to as equilibrium constant. It is a function of the temperature only. The temperature dependence of $K_\text{c}(T)$ is given by

$$\ln K_\text{c} = -\sum_{j\in\text{left}} \nu_j \left(\frac{\mu_j^0(T, P_0)}{RT} + \ln RT\right) + \sum_{j\in\text{right}} \nu_j \left(\frac{\mu_j^{(0)}(T, P_0)}{RT} + \ln RT\right). \tag{3.144}$$

Differentiating it with respect to $T$, we obtain

$$\frac{d\ln K_\text{c}}{dT} = \frac{\Delta H}{RT^2} - \sum_{j\in\text{left}} \left(\nu_j \frac{\Delta H_j}{RT^2} - \frac{\nu_j}{T}\right) + \sum_{k\in\text{right}} \left(\nu_k \frac{\Delta H_k}{RT^2} - \frac{\nu_k}{T}\right).$$

Using the relations $RT = p_j^{(0)} V_0^{(0)}$ and $U = G - PV$, this simplifies to

$$\frac{d\ln K_\text{c}}{dT} = \frac{\Delta U}{RT^2}, \tag{3.145}$$

where $\Delta U$ represents the change in internal energy during the reaction. Thus, the temperature dependence of $K_\text{c}(T)$ is given by the change of the internal energy $\Delta U$, which is commonly referred to as the "heat of reaction".

In the above arguments, we have assumed the behavior of an ideal gas. For the real gases, the activity $a_i$ is used instead of the molar fraction $x$. The chemical potential is expressed as

$$\mu_i(P, T) = \mu_i^{(0)}(T) + RT\ln x_i \rightarrow \mu_i(P, T) = \mu_i^{(0)}(T) + RT\ln a_i, \tag{3.146}$$

which modifies the law of mass action accordingly. The ratio:

$$\gamma_j = \frac{a_j}{x_j} \tag{3.147}$$

is known as the "activity coefficient". This coefficient quantifies deviations from ideal gas behavior and approaches 1 as $P \rightarrow 0$.

## 3.7 Other Topics Related to Entropy

### 3.7.1 Air Conditioner

A room can be heated either using an electric heater or an air conditioner. An electric heater converts electrical energy entirely into heat, achieving 100% energy conversion efficiency. In this case, the heat is produced directly by the conversion of electrical energy into thermal energy. In contrast, an air conditioner operates heat transfer between the inside and outside of the room. Remarkably, the amount of heat delivered into the room can be significantly greater than the electrical energy consumed. In other words, the efficiency of an air conditioner exceeds 100%. This process is known as a heat pump (Fig. 3.17).

As explained in Chap. 1, the heat engine extracts work $W$ from heat transferred from a high-temperature thermal bath to a low-temperature thermal bath (Fig. 2.1). Noting the heat flowing out of the high temperature bath as $Q_{\mathrm{H}}$, and the heat released into the low temperature bath as $Q_{\mathrm{L}}$, the relationship among these quantities is given by

$$W = Q_{\mathrm{H}} - Q_{\mathrm{L}}, \tag{3.148}$$

and the efficiency of the heat engine is defined as

$$\eta = \frac{Q_{\mathrm{H}} - Q_{\mathrm{L}}}{Q_{\mathrm{H}}}. \tag{3.149}$$

When the heat energy is utilized in the opposite direction, heat is transferred from the low-temperature bath to the high-temperature bath. In this case,

$$Q_{\mathrm{H}} = W + Q_{\mathrm{L}}, \quad Q_{\mathrm{L}} \geq 0, \tag{3.150}$$

where $W$ represents the work required to drive this process.

**Fig. 3.17** Heat transfer in an air conditioner

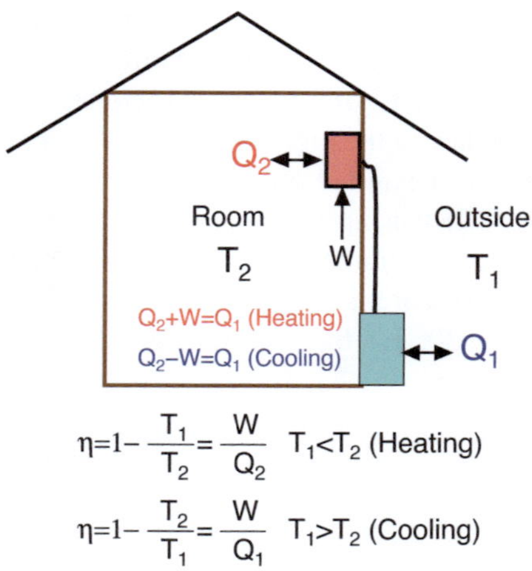

Although the efficiency of real engines is lower than that of the so-called reversible engine (e.g., the Carnot cycle), let us estimate the ratio of $Q_{\mathrm{H}}$ and $W$ for the reversible engine. In this case, the entropy change due to the heat flow are the same (2.24):

$$\frac{Q_{\mathrm{H}}}{T_{\mathrm{H}}} = \frac{Q_{\mathrm{L}}}{T_{\mathrm{L}}},$$

and the efficiency of the reversible engine is given by the ratio of absolute temperatures:

$$\eta = \frac{T_{\mathrm{H}} - T_{\mathrm{L}}}{T_{\mathrm{H}}}. \tag{3.151}$$

Using this relationship, (3.148) is written as

$$W = Q_{\mathrm{H}} - Q_{\mathrm{L}} = Q_{\mathrm{H}}\left(1 - \frac{T_{\mathrm{L}}}{T_{\mathrm{H}}}\right) = \frac{T_{\mathrm{H}} - T_{\mathrm{L}}}{T_{\mathrm{H}}} Q_{\mathrm{H}}. \tag{3.152}$$

Thus, the amount of heat brought into the high temperature bath is

$$Q_{\mathrm{H}} = \frac{T_{\mathrm{H}}}{T_{\mathrm{H}} - T_{\mathrm{L}}} W \geq W. \tag{3.153}$$

For example, if the outside temperature is 7°C ($T_1 = T_{\mathrm{L}}$ =280 K) and the inside temperature is 27 °C ($T_2 = T_{\mathrm{H}}$ =300 K), then

$$Q_{\mathrm{H}} = \frac{300}{300 - -280} W = 15W. \tag{3.154}$$

This means that the amount of heat supplied to the room is 15 times larger than the case of an electric heater. The reason why the ratio is so large comes to the fact that both temperatures 7 and 27 °C are close to each other when measured in absolute temperature, i.e., (7+273) K and (27+273) K. The efficiency of an air conditioner depends on the temperature difference between the inside and outside of the room. If $T_{\mathrm{L}} = 0$, the efficiency ratio becomes 1, which is the same as that of an electric heater. The key to the high efficiency of an air conditioner lies in its ability to transfer heat from an external source, rather than generating heat directly.

When using an air conditioner as a cooler, it removes heat from a low-temperature room and transfers it to the high-temperature outdoors. For instance, if the outside temperature is 32 °C ($T_2 = T_{\mathrm{H}}$ =305 K) and the inside temperature is 27 °C ($T_1 = T_{\mathrm{H}}$ =300 K), the heat transferred to the outside is

$$Q_{\mathrm{H}} = \frac{305}{305 - -300} W = 61W. \tag{3.155}$$

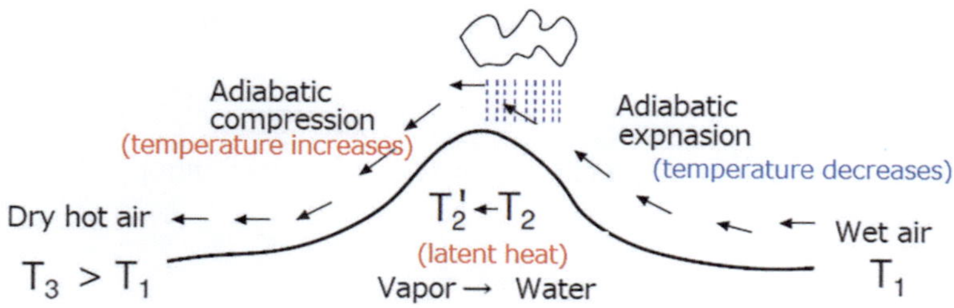


**Fig. 3.18** Fœhn phenomena

This result shows that the work required for cooling is significantly smaller than that for heating. Therefore, when purchasing an air conditioner, the machine's power capacity should be chosen to ensure adequate performance during heating.

### 3.7.2 *Fœhn Phenomena*

The Fœhn phenomena occurs as a combination of non-adiabatic processes caused by phase transitions (latent heat release) and adiabatic compression of air. As depicted in Fig. 3.18, when moist air at a temperature $T_1$ containing significant vapor rises over mountains, it expands adiabatically, causing its temperature to drop to $T_2$. At this point, the water vapor condenses into liquid water and precipitates as rain or snow. During this condensation process (vapor→water or ice), the latent heat is emitted to the air and the temperature increases from $T_2$ to $T_2'$.

Subsequently, the air descends on the leeward side of the mountain. During this descent, the air undergoes adiabatic compression, which raises its temperature further to $T_3$. This sequence of processes results in significantly higher temperatures in the downwind regions (Fig. 3.18).

### 3.7.3 *Various Heat-Engines*

Work can be extracted in various innovative ways from situations with temperature differences. For example, the temperature difference in seawater is utilized in the so-called Ocean Thermal Energy Conversion (OTEC) systems. Additionally, rubber heat engines [9] operate using the contraction force generated by the heating of rubber, such as through light irradiation. These methods have potential applications in achieving Sustainable Development Goals (SDGs) by promoting energy conservation and alternative energy sources.

### 3.7.4 Determination of Melting Entropy by Kinetic Coefficient

The latent heat $Q$ at a melting point $T_m$ gives the change of entropy (the Clausius entropy of fusion):

$$\Delta S_f = \frac{Q}{T_m}. \tag{3.156}$$

Recently, a new method to estimate the change of entropy by making use of the decay rate of the electron diffraction under variable voltage and temperature conditions in transmission electron microscopy (TEM) experiments was proposed by the Nakamura group [10]. By the Arrhenius law, the decay rate is given by

$$k = A \exp(-E_a/RT), \tag{3.157}$$

where $E_a$ is the energy of activation, and $R$ is the gas constant. Assuming that the coefficient $A$ is proportional to the number of disordered states $W_d$, they used the relation:

$$A = A_{INT} W_d, \tag{3.158}$$

from which the molar entropy of disorder ($\Delta S_d$) is defined as

$$\Delta S_d = R \ln W_d, \tag{3.159}$$

where $R$ is the gas constant. For flexible molecules, the activation energy $E_a$ is negligible, where the decay rate $k$ is directly related to $\Delta S_d$ as

$$k = A_{INT} e^{\Delta S_d/RT}. \tag{3.160}$$

They observed the decay rates of various materials and confirmed that the thus obtained values of $\Delta S_d$ agree with those obtained from the latent heat $\Delta S_f$. This method provides a new way to obtain the change of entropy for femtogram quantities of thermally unstable and biomolecular crystals.

## 3.8 Phase Transition: Entropy Versus Energy

In systems with interaction elements, e.g., magnetic systems composed of spins, gas-liquid systems with Lenard-Johns interaction, the interactions among the elements tend to lead the state to be ordered while entropy favors a disordered state which has larger entropy. This competition may induce a phase transition where the equilibrium state changes singularly. Normally, in the lowest energy state (the ground state), all local interactions take the lowest energy configuration, and the system exhibits a globally ordered state, such as the perfectly aligned spin state in ferromagnets and the Néel state in antiferromagnets.

As a typical example, consider the Ising model (3.18) with nearest neighbor ferromagnetic interaction $J$:

$$\mathcal{H} = -J \sum_{\langle ij \rangle} \sigma_i \sigma_j - H \sum_{i}^{N} \sigma_i, \quad \sigma_i = \pm 1. \tag{3.161}$$

The Ising model in two or more dimensions exhibits a phase transition between a disordered high-temperature phase (paramagnetic phase) and an ordered low-temperature phase (ferromagnetic phase). Properties of the phase transitions in two dimensions have been obtained exactly [11–22], which will be utilized in Part II. However, here, we analyze the model using the mean-field approximation.

### *3.8.1 Mean-Field Analysis on the Ferromagnetic Ising Model*

In the mean-field approximation, it is assumed that all spins in the system have a uniform magnetization, although each spin takes either value of $\pm 1$. Under this approximation, the energy of the system is given by

$$E = -\frac{zN}{2} J m^2, \tag{3.162}$$

where $z$ is the number of nearest-neighbor sites (e.g., $z = 4$ in the square lattice). Using the relations (3.24) and (3.27), the entropy is given by

$$S = k_{\mathrm{B}} \ln W(m) = -k_{\mathrm{B}} N \left[ \frac{1-m}{2} \ln \frac{1-m}{2} + \frac{1+m}{2} \ln \frac{1+m}{2} \right]. \tag{3.163}$$

Thus, the free energy is given by

$$F(m) = E - TS = -\frac{zN}{2} J m^2 + k_{\mathrm{B}} T N \left[ \frac{1-m}{2} \ln \frac{1-m}{2} + \frac{1+m}{2} \ln \frac{1+m}{2} \right], \tag{3.164}$$

which is known as Bragg-Williams free energy.

As depicted in Fig. 3.19, at high temperatures, the entropy term dominates, and $f(m)(= F(m)/N)$ has a minimum at $m = 0$, representing the disordered state. In contrast, at low temperatures $f(m)$ has minima at $m = \pm m_{\mathrm{S}} \neq 0$. Between them, at $T = zJ/k_{\mathrm{B}}$, $f(m)$ begins to develop minima at $\pm m_{\mathrm{S}}$, signaling the onset of the ordered state. This temperature is referred to as the critical temperature.

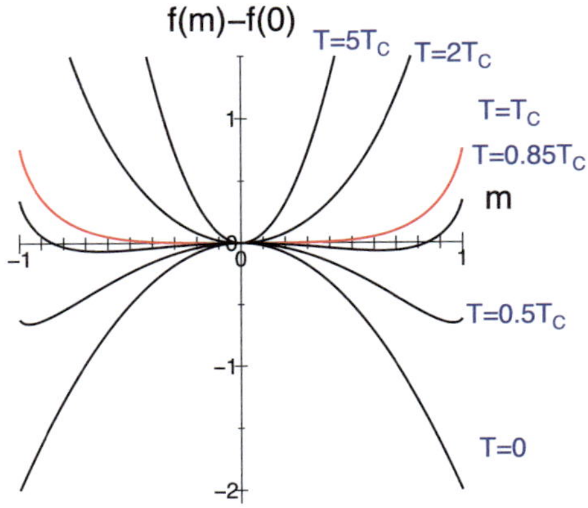


**Fig. 3.19** Bragg-Williams free energy as a function of magnetization $m$ at several temperatures. Here $T_C = zJ$

The minimum point of the free energy is given by solving $df(m)/dm = 0$:

$$\frac{df(m)}{dm} = -zJm + \frac{k_B T}{2} \ln \frac{1+m}{1-m} = 0. \tag{3.165}$$

This relation is rewritten as

$$m = \tanh(Jzm/k_B T), \tag{3.166}$$

which is known as the self-consistent equation in the mean-field analysis of the Ising model. From the change in the form of free energy, it is evident that the phase transition arises from the competition between energy (due to interactions) and entropy.

### 3.8.2 *Mean-Field Analysis on the Ferromagnetic Heisenberg Model*

The same idea can be applied to the Heisenberg model. The energy term remains the same ($E = -zNJm^2/2$), and the entropy (3.34) is given by

$$S = \ln Z = \ln \left[ (2\pi)^N \frac{1}{2\pi} \int_{-\infty}^{\infty} d\alpha \left[ \frac{2 \sin \alpha}{\alpha} e^{i\alpha m} \right]^N \right]. \tag{3.167}$$

Thus, the free energy ($F = E - TS$) for large $N$ is given by

$$F = -\frac{zN}{2} Jm^2 + Nk_B T \ln \left( \ln 2\pi + \ln \left[ \frac{2 \sinh x}{x} e^{-xm} \right] \right), \tag{3.168}$$

where the formula (3.37) is used. The minimum point of the free energy satisfies the relation:

$$\frac{df(m)}{dm} = -zJm - k_{\mathrm{B}}T\left[\left(\coth x - \frac{1}{x} - m\right)\frac{dx}{dm} - x\right] = 0. \tag{3.169}$$

Since $x$ satisfies the relation (3.36), we have

$$\frac{df(m)}{dm} = -zJm + k_{\mathrm{B}}Tx = 0. \tag{3.170}$$

This relation can be rewritten as

$$m = L\left(\frac{zJm}{k_{\mathrm{B}}T}\right), \tag{3.171}$$

which is known as the self-consistent equation of the mean-field analysis on the Heisenberg model.

For both the Ising and Heisenberg models, a non-zero solution for $m_{\mathrm{s}}$, corresponding to the spontaneous magnetization, appears below the critical point. Expanding the self-consistent equations (3.166) and (3.171) for small $m$ (up to the third order in $m$), we have

$$m = \tanh(\beta Jzm) \simeq \beta Jzm - \frac{1}{3}(\beta Jzm)^3 \Rightarrow m_{\mathrm{s}} \simeq \pm\sqrt{\frac{3(\beta Jz - 1)}{(\beta Jz)^3}} \propto \sqrt{T_{\mathrm{C}} - T}, \tag{3.172}$$

for the Ising model where $T_{\mathrm{C}} = zJ$, and

$$m = L\left(\frac{zJm}{k_{\mathrm{B}}T}\right) \simeq \frac{\beta Jzm}{3} - \frac{1}{45}(\beta Jzm)^3 \Rightarrow m_{\mathrm{s}} \simeq \pm\sqrt{\frac{\beta Jz - 3)}{15(\beta Jz)^3}} \propto \sqrt{T_{\mathrm{C}} - T}, \tag{3.173}$$

for the Heisenberg model where $T_{\mathrm{C}} = zJ/3$. In both cases, the spontaneous magnetization arises proportionally to $\sqrt{T_{\mathrm{C}} - T}$. It should be noted that difference of fluctuations in the models causes some quantitative differences in the ordering. For example, the critical temperature is lower in the Heisenberg model.

Additionally, the temperature dependence of $m_{\mathrm{s}}$ at low temperatures exhibits notable difference. For the Ising model, by setting $m_{\mathrm{s}} = 1 - \delta$, the self-consistent equation is expressed as

$$1 - \delta = \frac{e^{\beta Jz(1-\delta)} - e^{-\beta Jz(1-\delta)}}{e^{\beta Jz(1-\delta)} + e^{-\beta Jz(1-\delta)}} \simeq 1 - 2e^{-2\beta Jz(1-\delta)}, \tag{3.174}$$

which shows that $m_{\mathrm{s}}$ approaches 1 exponentially as temperature decreases. On the other hand, for the Heisenberg model,

$$1-\delta = \frac{e^{\beta Jz(1-\delta)} + e^{-\beta Jz(1-\delta)}}{e^{\beta Jz(1-\delta)} - e^{-\beta Jz(1-\delta)}} - \frac{1}{\beta Jz(1-\delta)} \simeq 1 - \frac{1}{\beta Jz}, \tag{3.175}$$

indicating that $m_s$ approaches 1 linearly as temperature decreases.

## 3.9 Roles of Entropy in Spin-Crossover Phenomena

In this section, we present the spin-crossover phenomenon as an example of systems where entropy plays a crucial role in phase transitions. This phenomenon involves a phase transition between an entropy-favored, highly degenerate phase and an energy-favored, low-energy phase. Extensive studies have been conducted on related materials [23].

Depending on the quantum states of electrons in atoms (or ions), an atom may exhibit bistable states known as the high-spin (HS) state and the low-spin (LS) state. For example, the ion $Fe^{3+}$ has five electrons in its 3d orbital. In the absence of an external force, the states of the 3d orbital are degenerate, and the five electrons form the maximum total spin state according to Hund's rule. In this case, the total spin is $S = 5/2$. On the other hand, if the ion is surrounded by ligands and experiences a crystal field, the 3d orbitals split into $e_g$ and $t_{2g}$ states. When the energy separation is large, the five electrons occupy the lower-energy levels. Due to the Pauli exclusion principle, electrons in the same orbital must have opposite spin directions, resulting in a total spin of $S = 1/2$. This situation is illustrated in Fig. 3.20.

The energy of the HS state, $E_{HS}$, is typically higher than that of the LS state, $E_{LS}$. The number of states (degeneracy) for the HS and LS states differ and are denoted as $n_{HS}$ and $n_{LS}$, respectively. Considering only the spin contributions, for $S = 5/2$, $n_{HS} = 6$, and for $S = 1/2$, $n_{LS} = 2$. However, in real systems, the primary contribution to degeneracy comes from the lattice degrees of freedom, and the ratio of the degeneracies, $n_{HS}/n_{LS}$, is often greater than 100.

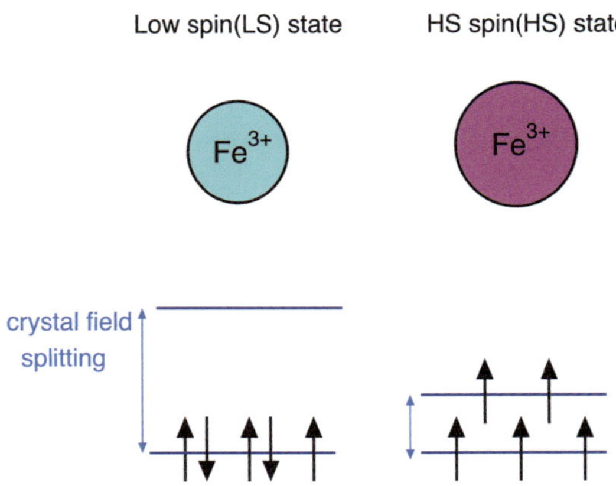


**Fig. 3.20** Low-spin state (LS) and high-spin state (HS) of $Fe^{3+}$

We denote HS and LS states by $s_i = \pm 1$, respectively. Introducing the following parameters:

$$E_0 = \frac{E_{HS} + E_{LS}}{2}, \quad D = \frac{E_{HS} - E_{LS}}{2}, \tag{3.176}$$

the Hamiltonian is written as

$$\mathcal{H} = -J \sum_{\langle ij \rangle} s_i s_j + D \sum_{i=1}^{N} s_i. \tag{3.177}$$

Here we must be careful about difference of the numbers of degeneracies, $n_{HS}$ and $n_{LS}$.

To analyze the model, it is convenient to express it by using the Ising variable $\sigma_i$ instead of $s_i$ [24, 25], where the states $\sigma_i = \pm 1$ ($n_+ = n_- = 1$). We now transform (3.177) using the Ising variable $\sigma_i$. The partition function, written in terms of the variable $s_i$ is given by

$$Z = \sum_{\text{all states}} e^{\beta\left(J \sum_{\langle ij \rangle} s_i s_j - D \sum_{i=1}^{N} s_i\right)}. \tag{3.178}$$

The Boltzmann factor of each state of $\{s_i\}$ can be expressed in terms of $\sigma_i = \pm 1$ as

$$e^{\beta\left(J \sum_{\langle ij \rangle} \sigma_i \sigma_j - D \sum_{i=1}^{N} \sigma_i\right)}. \tag{3.179}$$

Thus, taking degeneracies into account, the partition function is expressed as

$$Z = \sum_{\text{all states of} \sigma_i} \left( \prod_{k}^{N} n(\sigma_k) \right) e^{\beta\left(J \sum_{\langle ij \rangle} \sigma_i \sigma_j - D \sum_{i=1}^{N} \sigma_i\right)}. \tag{3.180}$$

Here, $n(\sigma_k)$ is the number of degeneracy for the state $s_k$ at the site $k$:

$$n(1) = n_{HS} = n_+, \quad n(-1) = n_{HS} = n_-, \tag{3.181}$$

which can be expressed as

$$n(\sigma_k) = \sqrt{n_+ n_-} e^{\frac{1}{2}\sigma_k \ln \frac{n_\pm}{n_-}} = \begin{cases} n_+ \text{ for } \sigma_k = 1 \\ n_- \text{ for } \sigma_k = -1 \end{cases}. \tag{3.182}$$

Using this notation, the partition function becomes

$$Z = \sum_{\{\sigma_i = \pm 1\}} e^{\beta\left(J \sum_{\langle ij \rangle} \sigma_i \sigma_j - D \sum_{i=1}^{N} \sigma_i\right)} (\sqrt{n_+ n_-})^N e^{\frac{1}{2} \ln \frac{n_\pm}{n_-} \sum_{i=1}^{N} \sigma_i}$$

$$= (\sqrt{n_+ n_-})^N \sum_{\{\sigma_i = \pm 1\}} e^{\beta\left(J \sum_{\langle ij \rangle} \sigma_i \sigma_j + \left(-D + \frac{1}{2} k_B T \ln \frac{n_+}{n_-}\right) \sum_{i=1}^N \sigma_i\right)}. \tag{3.183}$$

From this form, the model (3.177) is expressed by the Ising mode with the Hamiltonian:

$$\mathcal{H} = -J \sum_{\langle ij \rangle} \sigma_i \sigma_j + D \sum_{i=1}^{N} \sigma_i - \frac{1}{2} k_B T \ln \frac{n_+}{n_-} \sum_{i=1}^{N} \sigma_i + \text{const.}$$

$$= -J \sum_{\langle ij \rangle} \sigma_i \sigma_j - H_{\text{eff}} \left(-D + \frac{1}{2} k_B T \ln \frac{n_+}{n_-}\right) \sum_{i=1}^{N} \sigma_i + \text{const.} \tag{3.184}$$

Thus, the model is regarded as the Ising model with the temperature dependent external field:

$$H_{\text{eff}} = -D + \frac{1}{2} k_B T \ln \frac{n_+}{n_-}. \tag{3.185}$$

The critical properties of the spin-crossover (SC) model can be obtained from those of the Ising model in a field. The Ising model exhibits a phase transition at $T = T_C$, below which it undergoes a first-order phase transition as a function of the field $H$.

Let $T_0$ be the temperature at which $H_{\text{eff}} = 0$:

$$H_{\text{eff}} = 0 = -D + \frac{1}{2} k_B T_0 \ln \frac{n_+}{n_-} \implies T_0 = \frac{2D}{\ln \frac{n_+}{n_-}}. \tag{3.186}$$

Then, from the phase diagram $(T, H)$ of the Ising model, we know that the SC model shows a first-order phase transition when $T_0 < T_C$. On the other hand, if $T_0 > T_C$, the SC model undergoes a continuous, gradual change.

Which case occurs depends on the values of $D$ and $n_+/n_-$ [25–27]. Both cases have been observed in real SC materials. Indeed, various ordering processes occur in spin-crossover systems, including smooth crossover, first-order phase transitions, reentrant behavior leading to the metastable appearance of the HS state, and the presence of hidden LS states [25, 26]. Systematic changes in these ordering processes have been observed in experiments on diluted compounds such as $RbMn[Fe(CN)_6]$. For example, the reentrant appearance of the HS state has been identified in $Rb_{0.64}Mn[Fe(CN)_6]_{0.88}$ E1.7$H_2O$ [28], while the hidden LS state is observed in $Rb_{0.43}Mn[Fe(CN)_6]_{0.81}$ E3$H_2O$ [29]. Moreover, the reentrant appearance of the HS state provides a potential mechanism for the LIESST (Light-Induced Excited Spin State Trapping) phenomenon [30, 31].

### *3.9.1 Long-Range Interaction Due to Elastic Interaction*

A characteristic property of spin-crossover systems is the difference in degeneracy between HS and LS states, which is reflected in the temperature-dependent external field, as expressed in Eq. (3.185). In addition to this property, spin-crossover systems exhibit another unique feature due to the change in local lattice structure between HS and LS states, which is not typically considered in the standard Ising model.

In most cases, the HS state has a larger size than the LS state. As a result, inhomogeneity in $\sigma_i = \pm 1$ leads to lattice distortion, with which, elastic interaction is induced in the lattice. Figure 3.21 illustrates how the lattice distorts due to clusters of spin states. Such distortions incur significant energy costs, thereby preventing the formation of large clusters.

The elastic interaction creates an effective ferromagnetic long-range interaction among spins [32–34]. Consequently, the critical behavior of spin-crossover systems is expected to be of the mean-field type [35, 36]. The long-range nature of the interaction also gives rise to distinctive macroscopic ordering processes [37].

In the model described by (3.184), we assume a short-range interaction $J$ between neighboring sites to account for cooperative phenomena. This short-range interaction arises from local interactions that depend on configurations of neighboring spin, which has different role for the ordering from that of the long-range interaction due to the elastic effect [38]. Furthermore, when the short-range interaction is antiferromagnetic, its competition with the ferromagnetic long-range interaction leads to various interesting phase diagrams [39].

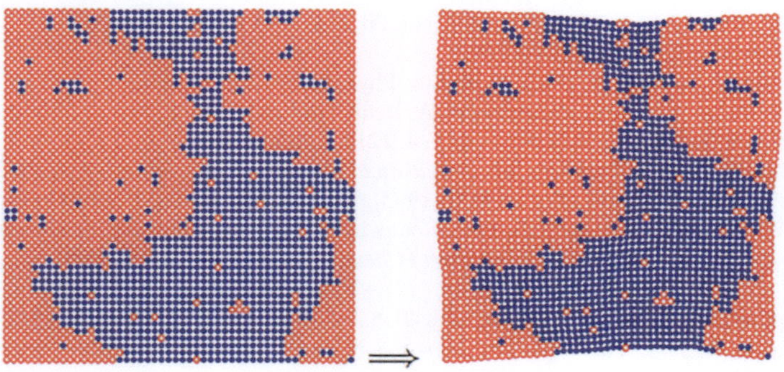

**Fig. 3.21** Deformation of the lattice due to differences in the sizes of HS and LS states, resulting in an effective long-range interaction among spins

## References


1. W. Lenz, Phys. Z., **21**, 613 (1920), E. Ising, Z. Physik, **31**, 253 (1925), S. G. Bruth, Rev. Mod. Phys. **39**, 883 (1967)
2. Y. Toga and S. Miyashita, unpublished
3. F. Wang, D.P. Landau, Phys. Rev. Lett. **86**. 2050 (2001); M. Troyer, S. Wessel, F. Alet, Phys. Rev. Lett. **90**, 120201 (2003)
4. S. Miyashita, *Yugen Ondo no Butsurigaku*, ed. by Parity Editorial Committee (Maruzen Publishing, 2004) (in Japanese)
5. R. Kubo, J. Phys. Soc. Jpn. **2**, 47 (1947), J. Phys. Soc. Jpn. **2**, 51 (1947), ibid., **3**, 119 (1948), **3**, 312 (1948), **4** 319 (1949). J. Phys. Soc. Jpn. **3**, 119 (1948). J. Phys. Soc. Jpn. **3**, 312 (1948), J. Phys. Soc. Jpn. **4** 319 (1949). R. Kubo, *GOMU DANSEI* in Japanese, (Shokabo 1996)
6. M. Madras and G. Slade, , *The Self-Avoiding Walk*, Birkhauser (Boston, 1993)
7. P.J. Flory, *Principles of Polymer Chemistry, Cornell University Press, Ithaca, NY (1953)* (P. J, Flory, Statistical Mechanics of Chain Molecules, Wiley, New York, 1969)
8. Y. Yoshikawa, N. Sakumichi, U. Chang. And T. Sasaki, Phys. Rev. X **11**,011045 (2021)
9. R.J. Farris, Rubber Chem. Tech, **52**, 159 (1979). W.B. Wiegand, Trans. Inst. Rubber Ind., **1**, 141 (1925); W.B. Wiegand, J.W. Snyder, Trans. Inst. Rubber Ind., **10**, 234 (1934). R. Hayward, Scientific American, **194**, 154 (1956); A. Furukawa, M. Eto, N. Hiraoks, T. Yokoyama, Nihon-Gomu-Kyoukai-shi (in Japanese), **64**, 255 (1991)
10. D. Liu, J. Fu, O. Elishav, M. Sakakibara, K. Yamanouchi, B. Hirshberg, T. Nakamuro, E. Nakamura, Science **384**, 1212 (2024)
11. L. Onsager, Phys. Rev. **65**, 117 (1944); G.H. Wannier, Rev. Mod. Phys., **17**, 50 (1945); B. Kaufman, Phys. Rev. **76**, 1232 (1949). B. Kaufman, L. Onsager, Phys. Rev. **76**, 1244 (1949); Y. Nambu, Prog. Thor. Phys. **1**, 5 (1950); C.N. Yang, Phys. Rev. **85**, 808 (1052)
12. R.M. MaCoy, T.T. Wu, *The Tow-dimensional Ising model* (Harvard Univ, Press, 1973)
13. T.D. Schultz, D.C. Mattis, E.H. Lieb, Rev. Mod. Phys. **36**, 856 (1964)
14. M. Kac, J.C. Ward, Phys. Rev. **88**, 1332 (1952)
15. C.A. Hurst, H.S. Green, J. Chem. Phys. **33**, 1059 (1960)
16. P.W. Kastelyn, J. Math. Phys. **4**, 287 (1963)
17. E.W. Montroll, R. Potts, J.C. Ward, J. Math. Phys. **4**, 308 (1963)
18. Soviet Physics JETP, **20**, 477 (1965)
19. V.G. Vaks, A.I. Larlin, N. Ovchinikov, Sov. Phys. JETP **22**, 820 (1966)
20. R P. Feynman, *Statistical Mecanics* (W. A. Benjamin Inc., 1972)
21. S. Sharman, J. Math. Phys. **1**,202 (1960), **4**, 1213 (1963)
22. R.J. Baxter, *Exact Solvable Models in Statistical Mechanics* (Academic Press, 1982)
23. O. Kahn, Molecular magnetism (VCII, 1993), in M.A. Halcrow(ed.), *Spin-Crossover Materials*(Wiley, 2013). G. Guetlich, H.E. Goodwin (eds.), *Spin Crossover in Transition Metal Compounds I, II, and III. Topics in Current Chemistry* (Springer, Berlin, 2004), vols. 233, 234, 235
24. J. Wajnflasz, J. (1970) Phys. Status Solidi **40**, 537-545 (1970); J. Wajnflasz, R. Pick, R.J. Phys. Colloq. **32**, C1 91-92 (1971); A. Bousseksou, J. Nasser, J. Linares, K. Boukheddaden, F. Varret, J. Phys. I France **2**, 1381–1403 (1992)
25. S. Miyashita, Y. Konishi, H. Tokoro, M. Nishino, K. Boukheddaden, F. Varret, Prog. Theor. Phys. **114**, 719 (2005)
26. S. Miyashita, Proc. Jpn Acad. Ser. B **86**, (643-666) (2010)
27. M. Sorai, S. Seki, J. Phys. Chem. Solid **35**, 555 (1972)
28. H. Tokoro, S. Miyashita, K. Hashimoto, S. Ohkoshi, Phys. Rev. B **73**, 172415 (2006)
29. H. Tokoro, S. Ohkoshi, Appl. Phys. Lett. **93**, 021906 (2008)
30. S. Decurtins et al., Inorg. Chem. **24**, 2174 (1985)
31. Y. Konishi, H. Tokoro, M. Nishino, S. Miyashita, J. Phys. Soc. Jpn. **75**, 114603 (2006)
32. M. Nishino, K. Boukheddaden, Y. Konishi, S. Miyashita, Phys. Rev. Letters **98**, 247203 (2007); K. Boukheddaden, M. Nishino, S. Miyashita, Phys. Rev. Lett. **100**, 177206 (2008). M. Nishino, K. Boukheddaden, S. Miyashita, Phys. Rev. B **79**, 012409 (1–4) (2009)

33. Y. Konishi, H. Tokoro, M. Nishino, S. Miyashita, Phys. Rev. Lett. **100**, 067206 (1-4) (2008)
34. C. Enachescu, L. Stoleriu, A. Stancu, A. Hauser, Phys. Rev. Lett. **102**, 257204 (2009); C. Enachescu, L. Stoleriu, A. Stancu, A. Hauser, Phys. Rev. B **82**, 104114 (2010); C. Enachescu, N. Nishino, S. Miyashita, in *Spin-Crosover Material* ed. by M.A. Halcrow (Wiley, 2013), pp. 455–474
35. S. Miyashita, Y. Konishi, M. Nishino, H. Tokoro, P.A. Rikvold, Phys. Rev. B **77**, 0144105 (2008)
36. T. Mori, S. Miyashita, P.A. Rikvold, Phys. Rev. E **81**, 011135 (2010)
37. M. Nishino, C. Enachescu, S. Miyashita, P.A. Rikvold, K. Boukheddaden, F. Varret , Sci. Rep.**1** 162, (2011)
38. T. Nakada, P. A. Rikvold, T. Mori, M. Nishino, S. Miyashita, Phys. Rev. B **84** , 054433 (2011). T. Nakada, T. Mori, S. Miyashita, M. Nishino, S. Todo, W. Nicolazzai, P.A. Rikvold, Phys. Rev. B **85**, 054408 (2012); M. Nishino, T. Nakada, C. Enachescu, K. Boukheddaden, S. Miyashita, Phys. Rev. B **88**, 094303 (2013)
39. M. Nishino, S. Miyashita, Phys. Rev. B**88** 014108 (2013). P.A. Rikvold, G. Brown, S. Miyashita, C. Omand, M. Nishino, Phys. Rev. B **93**, 064109 (2016); C.H. Chan, G. Brown, P.A. Rikvold, Phys. Rev. E **95**, 053302 (2017), Phys. Rev. B **96**, 174428 (2017); M. Nishino, S. Miyashita, P.A. Rikvold, Phys. Rev. B **96**, 144425 (2017); M. Nishino, P.A. Rikvold, C. Omand, S. Miyashita, Phys. Rev. B **98**, (2018) 144402

# Part II
# Frustration and Entropy

As mentioned in Sect. 3.8, thermal phase transitions occur due to competition between energy (interactions), which favors ordered states, and entropy, which favors disordered states. However, in some systems, conflicts among interactions arise, such that not all interactions can be satisfied simultaneously. In these cases, the ordering process involves not only the competition between energy and entropy but also competition among the interactions themselves. Determining how such systems order at low temperatures becomes a nontrivial problem, and the resulting ordering processes can be complex.

The effects of competing interactions have been studied in various contexts for many years. However, the explicit concept of frustration was introduced by Toulouse [1] in the study of spin glasses. Since then, the ordering processes of many uniformly frustrated systems have been extensively investigated, as discussed in sources such as Ref. [2].

In frustrated systems with discrete variables, such as Ising spins ($\sigma_{\rm i} = \pm 1$), macroscopic degeneracy may appear in the ground state. On the other hand, in systems with continuous spins (e.g., XY or Heisenberg spins), the order parameter, composed of several spins, is formed through a compromise among the frustrated interactions.

In Chap. 4, we study the effects of frustration on the ordering processes of systems with discrete variables. In Chap. 5, we examine systems with continuous spins. Quantum fluctuations also lead to interesting phenomena as studied in Chap. 6.

## References


1. G.Toulouse, Commun. Phys., **2**, 115, (1977).
2. *Magnetic systems with competing interactions (Frustrated spin systems)*, ed. by H. T. Diep, world scientific (1994) ISBN: 981-02-1715-3.

# Chapter 4
# Ground State Degeneracy in Frustrated Systems with Discrete Variables

**Abstract** In this chapter, we study the ground states and ordering processes of frustrated systems with discrete variables, such as Ising spins ($\sigma_i = \pm 1$):

$$\mathcal{H} = -\sum_{\langle ij \rangle} J_{ij}\sigma_i\sigma_j, \tag{4.1}$$

where $\langle ij \rangle$ denotes a pair of sites interacting with the coupling constant $J_{ij}$. Here, $J_{ij}$ can be positive for ferromagnetic couplings or negative for antiferromagnetic couplings, and not all interactions can be satisfied simultaneously. These models exhibit macroscopically degenerate ground states, resulting in residual entropy at $T = 0$. The nature of the degenerate ground state may lead to an order-by-disorder type phase transition. In frustrated systems, the ordering process is largely suppressed, and the nature of the phase transition is significantly modified, resulting in peculiar phenomena, such as partially disordered phases, reentrant phase transitions, and so on.

## 4.1 Residual Entropy: Macroscopically Degenerate Ground State

In frustrated systems with discrete variables, competing configurations can be energetically degenerate, resulting in macroscopic degeneracy $W$ of the ground state. When a macroscopic degeneracy remains in the ground state $W(0)$, the system exhibits the so-called residual entropy ($S_{\text{residual}} = k_{\text{B}} \ln W(0)$), i.e., the entropy at $T = 0$. In this section, we introduce some examples of such cases.

### *4.1.1 Ground State Degeneracy in Frustrated Lattices*

In Fig. 4.1, we depict lattices composed of antiferromagnetic bonds ($J_{ij} = J < 0$). In the non-frustrated case (Fig. 4.1a), the ground state is uniquely determined except

© The Author(s), under exclusive license to Springer Nature Singapore Pte Ltd. 2025

S. Miyashita, *Physics of Entropy*, Springer Series in Solid-State Sciences 212,
https://doi.org/10.1007/978-981-95-1758-9_4

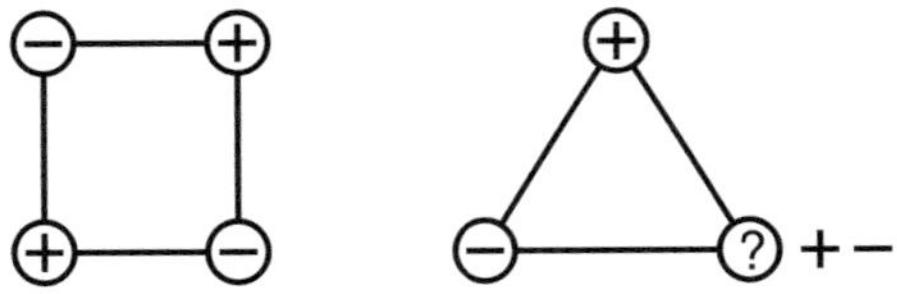


**Fig. 4.1** Spin configurations of antiferromagnetic Ising models **a** square lattice where antiferromagnetic order (Néel order) can be assigned without frustration, **b** triangular lattice where antiferromagnetic order (Néel order) cannot be assigned which causes frustration

for a global sign, resulting in two configurations: $(+, -, +, -)$ and $(-, +, -, +)$, which are known as Néel states. This degeneracy, caused by the global sign flip, does not depend on the size of the system.

In contrast, in the frustrated case (Fig. 4.1b), no configuration satisfies all the bond. Consequently, the lowest-energy states (the ground states) are degenerate, comprising six configurations such as $(+, -, +)$, $(+, -, -)$, …, and $(-, -, +)$. Unlike the non-frustrated case, the number of degenerate states increases exponentially with the size of the system.

### 4.1.2 Antiferromagnet on the Triangular Lattice (AFT) Ising Model

As mentioned above, there is no configuration that satisfies all bonds in the ground state of the antiferromagnet on a triangular lattice composed of triangles depicted in Fig. 4.1b. On this lattice, it is readily observed that many states are degenerate in the ground state.

Thanks to techniques enabling the exact solution of thermodynamic properties for the Ising model with non-crossing bonds on two-dimensional lattices (see Sect. 4.8), the thermal properties of the model can be obtained precisely even for the frustrated models.

The exact solution of the antiferromagnetic triangular Ising model (AFTI) was independently obtained in 1950 by several groups, including G. M. Wannier, H. Husimi and I. Syozi, and R. M. F. Houtappel [3–6]. These studies revealed that the system exhibits no phase transition at finite temperatures. The degeneracy $W$ of the ground state is calculated from the residual entropy at $T = 0$, which was determined by Wannier [4] as

$$\frac{S(0)}{Nk_B} = 0.323066\ldots, \tag{4.2}$$

indicating that a macroscopic degeneracy remains at $T = 0$:

$$W = e^{0.323066N}. \tag{4.3}$$

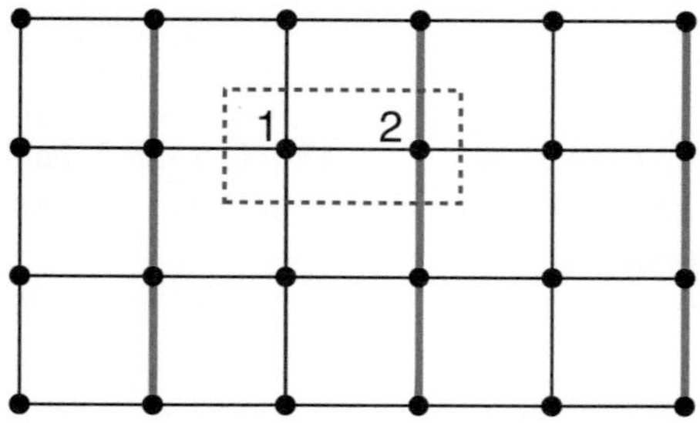


**Fig. 4.2** Villain lattice. Thin black lines represent ferromagnetic bonds ($J_{ij} = J$), and bold red lines represent antiferromagnetic bonds ($J_{ij} = -J$). The sites enclosed within the dashed block form the unit cell of the lattice, which will be used in Sect. 4.8.2

Not only the partition function but also the spin correlation function within each sublattice in the ground state was obtained by Stephenson [7] using Kasteleyn's method [8]. For a direction along the lattice, the spin correlation function is given by

$$\langle S(0)S(r)\rangle \simeq 0.63222608 \left(\frac{r}{a}\right)^{-1/2} \cos\left(\frac{2\pi r}{3a}\right), \tag{4.4}$$

where $a$ is the lattice constant. Although long-range order does not exist, it is notable that the correlation does not exhibit an exponential decay but rather follows a power-law decay. This behavior suggests development of sublattice ordering, and it may be appropriate to regard $T = 0$ as a kind of critical point.

As another example of a fully frustrated model, Villain proposed a model on a square lattice, as depicted in Fig. 4.2 which is called "Villain odd model" [9]. In this model, the interactions in every second row are assigned to be negative (antiferromagnetic), such that every unit loop of the square lattice contains one antiferromagnetic bond, making it frustrated, i.e., the product of interactions around each loop satisfies the relation:

$$\Pi_{\text{loop}} J_i = -J^4. \tag{4.5}$$

As explained in Sect. 4.8.2, the free energy of the model is given by

$$f = \frac{1}{N} \ln Z = -k_B T \ln\left(2\cosh 2J/k_B T\right)$$

$$-k_B T \frac{1}{4\pi^2} \int_0^{\pi} du \int_0^{\pi} dv \ln\left[(1+z^2)^2 - 2z^2 \cos u - 2z\, 2\cos v\right], \tag{4.6}$$

where

$$z = \tanh \frac{2J}{k_B T}. \tag{4.7}$$

This model shares common features with the antiferromagnetic Ising model on the triangular lattice. The ground-state entropy is

$$\frac{S(0)}{Nk_B} \simeq 0.246, \tag{4.8}$$

which will be obtained in Sect. 4.8.2, and the correlation function at $T = 0$ of this model exhibits a power-law decay with an exponent of $\eta = 1/2$ [10], which will be discussed in Sect. 4.2.

### 4.1.3 Kagome-AF Model

The ground-state entropy of the Ising kagome antiferromagnet has also been obtained as [11]:

$$S(0) = Nk_{\rm B} \times 0.502\ldots. \tag{4.9}$$

Using a dual lattice transformation with star-triangle decimation, it is known that the state of the AF-kagome-Ising model maps to a state of the honeycomb lattice at $T \geq 2/\ln 2 > T_{\rm C}$. Thus, states at all temperatures, including the ground state, are paramagnetic, where the spin correlation function decays exponentially [11, 12], in contrast to the case of the AFT-Ising model.

### 4.1.4 Frustration in Three Dimensions

So far, we have focused on models in two dimensions. The layered triangular lattice, i.e., the hexagonal lattice, is the simplest extension to three dimensions. It has been known that the model shows a phase transition at finite temperature below which the long-range order appears while one of the sublattice remains disorder [13, 14]. This ordered phase is the type of PD phase which will be introduce in Sect. 4.5.

Frustration occurs in plaquettes consisting of an odd number of antiferromagnetic bonds. Thus, all lattices with triangles as their fundamental units are frustrated. In three dimensions, tetrahedra are composed of triangles, and antiferromagnets on lattices formed by tetrahedra are also frustrated. The face-centered-cubic (fcc) is a lattice composed of tetrahedra and is highly frustrated. But it is known that the model exhibits a first-order phase transition at finite temperature [15].

The pyrochlore structure is a typical example, consisting of corner-sharing tetrahedra, as depicted in Fig. 4.3a. Here, the sites marked by green circles form a plane of the kagome lattice. The lattice can therefore be regarded as a layered kagome lattice, connected by the corner-sharing tetrahedron sites marked with red squares. As a result, the lattice is highly frustrated.

Within each tetrahedron, the four spins are mutually frustrated, resulting in the so-called two-in-two-out state, as illustrated in Fig. 4.3b. In the case of the Heisenberg spin model, defining the local axes along the directions of the arrows on the sites, the system is effectively described by an Ising model [16].

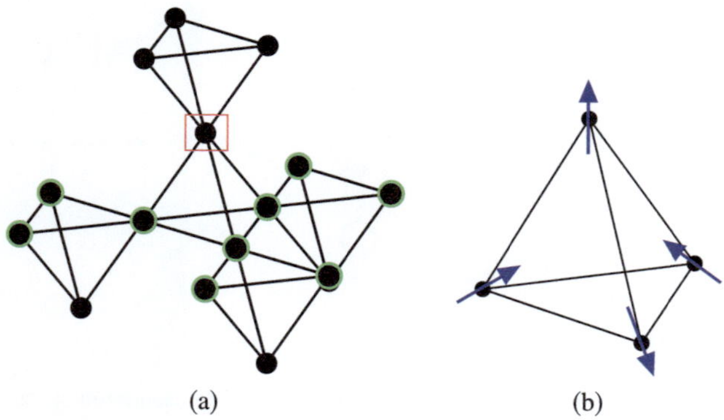


**Fig. 4.3** **a** Pyrochlore lattice. The sites marked by green circles form a plane of the kagome lattice. **b** Spin configuration of the antiferromagnetic Heisenberg model on a tetrahedron

When the two-in-two-out state is disrupted, local violations of the two-in-two-out rule occur in pairs, and each violation acts as a magnetic monopole [17]. The total configurations are macroscopically degenerate, and the entropy at $T = 0$ is approximately estimated as

$$S(0) \simeq Nk_{\mathrm{B}} \times 0.202\ldots. \tag{4.10}$$

Various interesting properties emerge, such as a novel type of first-order phase transition during the magnetization process [18]. The phenomena in this highly degenerate system are collectively referred to as "spin-ice", which has been extensively reviewed in a recently published book [19].

### 4.1.5 Residual Entropy of Ice

A famous example of macroscopic degeneracy in the ground state is the residual entropy of ice, which has been studied as the "ice model." In the ground state of ice, two hydrogen atoms tend to come close to an oxygen atom, while the other two move away from the oxygen atom, as depicted in Fig. 4.4a. This degeneracy was introduced by Linus Pauling [20] and is known as the "ice rule". There are macroscopically many states that satisfy this condition. Counting the number of such degenerate states is referred to as the ice problem. There are 16 $(= 2^4)$ possible configurations of hydrogen atoms on the four bonds around an oxygen atom, and 6 of them satisfy the ice rule, as will be shown in Fig. 4.28. Thus, the approximate number of degenerate states is estimated as

$$W_0 = 2^{2N}\left(\frac{6}{16}\right)^N \simeq 1.5^N, \tag{4.11}$$

where $2N$ is the number of bonds.

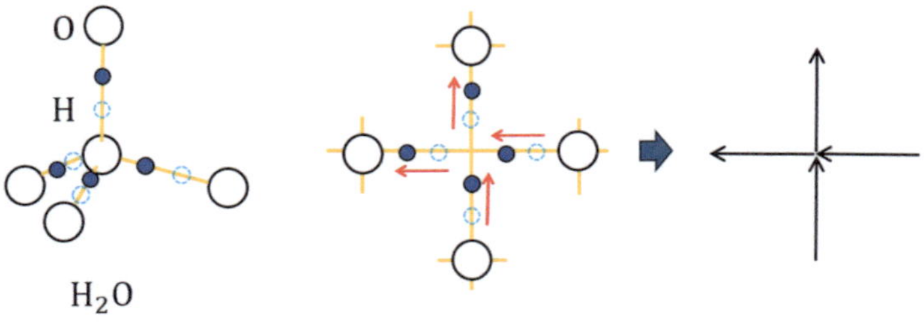


**Fig. 4.4** **a** Freedom of positions of hydrogen atoms in ice. **b** Representation of the ice rule configuration by a vertex composed of arrows corresponding to the hydrogen position

However, the configurations around oxygen atoms are not independent, and the precise count of allowed configurations must account for their mutual relationships. This problem in two dimensions has been studied using the so-called vertex model. The vertex model is a powerful method for studying frustrated systems, which we will briefly explain in Sect. 4.9.

The degeneracy of the ice model was calculated by Lieb [21, 22] using the transfer matrix method for the six-vertex model and the number of degenerate states was obtained as

$$W_{\text{2D-ice}} = \left(\frac{4}{3}\right)^{3N/2}, \quad \frac{S_{\text{2D-ice}}}{Nk_{\text{B}}} = 1.5396\ldots, \tag{4.12}$$

where $N$ is the number of vertices in the lattice.

### *4.1.6 Relation to the Third Law of Thermodynamics*

The third law of thermodynamics states that the entropy per degree of freedom approaches zero as the temperature approaches absolute zero. However, the above-mentioned systems exhibit macroscopic degeneracy, $W(0) \propto \alpha^N$, where $\alpha$ is a positive constant and $N$ is the number of degrees of freedom. In such cases, the entropy remains finite even in the ground state:

$$S(0) = k_{\text{B}} \ln W(0) = Nk_{\text{B}} \ln \alpha. \tag{4.13}$$

Some models, such as the ice model, exhibit this property, appearing to violate the third law of thermodynamics. In natural systems, however, a perfectly uniform system is rarely realized, and exact macroscopic degeneracy is typically lifted by local

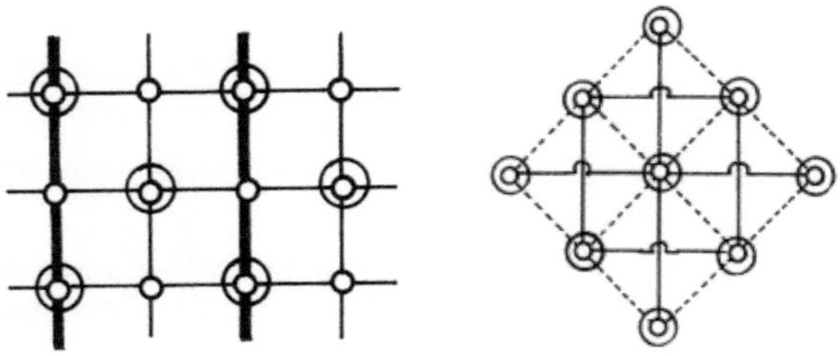


**Fig. 4.5** **a** Villain odd model. Thin lines represent ferromagnetic bonds ($J$), and bold lines represent antiferromagnetic bonds ($-J$). The spins on the sublattice denoted by large circles are traced out. **b** The effective lattice of the Ising model for the remaining spins after tracing out the sublattice spins. On this lattice, the effective nearest-neighbor (nn) interactions ($K_1$) are represented by dashed lines, which vanish in the Villain odd model. The effective next-nearest-neighbor (nnn) interaction $K_2$, and four-spin interaction $K_4$, are temperature-dependent. (These figures are reproduced from Ref. [23].©1983 The Physical Society of Japan)

randomness. Even in a perfectly uniform system, quantum mechanical interactions can resolve the degeneracy. Thus, in practice, the third law of thermodynamics holds. Nevertheless, even approximately macroscopic degeneracy can lead to interesting physical phenomena.

## 4.2 Forgacs Model

Forgacs studied the spin-correlation function of the Villain odd model, depicted in Fig. 4.2, using a dexterous method [10]. When the spins on one of the sublattices ($\{s_i\}$) (denoted by single circles in Fig. 4.5) are traced out, the model is transformed into a model of the remaining sublattice (denoted by double circles) with nearest-neighbor (nn) interaction $K_1$, next-nearest-neighbor (nnn) interaction $K_2$, and four-spin interaction $K_4$ on the remaining sublattice ($\{\sigma_i\}$):

$$\sum_{s=\pm 1} e^{\beta s(\sigma_1-\sigma_2+\sigma_3-\sigma_4)} = Ae^{\frac{1}{2}K_1(\sigma_1\sigma_2+\sigma_2\sigma_3+\sigma_3\sigma_4+\sigma_4\sigma_1)+K_2(\sigma_1\sigma_3+\sigma_2\sigma_4)+K_4\sigma_1\sigma_2\sigma_3\sigma_4}. \quad (4.14)$$

The effective interaction parameters are given as follows:

$$K_1 = 0, \quad K_2 = \frac{1}{8}\ln(\cosh 4K),$$
$$K_4 = -\frac{1}{8}\ln(\cosh 4K) - \frac{1}{2}\ln(\cosh 2K) = -K_2\left(1 - 4\frac{\ln\cosh 2K}{\ln\cosh 4K}\right), \quad (4.15)$$

and

$$A = e^{\left(\frac{1}{2}\ln(\cosh 2K)+\frac{1}{8}\ln(\cosh 4K)\right)}. \tag{4.16}$$

It should be noted that the nearest-neighbor interaction $K_1$ vanishes, reflecting the frustrating nature of the system.

As will be explained in Sect. 4.9, the Ising model with $K_2$ and $K_4$ is mapped to the 8-vertex model with the set of weights given in (4.151). For the present model, the parameters of the Baxter's model (4.152) in Sect. 4.9 are

$$a = b = Ae^{-K_4}, \quad c = Ae^{2K_2+K_4}, \quad d = Ae^{-2K_2+K_4}, \tag{4.17}$$

which yields

$$\frac{a}{c} = \frac{b}{c} = e^{-2K_2-2K_4}, \quad \frac{c}{c} = 1, \quad \frac{d}{c} = e^{-4K_2}. \tag{4.18}$$

Hereafter, we use the set of weights:

$$a = b = e^{-2K_2-2K_4}, \quad c = 1, \quad d = e^{-4K_2}, \tag{4.19}$$

because the common factor $A$ only contributes a non-critical factor in the partition function, and the phase diagram is independent of it.

Baxter obtained phase diagram of the present type of models is given in the space $(a, d)$ as depicted in Fig. 4.6. The bold line $(2a + d = 1)$ is the so-called Baxter's critical line, on which the critical exponent $\alpha$ changes continuously. The line $d = 0$ corresponds to the F model, on which the critical exponent of the spin correlation function changes continuously.

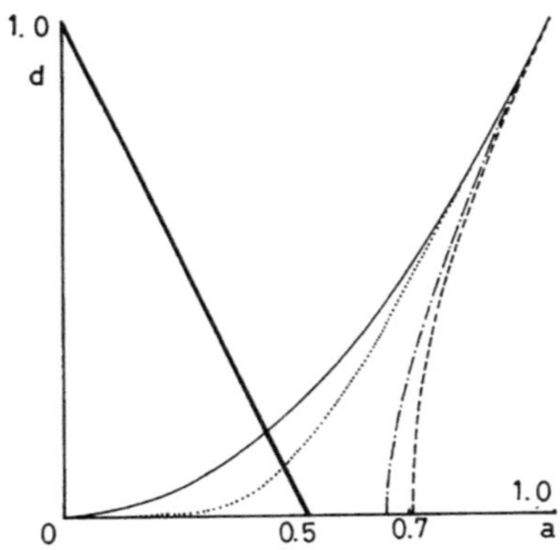


**Fig. 4.6** Phase diagram in $(a, d)$. The bold line $(2a + d = 1)$ is Baxter's critical line. The thin solid curve represents the trajectory of the Ising ferromagnetic model. The dashed curve represents that of the Villain odd model. The dotted line represents a model with next-nearest-neighbor interaction on the sublattice of the Villain odd model, and the dashed-dotted line represents a model with next-nearest-neighbor decorated interaction explained in Sect. 4.2.1. (This figure is reproduced from Ref. [23]. ©1983 The Physical Society of Japan)

The trajectory of the regular ferromagnetic model is also plotted by the solid curve. The Villain odd model (4.19) corresponds to the trajectory of the dashed line in the phase diagram (Fig. 4.6).

At $T = 0$ ($K = \infty$), we have

$$K_2 = \infty, \quad K_4 \simeq K_2, \tag{4.20}$$

and thus

$$a = b = 0, \quad c = \infty, \quad d = 0 \quad \Rightarrow \quad \frac{a}{c} = \frac{b}{c} = \frac{1}{\sqrt{2}}, \quad \frac{c}{c} = 1, \quad \frac{d}{c} = 0. \tag{4.21}$$

The present case ($d/c = 0$) corresponds to the F model. At the point $a/c = 1/\sqrt{2}$, the spin correlation function decays in a power law with $\eta = 1/2$ [10]:

$$\langle S(0)S(r)\rangle \propto \frac{1}{r^{1/2}}, \tag{4.22}$$

which exhibits the same behavior as the AFTI model.

### 4.2.1 Modified Forgacs Model

Along the Baxter critical line, the exponent $\eta$ is $1/4$. To study models that interpolate between the Villain odd model and the ferromagnetic Ising model, we propose a modification to the Forgacs model [23]. To enhance ordering, we introduced an additional ferromagnetic next-nearest-neighbor (nnn) interaction $J_2$ within a sublattice, as represented by the wavy bonds in Fig. 4.7a.

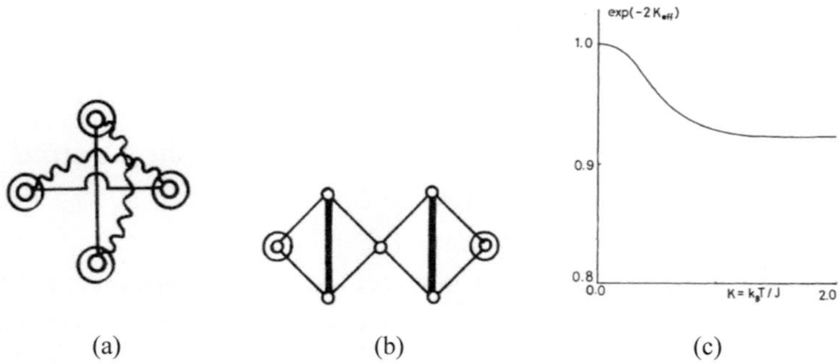


**Fig. 4.7** **a** Modified Forgacs model with next-nearest-neighbor (nnn) interactions within a sublattice. **b** A decorated bond used for the nnn interaction in the sublattice. **c** Temperature dependence of $\exp(-2K_{\text{eff}-\text{deco}})$. (These figures are reproduced from Ref. [23]. ©1983 The Physical Society of Japan)

The vertex weights for this modified model are given by

$$a = b = e^{-2(K_2-\beta J_2)-2K_4}, \quad c = 1, \quad d = e^{-4K_2+\beta J_2}. \tag{4.23}$$

The temperature dependence of the parameters $a$ and $d$ is shown as a dotted line in Fig. 4.6, which intersects the Baxter critical line. For $J_2 > 0$, the modified model exhibits a phase transition between a disordered phase and an ordered phase. The critical temperature decreases as $J_2$ becomes smaller, as expected. It is notable that while the critical exponent $\eta = 1/4$ remains constant, the critical exponent $\alpha$ varies continuously with $J_2$, as will be described by (4.154) in Sect. 4.9.2, which violates the concept of universality. Since the exponent $\eta = 1/4$ remains constant and the exponent $\nu$ changes along the Baxter critical line, the ratio of the critical exponents $\gamma/\beta$ remains constant. This behavior is referred to as weak universality [24].

When $J_2 = 0$, we recover the Villain odd model, and for $J_2 \neq 0$, the model exhibits a phase transition along the Baxter critical line. Thus, the dependence of the exponent $\eta$ on $J_2$ is singular. To realize a model whose trajectory lies between the limiting cases of $J_2 \to 0$ and the Villain odd model, the model is extended to include a decorated next-nearest-neighbor (nnn) interaction (Fig. 4.7b). The effective interaction for the decorated bond is given by

$$Ae^{K_{\rm eff-deco}\sigma_1\sigma_2} = \sum_{s_1=\pm1}\sum_{s_2=\pm1}\sum_{s_3=\pm1}\sum_{s_4=\pm1}\sum_{s_5=\pm1} e^{\beta(J\sigma_1(s_1+s_2)+Js_3(s_1+s_2+s_4+s_5)+J\sigma_2(s_3+s_4)-2J(s_1s_2+s_4s_5))}.$$

The temperature dependence of $K_{\rm eff-deco}$ is shown in Fig. 4.7c, where $K_{\rm eff-deco}$ is observed to remain finite even in the limit $T \to 0$.

Along the line where $d = 0$, the exponents change continuously [25], with the behavior resembling that of the Gaussian model [26–28]. The correspondence to the F-model (or BCSSOS model) has been clarified [29]. Thus, in this model, the exponent $\eta$ of the ground state spin correlation function changes continuously [23].

## 4.3 Order by Disorder: Long-Range Order in the Ground State of AFT-Ising with Large $S$

So far, we have studied the $S = 1/2$ Ising model ($\sigma_i = \pm1$), where it is well known that the phase at finite temperatures is paramagnetic. It is also established that the sublattice spin correlation length diverges as $T \to 0$, indicating that the system is marginally ordered within each sublattice.

For large values of $S$, where $S_z$ takes $2S+1$ discrete values $(S, S-1, \ldots, -S)$, it has been found that a long-range order exists in the ground state, even though the lattice remains fully frustrated. In this section, we study the ground state of the AFT-Ising model for large $S$. The Hamiltonian is given by

$$\mathcal{H} = -J \sum_{\langle ij \rangle} S_i^z S_j^z, \quad (J < 0) \quad \rightarrow \quad \mathcal{H} = |J| \sum_{\langle ij \rangle} S_i^z S_j^z. \tag{4.24}$$

In the case of $S = 1/2$, the spins at the disordered sites, at which internal fields from neighbors cancel out to zero, take either value of $\pm 1/2$. When $S > 1/2$, the spin at the disordered site can take any value from $S, S-1, \ldots, -S$. We refer such site as "free site", and the spin on it as a "free spin". Therefore, the degeneracy of the free site is $2S+1$, but not 2 for $\pm S$ only. As a result, configurations with a larger number of free sites are entropically favored. For a given configuration, let the number of free sites be $N_\mathrm{f}$. The total number of degenerate states in this configuration is then $(2S+1)^{N_\mathrm{f}}$.

Denoting the free spin by 0, there are six possible configurations for the three sublattices:

$$(-, 0, +), (-, +, 0), (0, +, -), (+, 0, -), (+, -, 0), (+, -, 0).$$

If the system adopts one of these configurations uniformly across the system, the value of $N_\mathrm{f}$ is maximized:

$$N_\mathrm{f} = \frac{N}{3}, \tag{4.25}$$

as shown in Fig. 4.8a.

On the other hand, if two different configurations coexist, a domain-wall-like structure appears, as shown by the dotted line in Fig. 4.8b. At the boundary (the domain wall region), free spins cannot be located, resulting in a decrease in $N_\mathrm{f}$. In Fig. 4.8b, $N_\mathrm{f}$ decreases by 3 due to the presence of the domain wall. Thus, from an entropic perspective, the configuration shown in Fig. 4.8a is more favorable.

The total degeneracy is given by the product of $(2S+1)^{N_\mathrm{f}}$ and the number of possible arrangements of the domain walls. The number of arrangements for domain walls of length $\ell$ is at most $3^\ell$. The loss in $N_\mathrm{f}$ caused by the domain wall is proportional to $\ell$, which reduces the degeneracy by a factor of $(2S+1)^{-c\ell}$, where $c$ is a positive constant of $O(1)$. Therefore, the total degeneracy is bound by

$$3^\ell \times (2S+1)^{-c\ell} = e^{\ell(\ln 3 - c \ln(2S+1))}. \tag{4.26}$$

For large values of $S$,

$$\ln 3 - c \ln(2S+1) < 0, \tag{4.27}$$

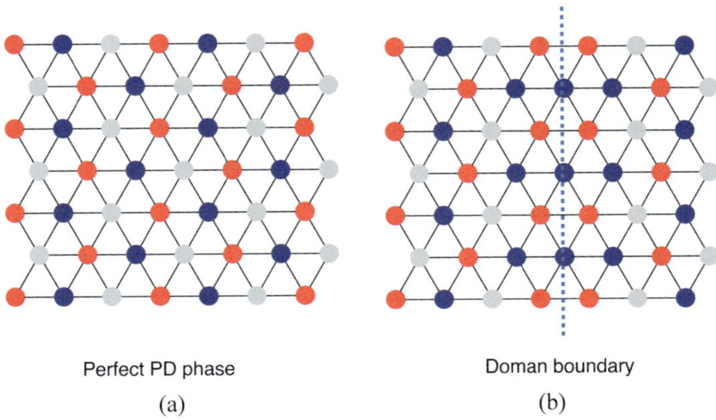


**Fig. 4.8** Configurations of the ground States of AFT-Ising model with large $S$. Configurations are characterized by $+S$ spins (red circles) and $-S$ spins (blue circles). Spins (gray circles) can take any value among $S, S-1, \ldots, -S$, and such sites are referred to as "free spins". **a** Perfect PD Phase (If all the spins on sublattice C are $\pm S$, the configuration becomes a 2-FR phase.) **b** Mixed PD Configurations: This configuration consists of two types of PD structures: the left side follows the $(+, -, 0)$ pattern, while the right side adopts the $(-, +, 0)$ pattern. This configuration also belongs to the degenerate ground states. The dashed line indicates a kind of domain wall separating different PD phases

and formation of large $\ell$ domain walls becomes unfavorable. This implies that, for large $S$, the ground state exhibits long-range order. The mechanism driving the emergence of long-range order comes from the difference in the number of states within the degenerate ground-state manifold. Thus, the long-range order arises as a result of an entropy effect [30], serving as an example of the phenomenon known as "order-by disorder" [31].

Regarding the critical value of $S$, extensive studies using the transfer matrix method have been conducted [32, 33]. These studies suggest that there is no long-range order for $S = 1/2, 1, \ldots, 5/2$, and the spin correlation function decays as a power law with $S$-dependent exponents $\eta(S)$:

$$\eta(1/2) = 0.5, \eta(1) \simeq 0.32, \eta(3/2) \simeq 0.24, \eta(2) \simeq 0.19, \text{ and } \eta(5/2) \simeq 0.15. \tag{4.28}$$

The transition from the critical phase with power-law spin correlation functions ($S < S_C$) to the phase with long-range order ($S > S_C$) has been discussed in the context of the Gaussian model with clock-model-type perturbations. The analysis indicates that long-range order emerges when $\eta(S) < 1/9$. Based on this observation, cases with $S > 3$ are most likely to exhibit long-range order [33].

The possibility of this entropy-driven effect causing an ordered state at finite temperatures has also been investigated included the classical spin case ($S = \infty$) [32], in which it was reported that a finite-temperature phase transition is highly plausible. If such a transition exists, understanding the type of criticality involved becomes an intriguing problem. At high temperatures, the system is in a disordered state dominated by entropy effects. At low temperatures, the model exhibits a long-range order, but this order is entropy driven. Thus, two distinct entropy-driven phases are present.

## 4.4 Effects of Entropy on Ordering Processes at Finite Temperatures

Frustration significantly affects not only the ground state but also the system's behavior at finite temperatures. Entropy plays a crucial role in determining which of the competing orders is realized. Since the entropic contribution to the free energy is proportional to the temperature, the preferred state can depend on temperature.

Frustration weakens spin correlations, which may lead to marginal ordering, as observed in spin glasses [34]. Additionally, frustration induces a nonmonotonic temperature dependence in the spin correlation function, reflecting the degeneracy of states (an entropy effect)[35]. By making use of such property, complex effective interactions can be engineered using the so-called "decorated bond" which was studied in detail by Syozi [36].

### *4.4.1 Non-monotonic Behavior of the Spin Correlation Function*

A simple example of such behavior can be demonstrated using the decorated bond model (Fig. 4.9). This decorated bond between $\sigma_1$ and $\sigma_2$ has two interaction paths, and its Hamiltonian is given by

$$\mathcal{H} = J_1\sigma_1\sigma_2 + J_2\left(\sigma_1\sigma_3 + \sigma_3\sigma_2\right), \quad J_2 > J_1 > 0. \tag{4.29}$$

Here, the direct interaction term $J_1\sigma_1\sigma_2$ induces an antiferromagnetic (antiparallel) correlation between $\sigma_1$ and $\sigma_2$, while the indirect interaction term through $\sigma_3$, $J_2\left(\sigma_1\sigma_3 + \sigma_3\sigma_2\right)$, induces a ferromagnetic (parallel) correlation. When $|J_2| > J_1 > 0$, these competing interactions result in a non-trivial temperature-dependent behav-

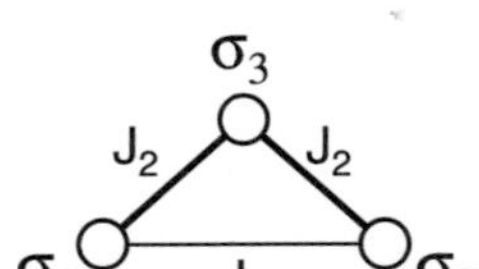


**Fig. 4.9** A decorated bond between $\sigma_1$ and $\sigma_2$

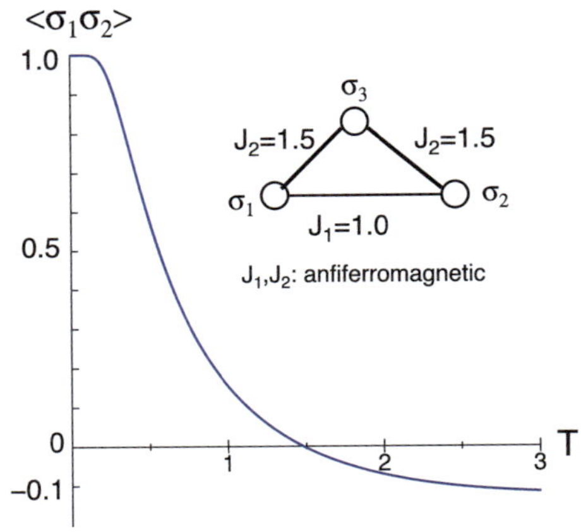


**Fig. 4.10** Temperature dependence of $\langle\sigma_1\sigma_2\rangle$ for $J_1 = 1.0$ and $J_2 = 1.5$

ior of the correlation function $\langle\sigma_1\sigma_2\rangle$. In Fig. 4.10, the temperature dependence of $\langle\sigma_1\sigma_2\rangle$ is shown for $J_1 = 1.0$ and $J_2 = 1.5$.

Since this system consists of only three spins, the partition function and correlation function can be calculated explicitly. For configurations where $\sigma_1 = \sigma_2$ (parallel alignment), the possible states are $(+\,+\,+)$, $(-\,-\,-)$, $(+\,-\,+)$, and $(-\,+\,-)$, with corresponding energies of $J_1 + 2J_2$, $J_1 + 2J_2$, $J_1 - 2J_2$, and $J_1 - 2J_2$, respectively. For configurations where $\sigma_1 = -\sigma_2$ (antiparallel alignment), the states $(+\,+\,-)$, $(+\,-\,-)$, $(-\,-\,+)$, and $(-\,+\,+)$ all have the same energy, $-J_1$. The partition function is therefore given by

$$Z = e^{-\beta J_1}\left(2e^{2\beta J_2} + 2e^{-2\beta J_2}\right) + 4e^{\beta J_1}, \tag{4.30}$$

and the spin correlation function is

$$\langle\sigma_1\sigma_2\rangle = \frac{e^{-\beta J_1}\left(2e^{2\beta J_2} + 2e^{-2\beta J_2}\right) - 4e^{\beta J_1}}{e^{-\beta J_1}\left(2e^{2\beta J_2} + 2e^{-2\beta J_2}\right) + 4e^{\beta J_1}}. \tag{4.31}$$

Since $J_2 > J_1$ at low temperatures (large $\beta$), the spin correlation function is approximately given by

$$\langle\sigma_1\sigma_2\rangle \simeq \frac{2e^{\beta J_2}e^{-\beta J_1}}{2e^{\beta J_2}e^{-\beta J_1}} = 1. \tag{4.32}$$

indicating a ferromagnetic correlation, while at high temperatures (small $\beta$), the spin correlation function becomes

$$\langle\sigma_1\sigma_2\rangle = \frac{e^{-\beta J_1}\left(2e^{2\beta J_2} + 2e^{-2\beta J_2}\right) - 4e^{\beta J_1}}{e^{-\beta J_1}\left(2e^{2\beta J_2} + 2e^{-2\beta J_2}\right) + 4e^{\beta J_1}} \simeq -\beta J_1 < 0, \tag{4.33}$$

indicating an antiferromagnetic correlation. The temperature at which the correlation becomes zero (disordered point [37]) is given by

$$e^{-\beta J_1}\left(2e^{2\beta J_2}+2e^{-2\beta J_2}\right)-4e^{\beta J_1}=0 \rightarrow e^{2\beta J_1}=\cosh(2\beta J_2). \tag{4.34}$$

This nonmonotonic behavior can be understood as follows. At $T=0$, the spins $\sigma_1$ and $\sigma_2$ align parallel because the interaction along the indirect path $J_2$ is stronger than the direct interaction ($J_1$). However, at high temperatures, thermal fluctuations weaken the effective interaction through the indirect path ($J_2$), which is interpreted as an entropy effect. As a result, the direct interaction ($J_1$) dominates, and an anti-ferromagnetic correlation emerges.

### 4.4.2 Effective Interaction

Here, we introduce the concept of "effective interaction" mediated by the decoration bond. The thermal weight for a given configuration of $\sigma_1$ and $\sigma_2$ through the indirect path is expressed as

$$\sum_{\sigma_2=\pm 1} e^{-\beta J_2(\sigma_1\sigma_3+\sigma_3\sigma_2)}=2\cosh(\beta J_2\sigma_1\sigma_2). \tag{4.35}$$

We define the effective interaction $K_{\text{eff indirect}}$ through the indirect path as

$$2\cosh(\beta J_2\sigma_1\sigma_2)=A(T)e^{K_{\text{eff indirect}}\sigma_1\sigma_2}, \tag{4.36}$$

where

$$A(T)=\sqrt{4\cosh(2\beta J_2}, \quad K_{\text{eff indirect}}=\frac{1}{2}\ln\left(\cosh 2\beta J_2)\right)=\tanh^{-1}\left[\tanh^2(\beta J_2)\right]. \tag{4.37}$$

Thus, the total interaction between $\sigma_1$ and $\sigma_2$ is given by

$$K_{\text{eff}}=-\beta J_1+K_{\text{eff indirect}}=\frac{1}{2}\ln\left(\cosh 2\beta J_2)\right)-\beta J_1. \tag{4.38}$$

To analyze the temperature dependence of $K_{\text{eff}}$, we consider its asymptotic behavior:

$$\frac{1}{2}\ln\left(\cosh 2\beta J_2)\right)-\beta J_1 \simeq \begin{cases}\simeq \beta(J_2-J_1) & \beta J_2 \gg 1\\ \simeq \beta^2 J_2^2-\beta J_1 & \beta J_2 \ll 1.\end{cases} \tag{4.39}$$

Here, we confirm the above observation in Fig. 4.10. This type of temperature-dependent spin correlation can give rise to distinct ordered phases at different temperatures, as will be discussed in the following sections. Additionally, it may lead to the so-called rejuvenation phenomena, where the ordering process observed at one temperature does not persist at another temperature. Such phenomena will be further explained in Sect. 4.7.1.1.

### 4.4.3 Reentrant Phase Transition with Decoration Bonds

So far, we considered the spin correlation function of the neighboring sites. Now, we consider phase transitions induced by the sign change of the effective interaction. For a square lattice, the critical temperature of the ferromagnetic Ising model is given by $1/k_B T_C = K_C = \sqrt{2} - 1$.

When the effective interaction between the nearest neighbor spins exceeds the critical value $K_C$, the system transitions to an ordered state. The model consisting of decoration bonds between neighboring lattice sites is equivalent to a uniform model with the coupling as an effective constant.

For the single decoration bond depicted in Fig. 4.9, the maximum value of $|K_{\rm eff}|$ in the antiferromagnetic side is less than $K_C$, although it exceeds $K_C$ in the ferromagnetic side at low temperatures. Thus, the model with the decoration bond depicted in Fig. 4.9 exhibits only a phase transition to a ferromagnetic phase. To enlarge $|K_{\rm eff}|$ in the antiferromagnetic side, we set $n$ decoration bonds between $\sigma_1$ and $\sigma_2$, leading to the effective interaction as

$$K_{\rm eff}^{(n)} = nK_{\rm eff}, \tag{4.40}$$

which exceeds the critical value in the antiferromagnetic side for a large $n$.

In this model, the following sequence of phase transitions can be observed. At very high temperatures, $|K_{\rm eff}^{(n)}| < K_C$, and the system remains in a paramagnetic (disordered) state. As the temperature decreases, $|K_{\rm eff}^{(n)}|$ increases, and in a certain range of temperature, amplitude of the effective coupling $|K_{\rm eff}^{(n)}|$ becomes larger than the critical value ($K_{\rm eff}^{(n)} < -K_C$), leading to an antiferromagnetic ordered state because $K_{\rm eff}^{(n)} < 0$. Upon further lowering the temperature, $|K_{\rm eff}^{(n)}|$ decreases and the antiferromagnetic order disappears. Then, the system reenters a paramagnetic state. At sufficiently low temperatures, the effective coupling reaches $K_{\rm eff}^{(n)} = K_C > 0$, and the system exhibits a transition to a ferromagnetic ordered state.

### 4.4.4 Reentrant Phase Transition in Regular Lattices

In the previous section, successive transitions between different types of ordered states were studied in a model with decoration bonds. The nonmonotonic temperature dependence of the spin correlation function can also be realized in regular lattices with appropriately arranged interaction constants. Here, we consider Ising models on a square lattice composed only of nearest-neighbor interactions:

$$\mathcal{H} = -\sum_{\langle ij \rangle} J_{ij}\sigma_i\sigma_j. \tag{4.41}$$

The Ising model on a two-dimensional lattice with periodically arranged bond structures (without crossing bonds) can be solved exactly, as explained in Sect. 4.8. Several

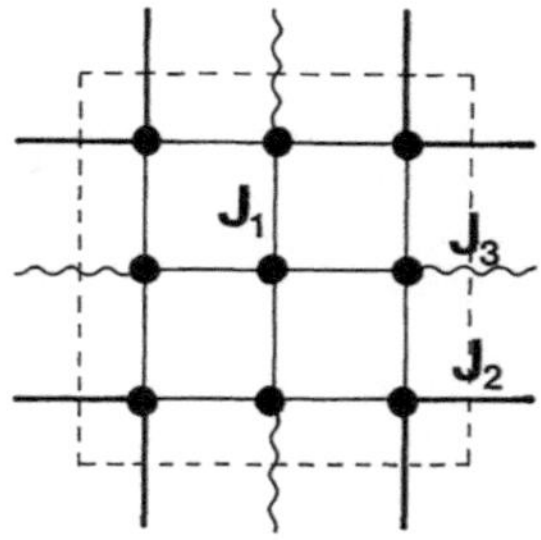


**Fig. 4.11** Unit cell of the model exhibiting reentrant phase transitions and successive transitions. (This figure is reproduced from Ref. [39]. ©1986 The Physical Society of Japan)

exactly solvable examples demonstrating reentrant and more complex sequences of phase transitions have been presented [38–40]. For example, a model with the distribution of $J_{ij}$ forming a $3 \times 3$ unit cell, as shown in Fig. 4.11, was studied.

Depending on the values of $J_1$, $J_2$, and $J_3$, the model exhibits various phase transitions, such as paramagnetic (P) to antiferromagnetic (AF), then back to paramagnetic, and finally to ferromagnetic (F) phases, as explained below.

Here, we consider the case of $J_1 > 0$, $J_2 < 0$ and $J_3 > 0$ and define the ratios:

$$x \equiv \frac{|J_2|}{J_1}, \quad y \equiv \frac{J_3}{|J_2|}. \tag{4.42}$$

The ratio $y$ plays a crucial role in determining whether the $3 \times 3$ unit cells align ferromagnetically ($y < 2$) or antiferromagnetically ($y > 2$). Depending on the values of $x$ and $y$, the system exhibits various ordered states. Representative configurations at $T = 0$ are shown in Fig. 4.12. For the case, $x < 1$, $y > 2$, the system exhibits a ferromagnetic order (Fig. 4.12a), and for $xy < 2$, $y < 2$, an antiferromagnetic order of the $3 \times 3$ unit cell appears (Fig. 4.12b). For $x > 0$, $xy > 2$, a ferromagnetic order with fluctuation plaquettes emerges (Fig. 4.12c). In the case of Fig. 4.12c, the spins within each plaquette connected by thin lines are aligned (i.e., parallel), but the overall spin direction (sign) can differ from plaquette to plaquette, i.e., the system has parts behaving paramagnetically.

(a) (b) (c)

**Fig. 4.12** Spin configuration of the unit cell. **a** $x < 1$, $y > 2$, **b** $xy < 2$, $y < 2$, **c** $x > 0$, $xy > 2$. In **c**, the four spins connected by the lines align ferromagnetically within each group, but their orientations remain independent of one another. (This figure is reproduced from Ref. [39].©1986 The Physical Society of Japan)

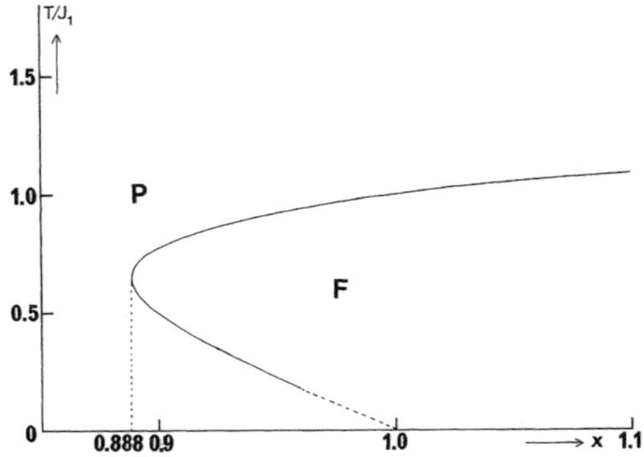


**Fig. 4.13** Phase diagram $(x, T/J_1)$ showing the reentrant phase transition at $y = 2$. (This figure is reproduced from Ref. [39]. ©1986 The Physical Society of Japan)

Typical phase diagrams are shown in Figs. 4.13 and 4.14. The phase boundaries are calculated by using the graphical method explained in Appendix 4.8. The free energy per site is given by

$$F/Nk_BT = -\ln 2 + \ln\left(\cosh^2\beta\right) - \frac{1}{2N}\sum_{p,q}\ln\left(\det(1-\Lambda_{p,q})\right), \tag{4.43}$$

where $p$ and $q$ represent the wave number, and $N$ is the number of lattice sites. The matrix of $(\Lambda_{p,q})$ is a $36 \times 36$ matrix, constructed based on the bond structure $(J_{ij})$ depicted in Fig. 4.11.

To determine the phase transition points, the following criteria are used. For the ferromagnetic order, the phase transition point is obtained by

$$\det(1-\Lambda_{0,0}) = 0, \tag{4.44}$$

while the antiferromagnetic phase transition point is detected by

$$\det(1-\Lambda_{\pi,\pi}) = 0. \tag{4.45}$$

At $y = 2$, the system exhibits a reentrant phase transition (P-F-P) for $0.888 < x < 1$. The corresponding phase diagram in $(x, T/J_1)$ plane is depicted in Fig. 4.13. For $y \neq 2$, the system exhibits more complex phase diagrams, as shown in Figs. 4.14. For $y < 2$, an antiferromagnetically ordered state appears at low temperatures. There, sequential phase transitions changes P-F-P-AF appears. For $y > 2$, the phase sequence P-F-P-F appears. In this way, various sequences of different ordered states are possible with tactful arrangements of frustrated interactions, tuning entropic effects.

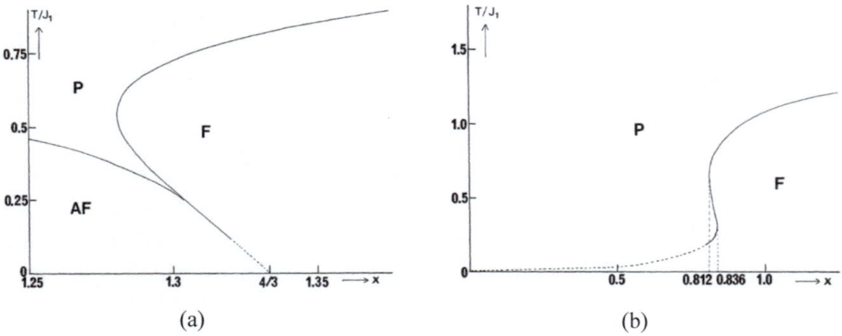


**Fig. 4.14** Phase diagram $(x, T/J_1)$ showing the reentrant phase transitions: **a** $y = 1.5$, and **b** $y = 2.1$. (These figures are reproduced from Ref. [39]. ©1986 The Physical Society of Japan)

## 4.5 Mekata Model: Antiferromagnetic Triangular Ising Model (AFTI) with Next-Nearest-Neighbor Interaction

Experimentally, successive phase transitions have been observed in $CsCoCl_3$ [41, 42]. The lattice structure and magnetic interactions are depicted in Fig. 4.15. In this compound, the Co ions behave as Ising spins with $S = 1/2$.

The system can be described by the following Hamiltonian:

$$\mathcal{H} = J_1 \sum_{\langle ij \rangle \mathrm{nn}, n} \sigma_{i,n}\sigma_{j,n} - J_2 \sum_{\langle ik \rangle \mathrm{nnn}, n} \sigma_{i,n}\sigma_{k,n} - J_0 \sum_{i,n} \sigma_{i,n}\sigma_{i,n+1}, \tag{4.46}$$

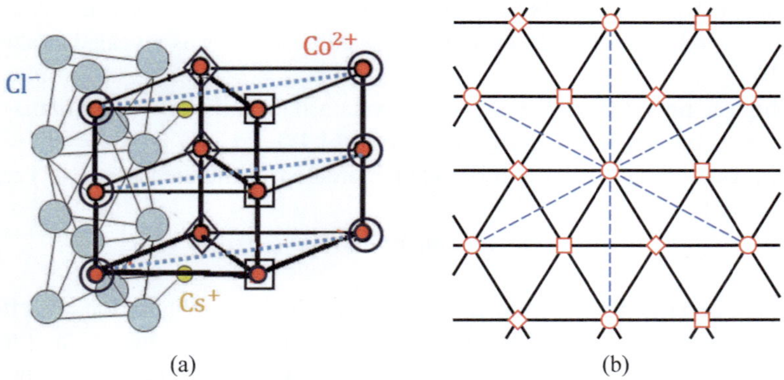


**Fig. 4.15** Layered triangular lattice ($ABX_3$-type structure). **a** Lattice structure of $CsCoCl_3$ and magnetic interactions: nearest-neighbor interaction ($J_1$, solid lines in the ab-plane), next-nearest-neighbor interaction ($J_2$, dotted lines), and interlayer interaction ($J_0$, solid vertical lines). **b** Top view of the lattice is a triangular lattice (in the ab-plane). Co-ions connected by $J_0$ along the c-axis are represented by points. The region enclosed by the dotted line corresponds to the detailed structure shown in **a**. The circles, squares, and diamonds represent the sublattices of the triangular lattice

where $i$, $j$, $k$ are sites in the $n$-th layer of the lattice, $\langle ij \rangle_{\mathrm{nn}}$ denotes the nearest neighbor sites and $\langle ik \rangle_{\mathrm{nnn}}$ denotes the next nearest neighbor sites. The interlayer interaction $J_0$ is estimated to be approximately $-75$K, while the nearest-neighbor interaction within a layer $J_1$ is about 1K. Consequently, the system can be modeled as a triangular lattice composed of strongly coupled one-dimensional spin chains.

Mekata and Adachi [42] found that the material $CsCoCl_3$ shows interesting successive phase transitions by neutron diffraction measurement. Around $T \simeq 20$ K, spins on two sublattices become ordered, while the spins on the remaining sublattice remain disordered. At lower temperatures, all three sublattices exhibit ordering.

Mekata [41] analyzed these phase transitions by a mean-field method using a model including the next-nearest-neighbor (nnn) interaction $J_2$ (Mekata model) and explained the observed successive transitions.

The lattice is divided into three sublattices, A, B, and C, as shown in Fig. 4.15b. The average magnetization of the sublattices is denoted as $m_{\mathrm{A}}, m_{\mathrm{B}}, m_{\mathrm{C}}$, respectively. On a spin in sublattice A, an effective field $-J_1(3m_{\mathrm{B}} + 3m_{\mathrm{C}})$ arises from the spins in the nearest-neighboring sublattices (B and C) and $6J_2m_{\mathrm{A}}$ from the next-nearest-neighboring sublattice (A). Thus, the self-consistent equation for $m_{\mathrm{A}}$ is given by

$$m_{\mathrm{A}} = \langle \sigma_{\mathrm{A}} \rangle = \frac{\sum_{\sigma_{\mathrm{A}}=\pm 1} \sigma_{\mathrm{A}} e^{-\beta\sigma_{\mathrm{A}}(J_1(3m_{\mathrm{B}}+3m_{\mathrm{C}})-6J_2m_{\mathrm{A}})}}{\sum_{\sigma_A=\pm 1} e^{-\beta\sigma_{\mathrm{A}}(J_1(3m_{\mathrm{B}}+3m_{\mathrm{C}})-6J_2m_{\mathrm{A}})}}. \tag{4.47}$$

Similar equations are derived for $m_{\mathrm{B}}$ and $m_{\mathrm{C}}$. The set of self-consistent equations is given by

$$\begin{cases} m_{\mathrm{A}} = \tanh\left(-\beta\left(J_1(3m_{\mathrm{B}} + 3m_{\mathrm{C}}) - 6J_2m_{\mathrm{A}}\right)\right) \\ m_{\mathrm{B}} = \tanh\left(-\beta\left(J_1(3m_{\mathrm{C}} + 3m_{\mathrm{A}}) - 6J_2m_{\mathrm{B}}\right)\right) \\ m_{\mathrm{C}} = \tanh\left(-\beta\left(J_1(3m_{\mathrm{A}} + 3m_{\mathrm{B}}) - 6J_2m_{\mathrm{C}}\right)\right). \end{cases} \tag{4.48}$$

As a characteristic parameter, we define the ratio $\alpha = J_2/J_1$. Solving this set of equations (4.48) for $\alpha = 0.2$, a series of phase transitions was revealed as shown in Fig. 4.16.

At high temperatures ($T > T_{\mathrm{C1}}$), the system is in a disordered (paramagnetic) state, corresponding to the trivial solution of (4.48), i.e., $m_{\mathrm{A}} = m_{\mathrm{B}} = m_{\mathrm{C}} = 0$. At $T = T_{\mathrm{C1}}$, two sublattices (A and B) become ordered, while the last one (C) remains in disordered:

$$(m_{\mathrm{A}}, m_{\mathrm{B}}, m_{\mathrm{C}}) = (m, -m, 0). \tag{4.49}$$

This phase persists in the period $T_{\mathrm{C1}} > T > T_{\mathrm{C2}}$. In this configuration, the effective fields from sublattices A and B cancel out at the sites in sublattice C, and the nnn interaction $J_2$ is not significant enough to make them ordered at this temperature, leading to $m_{\mathrm{C}} = 0$. Because only two of sublattices are ordered in this phase, it is called "partially disordered (PD) phase", which is a key feature of frustration in the system.

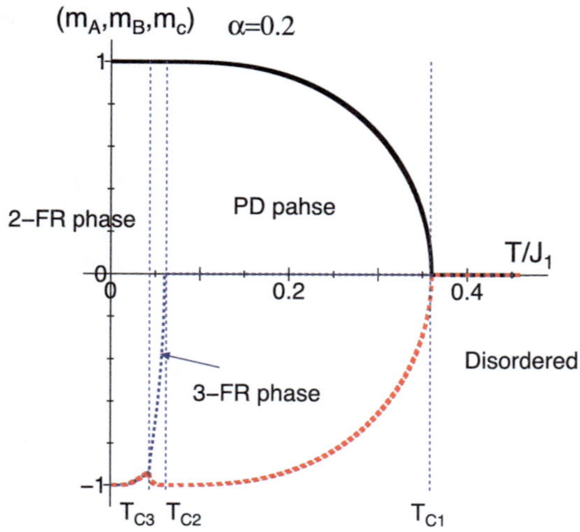


**Fig. 4.16** Successive phase transitions of the Mekata model for $\alpha = 0.2$, The temperature dependence of $(m_{\rm A}, m_{\rm B}, m_{\rm C})$ is shown

As the temperature decreases, the next-nearest-neighbor interaction $J_2$ becomes relevant, which allows the remaining sublattice to be also ordered (blue dotted curve). In the period $T_{\rm C2} > T > T_{\rm C3}$, a phase composed of

$$m_{\rm A} = m_1 > 0, \quad m_{\rm B} = m_2 < 0, \quad m_{\rm C} = m_3 < 0. \tag{4.50}$$

appears, where all three sublattices become ordered with different magnetizations. This phase is called the 3-sublattice ferrimagnetic (3-FR) phase.

At $T_{\rm C3} > T$, two sublattices are equally ordered and the remaining one is ordered in the opposite direction:

$$(m_{\rm A}, m_{\rm B}, m_{\rm C}) = (m_1, -m_2, -m_2), \quad (|m_1| > |m_2|), \tag{4.51}$$

This is called the 2-sublattice ferrimagnetic (2-FR) state. At zero temperature, all the spins in each sublattice are ordered, and two of them are parallel while the last one is antiparallel. There are six possible configurations, one of which is $(m_{\rm A}, m_{\rm B}, m_{\rm C}) = (1, -1, -1)$.

However, it is not clear whether the 3-FR phase exists in real systems, because the delicate balance between the phases in the mean-field analysis can be ambiguous. If the system transitions directly from the PD phase $(m, -m, 0)$ to the 2-FR phase $(m_1, -m_2, -m_2)$, the transition must be discontinuous because the order parameters cannot be transferred smoothly. In this case, we expect a first-order phase transition instead. The transition from the PD phase to the 2-FR phase has been a subject of long-standing discussion. Additionally, the existence of the PD phase, where 1/3 of the sites remain disordered, has also been debated. Due to concerns regarding the mean-field approach, as explained in the following sections, Monte Carlo simulations have been performed to investigate the thermodynamic properties of the system more directly and accurately.

### 4.5.1 MC Studies in Two-Dimensional Models

Historically, Monte Carlo (MC) studies of the Mekata model were first conducted for two-dimensional systems:

$$\mathcal{H} = J_1 \sum_{\langle ij \rangle \mathrm{nn}} \sigma_i \sigma_j - J_2 \sum_{\langle ik \rangle \mathrm{nnn}} \sigma_i \sigma_k. \tag{4.52}$$

One naturally expects significant differences in the characteristics of phase transitions depending on the dimensionality in highly frustrated models. However, historically, studies began with two-dimensional models to gain insight into phase transitions in such systems, and to explore whether the PD phase truly exists in the thermodynamic limit and how the transition between the PD phase and the 2-FR phase occurs.

Although the antiferromagnetic Ising model on a triangular lattice (AFTI) does not exhibit a phase transition at finite temperatures, the introduction of a ferromagnetic next-nearest-neighbor interaction $J_2$ enhances sublattice ordering, leading to long-range order at low temperatures. Furthermore, it was discovered that the model exhibits successive phase transitions [43].

In Fig. 4.17a, the temperature dependence of the specific heat $C$, sublattice magnetization:

$$M_{\mathrm{sub}}^2 = \frac{1}{(N/3)^2} \left\langle \left( \sum_{i \in \mathrm{A}} \sigma_i \right)^2 + \left( \sum_{i \in \mathrm{B}} \sigma_i \right)^2 + \left( \sum_{i \in \mathrm{C}} \sigma_i \right)^2 \right\rangle, \tag{4.53}$$

and the total magnetization:

$$M_{\mathrm{all}} = \frac{1}{N^2} \left\langle \left( \sum_{i \in \mathrm{all}} \sigma_i \right)^2 \right\rangle, \tag{4.54}$$

are shown. The specific heat exhibits two peaks, indicating the occurrence of two distinct ordering processes.

In Fig. 4.17b, the size dependence of the sublattice order parameter $M_{\mathrm{sub}}^2$ is plotted. Both $C$ and $M_{\mathrm{all}}$ exhibit little size dependence (not shown). From the size dependence of $M_{\mathrm{sub}}^2$, it is revealed that the system has three distinct phases:

(1) the disordered (paramagnetic) phase, where $M_{\mathrm{sub}}^2 \propto O(N^{-1})$,
(2) the critical (KT) phase, where $M_{\mathrm{sub}}^2 \propto O(N^{-\eta}), \quad \eta > 0$
(3) the ordered (2-FR) phase, where $M_{\mathrm{sub}}^2 \propto O(N^0)$.

This observation indicates that the model exhibits an intermediate phase in which the correlation function decays as a power law and the correlation length diverges.

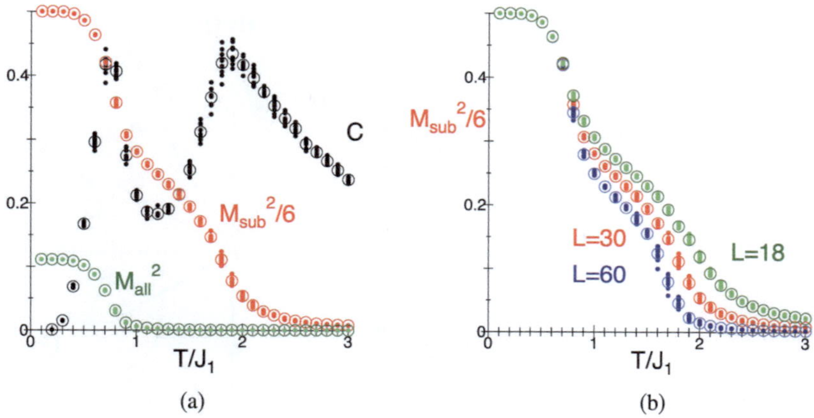


**Fig. 4.17** **a** Temperature dependence of the specific heat $C$, $M_{\text{sub}}$, and $M_{\text{all}}$ for AFTI with the nnn interaction $J_2 = 0.2J_1$ (Eq. 4.52). The system size is $L = 30$. **b** Size dependence of $M_{\text{sub}}$ for $L = 18$, 30 and 60. Simulations were performed on 10 samples, each with 10,000 MCS (plus 10,000 MCS for thermalization) at each temperature. Small dots represent individual sample data, while large circles indicate the average values

This implies that all temperatures within this phase are effectively critical, forming a critical line that is a hallmark of the Kosterlitz-Thouless (KT) transition[44]. Such intermediate phases are also found in non-frustrated systems, such as clock models[43, 45].

Indeed, the 6-fold degenerate ground states of the AFT-Ising model:

$$(-,-,+), (-,+,+), (-,+,-), (+,+,-), (+,-,-), (+,-,+) \tag{4.55}$$

can be mapped onto the 6-state clock model:

$$\mathcal{H}_{6\text{SC}} = -J \sum_{\langle ij \rangle} \cos(\theta_i - \theta_j), \quad \theta_i = \frac{2\pi}{6}k, k = 0, 1, \ldots, 5, \tag{4.56}$$

as depicted in Fig. 4.18.

José et al. [45] studied the 6-state clock model and found that it exhibits two successive phase transitions, both of which are KT transitions. These two transitions are related by a dual transformation [46, 47]. The nature of the ordering in these double KT transitions has also been explored in other contexts [48].

While the intermediate phase in these models is a KT phase with a diverging correlation length, the PD phase of the Mekata model features long-range order in two sublattices and a disordered third sublattice. Thus, it is clarified that the two-dimensional model is incompatible with the Mekata solution obtained through mean-field analysis. To avoid the emergence of the KT phase, which is characteristic of two-dimensional systems with significant fluctuations caused by the infrared divergence,

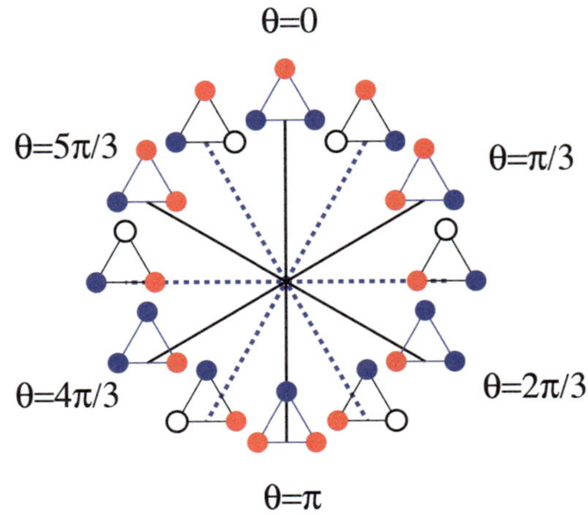


**Fig. 4.18** Mapping between the ground-state configurations on a triangular lattice and the 6-state clock model

it is necessary to focus on three-dimensional models. This will help elucidate the behavior of the original model (4.46).

### *4.5.2 Three-Dimensional Six-State Clock Model*

Extensive Monte Carlo simulations of the three-dimensional Mekata model on a hexagonal lattice have been conducted [14, 49]. These studies observed behavior of correlation functions similar to those found in experimental observations. However, the nature of the PD phase remained unclear due to significant fluctuations in the data. Additionally, the transition between the intermediate-temperature phase and the 2-FR phase was not fully understood because no specific heat anomaly was detected.

It is natural to hypothesize that the six-state clock model on a hexagonal lattice exhibits a similar ordering process to the Mekata model. Indeed, thermodynamic quantities such as the specific heat (Fig. 4.19a) and magnetization (Fig. 4.19b) show remarkably similar behavior [50].

Specifically, the specific heat exhibits one sharp divergence at ($T \simeq 3J$) and a broad peak around $T = 1.5J$, and the spontaneous magnetization appears at the temperature where the specific heat diverges. The quantity $\langle M^2 \rangle / N^2$ increases almost linearly with temperature over a wide range ($1 < T/J < 2.5$) and approaches its saturation value exponentially at low temperatures ($T < 0.5J$). These observations suggest the existence of two distinct phases [50]: (1) an intermediate phase with characteristics of the gapless low-temperature phase of the ferromagnetic XY model and (2) a low-temperature phase that is gapful due to the discreteness of the six-state clock model.

However, a careful analysis of size dependence of the six-fold anisotropy revealed that no true phase transition occurs in the low-temperature phase [50]. This property

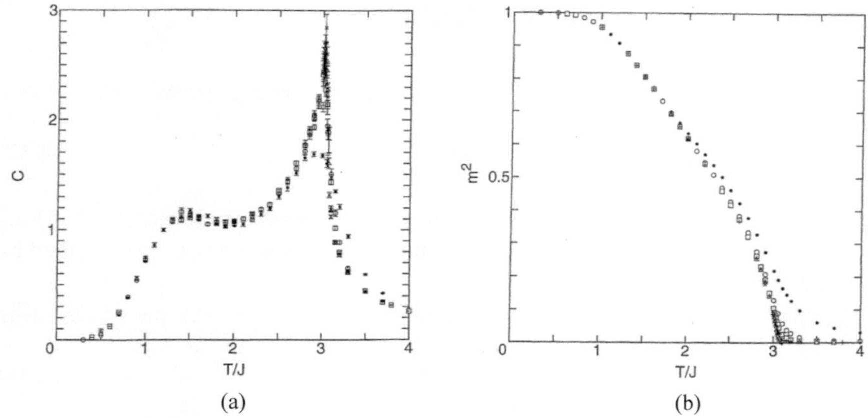


**Fig. 4.19** **a** Temperature dependence of the specific heat. **b** Temperature dependence of $\langle M^2/N^2 \rangle$ for 3D 6-state clock model. Closed circles, open circles, squares, plus sings, triangles, crosses, and inverted triangles represent system sizes $L = 6, 12, 18, 24, 30, 42$ and 60, respectively. (These figures are reproduced from Ref. [50]. ©1997 The Physical Society of Japan)

was also proposed by a renormalization group (RG) analysis [51]. Therefore, the existence of the intermediate phase is confirmed to be merely apparent. This finding implies that the PD phase is not a genuine thermodynamic phase but an apparent phenomenon. Consequently, the very existence of the PD phase has been called into question based on this analysis.

Nevertheless, the experiments suggest the existence of a PD-type intermediate phase, which raises the question of whether the three-dimensional six-state clock model accurately represents the Mekata model. According to the general principle of universality in phase transitions, models with the same symmetry are expected to exhibit the same type of phase transition. However, in frustrated systems, the definition of "same symmetry" requires careful consideration. Even subtle differences can significantly alter the behavior of the system. In this context, the so-called generalized six-state clock model has been introduced [52, 53].

## 4.6 Generalized 6-State Clock ($Z_6$) Model

In the standard 6-state clock ($Z_6$) model, the interaction energy between the $i$-th and $j$-th states is given by

$$E_{ij} = -J\cos((i-j)\pi/3), \tag{4.57}$$

where $i - j$ is defined in modulo 6. However, the $Z_6$ symmetry is preserved as long as $E_{ij}$ is periodic with a period of 6. The interaction is not necessarily described by a simple cosine function.

In this section, we study a generalized model that retains $Z_6$ symmetry. Denoting the six states by

$$S_i = 0, 1, 2, \ldots, 5, \tag{4.58}$$

the Hamiltonian of the generalized $Z_6$ model is expressed as

$$\mathcal{H}_{GZ_6} = \sum \langle ij \rangle \left(\varepsilon_0 \delta_{S_i,S_j} + \varepsilon_1 \delta_{S_i,S_j \pm 1} + \varepsilon_2 \delta_{S_i,S_j \pm 2} + \varepsilon_3 \delta_{S_i,S_j \pm 3}\right), \quad \text{mod}(6). \tag{4.59}$$

In the generalized $Z_6$ model, the parameters $\varepsilon_i$ $(i = 0, 1, 2, 3)$ can take arbitrary values. However, in this study, we focus on a natural ordering of the parameters:

$$\varepsilon_0 \leq \varepsilon_1 \leq \varepsilon_2 \leq \varepsilon_3. \tag{4.60}$$

It has been found that, depending on the choice of these parameters, non-uniformly ordered phases can emerge [52]. As long as $\varepsilon_1 > \varepsilon_0$, the ground state selects a single state. However, when

$$\varepsilon_0 \simeq \varepsilon_1 \ll \varepsilon_2, \tag{4.61}$$

the energy cost associated with the mixing of two neighboring states becomes small. In such cases, within an intermediate temperature range,

$$|\varepsilon_1 - \varepsilon_0| < T \ll \varepsilon_2, \tag{4.62}$$

a mixed state consisting of two neighboring states can emerge (entropy-induced mixed state). In this state, a specific pair of neighboring states is spontaneously selected [52, 53].

It should be emphasized that this mixing is not simply due to fluctuation, but the mixed state gives a definite phase characterized by symmetry breaking. The two selected states mix in a fixed ratio of 1 : 1. Between regions of two different mixed ordered states, a well-defined domain wall is observed [53].

At high temperatures ($T > \varepsilon_2$), the system becomes disordered. Thus, the model exhibits successive phase transitions: from a disordered phase to a phase consisting of a pair of states (referred to as the "incompletely ordered phase (IOP1)") and finally to a uniformly ordered state. This sequence of phases closely resembles the transitions observed in the Mekata model, which progresses through disordered, partially disordered (PD), and 2-FR phases.

The phase transition between the disordered state and the mixed state belongs to the 3D XY universality class, characterized by critical exponents $\beta = 0.34(1)$ and $\nu = 0.66(4)$. However, the phase transition between the mixed state and the uniformly ordered state is of the first order. This behavior differs from the results of mean-field (MF) analysis, which predicts the existence of a 3-FR phase that connects the PD phase and the 2-FR phase continuously.

In experiments, no latent heat or divergence is observed at the transition between the PD and 2-FR phases. However, other quantities, such as the sublattice magnetization, exhibit discontinuous changes. These observations strongly suggest that a first-order phase transition occurs.

Thus, the remaining question is why the latent heat is absent. A key factor behind this discrepancy lies in the fact that the original model consists of highly ordered one-dimensional chains, each of which effectively represents a single spin in a triangular lattice:

$$\left\langle \frac{\sum_{i \text{ along chain}} \sigma_i}{L} \right\rangle = \pm 1, \tag{4.63}$$

where $L$ is the length of the chain. On the other hand, when the spins within a chain divide into two regions with opposite orientations, a domain wall forms within the chain. In this case, the average magnetization of the chain becomes small:

$$\left\langle \frac{\sum_{i \text{ along chain}} \sigma_i}{L} \right\rangle \simeq 0. \tag{4.64}$$

As a result, the energy difference between the configurations in the 2-FR phase and the PD phase becomes very small. This explains why the energy change associated with the transition between the PD and 2-FR phases is negligible.

To study this effect, a model with $|J_0| \gg |J_1|$ was investigated through Monte Carlo simulations [54]. In Fig. 4.20a, the temperature dependence of the specific heat for systems of different sizes is shown. A sharp peak is observed around $T = T_{C1} \simeq 3.7J$, but no significant anomaly is detected at lower temperatures. On the other hand, Fig. 4.20b shows the temperature dependence of the domain wall densities of the three sublattices ($\rho_1, \rho_2, \rho_3$). These densities clearly exhibit discontinuous changes at $T_{C2}$.

An important relation for first-order phase transitions is the Clausius-Clapeyron relation:

$$\frac{\Delta Q}{\Delta M} = -T \frac{d H_C(T)}{dT}. \tag{4.65}$$

In the present model, the PD phase transitions to the 2-FR phase with only a very small change in energy ($\Delta Q$). This is because the energy difference between these phases originates solely from a single domain wall within a chain. Consequently,

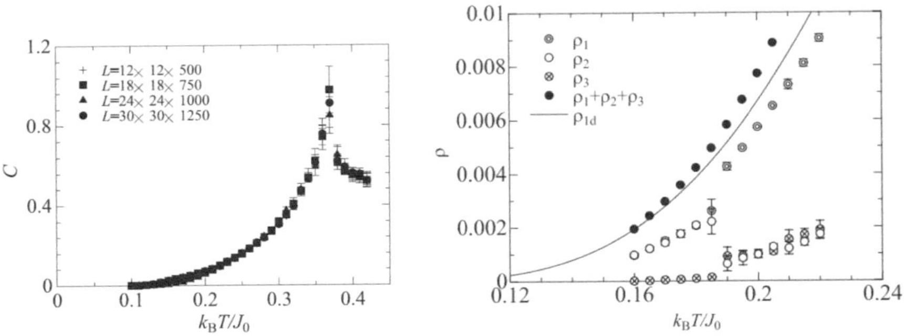


(a) (b)

**Fig. 4.20** **a** Specific heat and **b** domain wall densities of chains in the three sublattices of the generalized six-state clock model with $|J_0| \gg |J_1|$. A discontinuous change at $T_{C2}$ is clearly found. (These figures are reproduced from Ref. [54]. ©2004 The Physical Society of Japan)

$$\frac{dH_C(T)}{dT} \simeq 0. \tag{4.66}$$

As a result, $\Delta Q \simeq 0$, which explains the absence of any observable anomaly in the specific heat shown in Fig. 4.20a.

This successive phase transition can be understood through an analysis of the RG flow chart shown in Fig. 4.21 [54]. In the figure, $t$ denotes the temperature, and $\lambda_6$ represents the strength of the six-fold anisotropy, described by the term $\lambda_6 \cos(6\theta)$. In the region where $\lambda_6 \cos(6\theta) < 0$, the configurations depicted in Fig. 4.18 for $\theta = 2\pi/6 \times k$, $k = 0, 1, \ldots, 5$ showing 2FR states are selected. On the other hand, for $\lambda_6 \cos(6\theta) > 0$, the configurations $\theta = 2\pi/6 \times k + \pi/6$, $k = 0, 1, \ldots, 5$ showing the PD states are selected.

In the observation of the three-dimensional six-state clock model studied in Sect. 4.5.2, the PD phase does not appear as a stable phase, and only the 2-FR phase is realized. This situation corresponds to the lower half of Fig. 4.21, where the RG flow is confined to the region with negative $\lambda_6$ [51]. There, the flow along the $t$ axis corresponds to an apparent PD phase, but the PD phase does not emerge as a stable phase. The Mekata model, represented by the generalized $Z_6$ model, corresponds to the curve that crosses the line $\lambda_6 = 0$ (schematically shown by a bold line).

## 4.7 Effects of Randomness on Ordering Processes

Randomness in a system can significantly affect the nature of ordering, even in non-frustrated cases. In systems with inhomogeneous interaction strengths, such as diluted magnets, ordering begins in strongly connected regions at higher temperatures, forming clusters. These clusters may orient in different directions, aligning them incurs an energy cost.

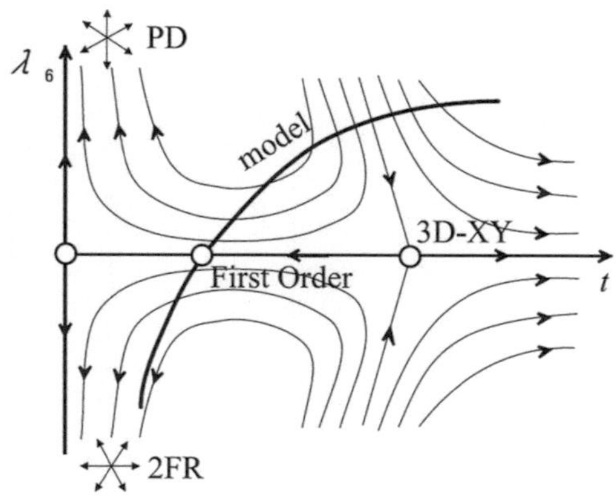


**Fig. 4.21** RG flow chart of a 3D generalized six-state clock model corresponding to the Mekata model. The thin curves represent the RG flow, while the bold line indicates the position of the specific model. The circle labeled "3D-XY" marks the fixed point for the phase transition between the disordered and PD phases. The line labeled "First Order" denotes the phase boundary for the first-order phase transition between the PD and 2-FR phases. (This figure is reproduced from Ref. [54]. ©2004 The Physical Society of Japan)

In systems with discrete variables, such as the Ising model, domain walls must pass through tightly ordered regions, creating energy barriers [55], and in continuous spin models, vortices (topological defects) are pinned by the randomness. These effects lead to a multi-valley structure in the energy landscape, which gives rise to unique entropy properties.

Understanding thermal phenomena in systems with multi-valley energy structures is an intriguing problem. Spatial randomness in strength of interactions tends to smear out critical phenomena [56–61]. It has been pointed out that, if the specific heat of the original pure model diverges with an exponent $\alpha > 0$, the exponent is renormalized in the presence of randomness to $\alpha_X = -\alpha/(1-\alpha)$ [56]. It has been shown that in the Ising model on a square lattice with randomly distributed vertical interactions while maintaining uniform horizontal interactions, the specific heat exhibits an essential singularity. In this case, all derivatives of the free energy remain continuous [57].

Ordering processes in diluted ferromagnets is a significant problem in statistical physics. The connectivity of sites in a diluted lattice is often analyzed as a percolation problem (Fig. 4.22).

Consider a ferromagnetic system on a lattice. Let the probability of removing a spin at a site be $p$. If $p$ exceeds the critical percolation threshold $p_C$ [62], the lattice becomes fragmented into isolated finite clusters. In such cases, the system is not expected to have a phase transition leading to long-range order.

However, there remains a non-zero probability of forming a perfect lattice, which is given by $(1-p)^N$, where $N$ is the total number of sites in the lattice. This probability decreases exponentially with increasing $N$, making its contribution negligible in large systems, so it can be safely ignored when studying system's thermal properties.

**Fig. 4.22** A diluted configuration, with site occupation probability $p = 0.5$, which is below the critical probability of the site percolation

However, when considering cases where the probability is very small, but the effects are large, we must be cautious. Griffiths pointed out that significant effects can arise from thermodynamically negligible states in dynamic problems, a phenomenon known as the Griffiths singularity[63]. This problem has been studied numerically, and peculiar types of relaxation have been observed [64].

In diluted systems, the distribution of cluster size is determined by the probability $p$. In the ordered state of Ising ferromagnets, spin domains with opposite magnetization exist due to thermal fluctuations. The probability of a large domain decreases exponentially with its size and does not seriously affect static properties. However, the averaged autocorrelation of a spin shows a stretched exponential decay (rather than a simple exponential decay) due to the long relaxation time of large clusters, which can be regarded as a Griffiths singularity [65, 66].

The question of how a random field applied to each site destroys long-range uniform order is also an important problem. The dependence on the dimensionality of the spin variable and on the spatial dimension has been studied [67]. Moreover, it has been pointed out that even in cases where the random field has negligible effects on equilibrium properties, it can cause significant effects on spin relaxation dynamics [68].

### *4.7.1 Spin Glasses*

In spin glasses, both frustration and randomness play crucial roles. The interplay between these two factors results in highly complex ordering processes. Over recent decades, properties of spin glasses have been extensively studied [69–73]. Experimentally, numerous intriguing phenomena have been observed, including the divergence of non-linear susceptibility [74, 75], memory effects, rejuvenation phenomena, and more [76].

To study random systems, two primary cases are generally considered: quenched randomness and annealed randomness. In the case of quenched randomness, the random parameters are fixed during the calculation of the free energy. The average is then taken over the distribution of randomness in the calculated free energy:

$$F_{\text{quenched random}} = -k_{\text{B}}T \, \overline{\ln Z_{\text{fixed random configuration}}} \,, \tag{4.67}$$

where the overline denotes averaging over the distribution of randomness. This process is referred to as the quenched average. Most significant cases involving randomness belong to this category.

In contrast, for annealed randomness, the random variables fluctuate due to thermal contact, and the average is taken over the partition function:

$$F_{\text{annealed random}} = -k_{\text{B}}T \ln \overline{Z_{\text{random configuration}}} \,, \tag{4.68}$$

which is referred to as the "annealed average".

The quenched average is generally difficult to calculate, except in special cases, such as along the so-called Nishimori line [77]. To overcome the difficulty in quenched average, the replica trick was introduced [70]. The key idea behind the replica method relies on the relation:

$$\ln(x) = \lim_{n \to 0} \frac{x^n - 1}{n}. \tag{4.69}$$

By applying this relation to (4.67), we obtain the formal expression:

$$F_{\text{quenched random}} = -k_{\text{B}}T \lim_{n \to 0} \frac{\overline{Z^n_{\text{fixed configuration}}} - 1}{n}. \tag{4.70}$$

Here, $\overline{Z^n_{\text{fixed configuration}}}$ represents an annealed average over $n$ independent replica systems, which can be computed without difficulty.

For a model with random bonds following a Gaussian distribution, an explicit calculation was carried out within the mean-field approximation by Sherrington and Kirkpatrick [71]. In this approach, although the number of replicas $n$ is integer, it is treated as real number in the limit $n \to 0$ in (4.69). This discrepancy raises mathematical concerns about the validity of the formula. Nonetheless, by assuming specific correlations among replicas, meaningful insights into spin freezing phenomena can still be extracted.

However, in the replica trick, the calculated entropy becomes negative [71]. In this calculation, all ordered states of the replicas are assumed to be identical, a condition known as the replica symmetric state. For this problem, Parisi proposed a groundbreaking idea, suggesting that the ordered states in different replicas do not need to be identical but can vary [78]. This variability reflects the highly degenerate multi-valley energy landscape caused by randomness. With this approach, the entropy is restored to a positive value. This concept, known as replica symmetry breaking (RSB), offers a new perspective on the nature of ordered states in random systems. Particularly, RSB has been used to study the structure of metastable states in spin glasses. It has been shown that the number of metastable states increases exponentially with the system size. This intriguing property is associated with a unique hierarchical organization of these states, referred to as ultrametricity [79–81].

As an alternative interpretation of spin glass order, several other models have been proposed. A naive picture of spin glass order is an ordered state with random spin directions as an extension of the Néel order for antiferromagnets. In the absence of frustration among the interactions, this view holds true and is formalized in the Mattis model [82]. In this model, randomness leads to randomly distributed spin orientations in the ordered state, determined by the signs of the coupling constants between spins. This behavior can be represented through a local gauge transformation:

$$\sigma_i \to -\sigma_i, \quad J_{ij} \to -J_{ij}. \tag{4.71}$$

However, spin glasses exhibit unique properties inherent to randomly frustrated systems, such as the divergence of the non-linear susceptibility [74, 75], that cannot be explained by the Mattis model.

In general, frustration among interactions results in a distribution of strength of correlations [83]. Within such frustrated systems, one can define domains where spins interact strongly. The structure and dynamics of these domains play a crucial role in the behavior of spin glasses. Specifically, large clusters relax more slowly, contributing to the characteristic slow relaxation observed in the average spin autocorrelation function. This domain picture with weak stiffness has been proposed to explain certain ordering phenomena [83–85]. Furthermore, a scenario involving chirality ordering in spin glasses has also been proposed, offering another avenue for understanding the complex ordering processes in these systems [86].

#### 4.7.1.1 Rejuvenation

Because of the peculiar properties of spin glasses, the ordering process below the critical temperature is not straightforward. In experiments, various interesting observations have been reported [87]. One notable example is the history-dependent ordering dynamics. The spin-glass ordering can be monitored by the imaginary part of the ac-susceptibility, $\chi''$.

When the system is held at a temperature below the critical temperature, $\chi''$ decreases over time, indicating ongoing ordering; this process is referred to as "aging". If the temperature is lowered slightly further, one might expect $\chi''$ to continue decreasing. However, surprisingly, experiments show that $\chi''$ instead increases abruptly to a higher value. An example is shown in Fig. 4.23a for an experiment on $\chi''$ of $CdCr_{1.7}In_{0.3}S_4$.

The spin-glass critical temperature of this material is about 16.7 K. In the figure, when the temperature is decreased from 12 to 9 K, $\chi''$ jumps upward, suggesting that the ordered structure established at 12 K is incompatible with the ordering at 9 K. This

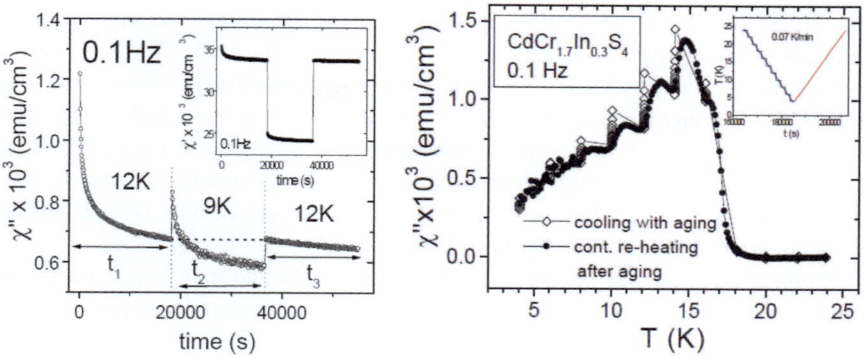

(a) (b)

**Fig. 4.23** **a** Time dependence of $\chi''$ of $CdCr_{1.7}In_{0.3}S_4$ under a stepwise temperature change (12K→ 9K→ 12K, $T_g \simeq 16.7$K). The inset shows the real part $\chi'$. **b** Rejuvenation and memory. The open diamonds are data under the stepwise cooling process and the closed circles show data under the subsequent continuous reheating. (These figures are reproduced from S. Miyashita and E. Vincent, Eur. Phys. J. B **22**, 203 (2001) [89], ©Springer-Verlag 2001, with permission

observation indicates that ordering at one temperature does not necessarily support ordering at another temperature. This phenomenon is called "rejuvenation" [88]. The non-compatibility of ordering at different temperatures indicates that the mechanism of temperature-dependent spin correlation function studied in Sect. 4.4.1 may provide a possible microscopic interpretation of this phenomenon [89, 90].

Another surprising phenomenon was observed in Fig. 4.23a, when the system was brought back to 12 K after being cooled to 9 K. Despite the new aging process that occurred at 9 K, the system retrieved the aging state previously reached at 12 K. This demonstrates that the memory of the prior aging at 12 K was preserved, even after the rejuvenation at 9 K.

In Fig. 4.23b, the open diamond symbols illustrate the behavior of the imaginary part of the ac susceptibility, $\chi''$, during a stepwise cooling process [89, 91]. As the temperature drops below the spin-glass critical temperature (16.7K), the relaxation time increases significantly, and slow relaxation becomes evident. In the experiment, the temperature was reduced stepwise, as shown in the inset. At each fixed temperature, $\chi''$ gradually decreased over time, indicating ongoing relaxation (aging). However, when the temperature was suddenly lowered by 2 K, $\chi''$ increased, signifying rejuvenation at the new temperature. This phenomenon, known as "temperature chaos" [92, 93], has been supported by experimental findings [88, 94].

The figure also includes $\chi''$ data collected during the subsequent warming process, represented by closed circles. Remarkably, the system appeared to "remember" the aging processes that occurred at specific temperatures during the cooling phase. This behavior, termed the "memory effect" [88, 95], has been observed in various temperature cycling experiments. Such results suggest that complex systems like spin glasses possess an intrinsic mechanism for encoding and retaining information about their history, providing valuable insights into their unique ordering dynamics and memory properties.

Regarding the mechanism of memory and rejuvenation, many theories have been proposed, including the idea of the slow relaxation of less frustrated clusters [89, 90] and also an idea of the complicated valley structure of free energy [96]. These theories suggest that a highly inhomogeneous structure is essential. Recently, very extensive numerical simulations have been conducted, which may provide further insights into these intriguing aging properties [97].

Although spin glasses have been studied extensively and many more intriguing properties have been uncovered, we will not delve into the detailed aspects of their behavior here.

## 4.8 Appendix A: Exact Solution of the Two-Dimensional Ising Model (Combinatorial Method)

The exact solution of the Ising model has been a central topic in statistical physics [98]. The exact formula for the free energy of two-dimensional Ising models without crossing bonds has been derived using various methods [99–110]. For frustrated models, where the bond distribution varies spatially, the combinatorial method [102, 104–109, 111] proves particularly convenient.

The general form of the Hamiltonian for two-dimensional Ising models is expressed as

$$\mathcal{H} = -\sum_{\langle ij \rangle} J_{ij}\sigma_i\sigma_j, \tag{4.72}$$

where $i$ and $j$ are lattice points connected by the bond $J_{ij}$. The partition function is given by

$$\begin{aligned} Z &= \sum_{\{\sigma_i=\pm 1\}} e^{-\beta\mathcal{H}} = \sum_{\{\sigma_i=\pm 1\}} \prod_{\langle ij \rangle} e^{\beta J_{ij}\sigma_i\sigma_j} \\ &= \left(\prod_{\langle ij \rangle} \cosh(\beta J_{ij})\right) \sum_{\{\sigma_i=\pm 1\}} \prod_{\langle ij \rangle} (1 + t_{ij}\sigma_i\sigma_j), \end{aligned} \tag{4.73}$$

where

$$t_{ij} = \tanh(\beta J_{ij}). \tag{4.74}$$

When we expand the product part, there appear various terms:

$$\prod_{\langle ij \rangle} (1 + t_{ij}\sigma_i\sigma_j) = 1 + \sum_{\langle ij \rangle} t_{ij}\sigma_i\sigma_j + \sum_{\langle ij \rangle} t_{ij}\sigma_i\sigma_j \sum_{<kl>} t_{kl}\sigma_k\sigma_l + \cdots. \tag{4.75}$$

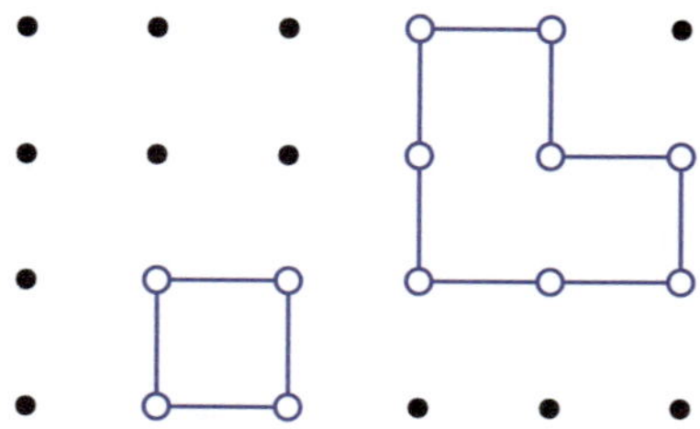

**Fig. 4.24** Closed loops formed by $t_{ij}$s

Since

$$\sum_{\sigma_i=\pm 1} \sigma_i = 0, \quad \sum_{\sigma_i=\pm 1} \sigma_i^2 = 2, \tag{4.76}$$

the terms containing odd product of $\sigma_i$ have no contribution to the sum. The remaining terms consist of closed loops of $t_{ij}$, as depicted in Fig. 4.24. Thus, the partition function is given by the sum over all possible closed loops. Since the sum is given as a series of the power of $t_{ij}$ which is small at high temperatures, it provides a high-temperature series expansion of the partition function. The free energy, $-k_{\mathrm{B}}T \ln Z$, is given by all the connected loops.

The number of paths on the lattice can be counted using a matrix $A$ whose matrix element $A(i, j|i'j')$ is 1 if the motion from $(i, j)$ to $(i'j')$ is possible, and 0 otherwise. This matrix is known as the "adjacency matrix", and $(A)_{(ij),(i^2j^2)}(A)_{(i^2j^2),(i^3j^3)} \cdots (A)_{(i^nj^n),(kl)}$ represents the paths from the site $(i, j)$ to $(k, l)$ in $n$ steps. The matrix element $(A^n)_{(ij),(kl)}$ gives the number of paths starting $(i, j)$ and ending at $(k, l)$ in $n$ steps. The diagonal elements $(A^n)_{(ij),(ij)}$ represent the number of paths that return to the original point.

But, in the sum (4.75), each bond $t_{ij}$ must appears only once, whereas the simple adjacency matrix method counts loops in which the same bond is used multiple times. To obtain the exact partition function, we need a method to correct these overcounts. This correction is achieved through the so-called combinatory method, which uses a modified adjacency matrix [102, 104–106, 108, 109]. In the method, we introduce a weight for the path entering a site $(i, j)$ from a direction $\nu$ and leaving to another direction $\nu'$, where the weight depends on the directions $\nu$ and $\nu'$. The weight is set to 1 if the path enters the site $(i, j)$ from direction $\nu$ and exits in the same direction. If the path exits in a different direction $\nu'$, with a change of angle $\phi$, the weight is set to $e^{-i\phi/2}$. The motion that returns to the original point is not considered, setting the weight 0. For a single loop with $m$ steps with the angle change $\theta_m$, the product of the weights is given by $(e^{-i\sum_m \phi_m/2}) = e^{i\pi} = -1$. Brief explanation of this method will be given at the end of this section (Sect. 4.8.5).

### 4.8.1 Uniform Model on the Square Lattice

First, we study the Ising model on the square lattice with uniform ferromagnetic interactions. In this case, all interactions are identical:

$$t_{ij} = t = \tanh(\beta J). \tag{4.77}$$

Let us consider how the weights for the paths at the $(n+1)$-th step, $W_{n+1}(i, j, \nu)$, are expressed in terms of those at the $n$th step, $W_n(i, j, \nu)$. Here $\nu$ denotes the direction of entry to the site $(i, j)$.

For the square lattice, there are four possible directions: $\nu = 1, 2, 3$ or 4. Noting that

$$\alpha = e^{-\pi/4}, \tag{4.78}$$

the weights for the paths of $(n+1)$-th step entering from the direction $\nu$ are given by

$$\begin{cases} W_{n+1}(i,j,1) = tW_n(i,j,1) + t\alpha^{-1}W_n(i,j-1,2) \\ \qquad +0\times W_n(i+1,j,3) + t\alpha W_n(i,j+1,3), \\ W_{n+1}(i,j,2) = t\alpha W_n(i,j,1) + tW_n(i,j-1,2) \\ \qquad +\alpha^{-1}W_n(i+1,j,3) + 0\times W_n(i,j+1,3), \\ W_{n+1}(i,j,3) = 0\times W_n(i,j,1) + t\alpha W_n(i,j-1,2) \\ \qquad +tW_n(i+1,j,3) + t\alpha^{-1}W_n(i,j+1,3), \\ W_{n+1}(i,j,4) = t\alpha^{-1}W_n(i,j,1) + 0\times W_n(i,j-1,2) \\ \qquad +t\alpha W_n(i+1,j,3) + tW_n(i,j+1,3). \end{cases} \tag{4.79}$$

This process is expressed by a $(4N \times 4N)$ matrix $\Lambda$:

$$\mathbf{W}_{n+1} = \Lambda \mathbf{W}_n, \tag{4.80}$$

where $\mathbf{W}_n$ is a $4N$ dimensional vector consisting of $W_n(i, j, \nu)$. The free energy per spin, $f$, is given by

$$\beta f = \frac{1}{N}\ln Z = -\ln 2 + \ln(1-t^2) - \frac{1}{2N}\ln \mathrm{Det}(I - \Lambda). \tag{4.81}$$

Since the lattice exhibits translational symmetry, we can introduce the Fourier transformation with respect to the unit cell. For a regular square lattice, the unit cell corresponds to a lattice point, and the Fourier transformation is given by

$$\hat{W}_n(p,q) = \sum_{k=1}^{L}\sum_{l=1}^{L} e^{-ipk}e^{-iql}W_n(k,l), \quad p = \frac{2\pi}{L}p', \quad q = \frac{2\pi}{L}q' \tag{4.82}$$

and the inverse Fourier transformation is

$$W_n(k,l) = \frac{1}{N} \sum_{p'=1}^{L} \sum_{q'=1}^{L} e^{-ipk} e^{-iql} \hat{W}_n(p,q). \tag{4.83}$$

The matrix $\Lambda$ for the operation of a step (4.79) is block diagonalized by $\hat{W}_n(p,q)$ given by

$$\Lambda(p,q) = \begin{pmatrix} te^{-ip} & t\alpha^{-1}e^{-iq} & 0 & t\alpha e^{iq} \\ t\alpha e^{-ip} & te^{-iq} & t\alpha^{-1}e^{ip} & 0 \\ 0 & t\alpha e^{-iq} & te^{ip} & t\alpha^{-1}e^{iq} \\ t\alpha^{-1}e^{-ip} & 0 & t\alpha e^{ip} & te^{iq} \end{pmatrix}. \tag{4.84}$$

Then, the free energy per spin (4.81) is given by

$$\beta f = -\ln 2 + \ln(1-t^2) - \frac{1}{2N} \sum_{p'=1}^{L} \sum_{q'=1}^{L} \ln \mathrm{Det}(I - \Lambda(p,q)). \tag{4.85}$$

The determinant of $I - \Lambda(p,q)$ is explicitly calculated as

$$\mathrm{Det}(I - \Lambda(p,q)) = (1+t^2)^2 - 2t(1-t^2)(\cos p + \cos q). \tag{4.86}$$

In the limit $L \to \infty$, the sum is expressed by an integral:

$$\frac{1}{2N} \sum_{p'=1}^{L} \sum_{q'=1}^{L} \to \frac{L}{2\pi} \int_0^{2\pi} dp \frac{L}{2\pi} \int_0^{2\pi} dq. \tag{4.87}$$

leading to the expression for free energy:

$$\beta f = -\ln 2 + \ln(1-t^2)$$

$$- \frac{1}{2} \frac{1}{(2\pi)^2} \int_0^{2\pi} \int_0^{2\pi} dp dq \ln\left((1+t^2)^2 - 2t(1-t^2)(\cos p + \cos q)\right). \tag{4.88}$$

The critical point is obtained from the singular point of the integrand:

$$(1+t^2)^2 - 2t(1-t^2)(\cos p + \cos q) = 0, \tag{4.89}$$

which has a solution for $p = q = 0$. The solution of

$$(1+t^2)^2 - 4t(1-t^2) = (t^2 + 2x - 1)^2 = 0 \tag{4.90}$$

is given by

$$x_C = \sqrt{2} - 1, \quad T_C = \frac{J}{k_B \ln(\sqrt{2} - 1)} \simeq 2,269J. \tag{4.91}$$

The free energy itself is obtained by the integral (4.88).

### 4.8.2 *Villain Lattice*

The unit cell of the Villain odd model consists of two sites, enclosed by the dashed line in Fig. 4.2. The critical temperature of this model has been derived in previous studies [112–116]. Here, we use a vector $W_{n+1}(i, j, k, \nu)$, where $i, j$ denotes the position of the unit cell, $k$ is the number of site index in the cell (i.e., $k = 1, 2$), and $\nu$ indicates the direction. First, we consider the general case in which

$$x = \tanh(\beta J), \quad y = \tanh(\beta J'). \tag{4.92}$$

The weight for the path at the $n + 1$ steps, $W_{n+1}(i, j, k, \nu)$ are given as:

$$\begin{cases} W_{n+1}(i, j, 1, 1) = yW_n(i-1, j, 1, 1) + x\alpha^{-1}W_n(i, j-1, 2, 2) \\ \qquad +x\alpha W_n(i, j, 2, 3), \\ W_{n+1}(i, j, 1, 2) = y\alpha W_n(i-1, j, 1, 1) + y\alpha^{-1}W_n(i, j-1, 1, 3) \\ \qquad +xW_n(i, j-1, 2, 2) \\ W_{n+1}(i, j, 1, 3) = yW_n(i+1, j, 1, 3) + x\alpha W_n(i, j-1, 2, 2) \\ \qquad +x\alpha^{-1}W_n(i, j, 2, 4), \\ W_{n+1}(i, j, 1, 4) = x\alpha^{-1}W_n(i-1, j, 1, 1) + y\alpha W_n(i+1, j, 1, 3) \\ \qquad +xW_n(i, j, 2, 4) \\ W_{n+1}(i, j, 2, 1) = x\alpha^{-1}W_n(i, j, 1, 2) + x\alpha W_n(i, j+1, 1, 4) \\ \qquad +x\alpha W_n(i-1, j, 2, 1), \\ W_{n+1}(i, j, 2, 2) = xW_n(i, j, 1, 2) + x\alpha^{-1}W_n(i-1, j, 2, 1) \\ \qquad +x\alpha W_n(i+1, j, 2, 3) \\ W_{n+1}(i, j, 2, 3) = x\alpha W_n(i, j, 1, 2) + x\alpha^{-1}W_n(i, j+1, 1, 4) \\ \qquad +xW_n(i+1, j, 2, 3), \\ W_{n+1}(i, j, 2, 4) = xW_n(i, j+1, 1, 4) + x\alpha^{-1}\alpha W_n(i-1, j, 2, 1) \\ \qquad +x\alpha W_n(i+1, j, 2, 3) \end{cases} \tag{4.93}$$

Using the Fourier transformation, the vector is transformed as

$$W_n(p, q, k, \nu) = \sum_{l=1}^{L} \sum_{m=1}^{L/2} e^{ipk+iql} W_n(l, m, k, \nu). \tag{4.94}$$

Then, (4.93) is expressed as

$$\boldsymbol{W}_{n+1} = \Lambda(p, q)\boldsymbol{W}_n, \tag{4.95}$$

where

$$\Lambda(p,q) = \begin{pmatrix} ye^{-ip} & 0 & 0 & 0 & 0 & x\alpha^{-1}e^{-iq} & 0 & x\alpha \\ y\alpha e^{-ip} & 0 & ye^{-iq}0 & 0 & te^{-iq} & 0 & 0 & \\ 0 & 0 & ye^{ip} & 0 & 0 & x\alpha e^{-iq} & 0 & x\alpha^{-1} \\ y\alpha^{-1}0 & \alpha e^{ip} & 0 & 0 & 0 & 0 & x & \\ 0 & x\alpha^{-1} & 0 & x\alpha e^{-ip} & 0 & 0 & 0 & 0 \\ 0 & x & 0 & 0 & x\alpha e^{-ip} & 0 & x\alpha^{-1}e^{ip} & 0 \\ 0 & x\alpha & 0 & x\alpha^{-1}e^{iq} & 0 & 0 & xe^{ip} & 0 \\ 0 & 0 & 0 & xe^{iq} & x\alpha^{-1}e^{-ip} & 0 & x\alpha e^{ip} & 0 \end{pmatrix}. \tag{4.96}$$

The free energy per spin is then given by

$$\beta f = \frac{1}{N}\ln Z = -\ln 2 + \frac{3}{4}\ln(1-x^2) + \frac{1}{4}\ln(1-y^2) - \frac{1}{2N}\ln \mathrm{Det}(I-\Lambda)$$

$$= -\ln 2 + \frac{3}{4}\ln(1-x^2) + \frac{1}{4}\ln(1-y^2) - \frac{1}{2N}\sum_{p'=1}^{L}\sum_{q'=1}^{L/2}\ln \mathrm{Det}(I-\Lambda(p,q)). \tag{4.97}$$

In the limit $L \to \infty$, the sum is expressed by an integral:

$$\frac{1}{2N}\sum_{p'=1}^{L}\sum_{q'=1}^{L/2} \to \frac{1}{2\pi}\int_0^{2\pi} dp \frac{1}{4\pi}\int_0^{2\pi} dq, \tag{4.98}$$

yielding the free energy expression:

$$\beta f = -\ln 2 + \frac{3}{4}\ln(1-x^2) + \frac{1}{4}\ln(1-y^2)$$
$$-\frac{1}{4}\frac{1}{(2\pi)^2}\int_0^{2\pi}\int_0^{2\pi} dp dq \ln \mathrm{Det}(I-\Lambda(p,q)). \tag{4.99}$$

The determinant is explicitly given by

$$\begin{aligned} &\mathrm{Det}(I-\Lambda(p,q)) \\ &= (1+x^2)\left(1+x^4+2x(1+x^2)y+(1+x^4)y^2\right) \\ &+2(x^4-1)(x+y)(1+xy)\cos(p) \\ &+2x(x^2-1)((x^2-1)y\cos(2p) - x(y^2-1)\cos q). \end{aligned} \tag{4.100}$$

The critical point is given by the condition $\mathrm{Det}(I - \Lambda(p,q)) = 0$. At $p = q = 0$,

$$(1+x^2)\left(1+x^4+2x(1+x^2)y+(1+x^4)y^2\right)+2(x^2-1)(x+y)(1+xy)$$

$$+2x(x^2-1)(x^2-1)y-x(y^2-1))=0. \tag{4.101}$$

For the special case $J = -J'$, i.e., $x = -y$, the critical temperature is zero:

$$T_C = 0. \tag{4.102}$$

The free energy is given by

$$\begin{aligned}\beta f =& -\ln 2 + \frac{1}{2}\ln(1-x^2)\\ &- \frac{1}{16\pi^2}\int_0^{2\pi}\int_0^{2\pi} dpdq \ln\left((1+x^2)^2 - 2x^2(\cos(p)+\cos(q))\right).\end{aligned} \tag{4.103}$$

The entropy at $T = 0$ is given by

$$\frac{S}{N} = \frac{E-F}{NT} = -\frac{J}{T} - \frac{f}{T}. \tag{4.104}$$

At $T \to 0$,

$$1 - x^2 = \frac{1}{\cosh^2(\beta J)} \simeq 4e^{2\beta J}. \tag{4.105}$$

$$\begin{aligned}\frac{f}{T} &\simeq -\frac{J}{T} - \frac{k_B}{16\pi^2}\int_0^{2\pi}\int_0^{2\pi} dpdq\ \ln\left((4 - 2(\cos(p)+\cos(q))\right)\\ &\simeq -\frac{J}{T} - 0.291561 k_B.\end{aligned} \tag{4.106}$$

Thus, the residual entropy per spin is

$$\frac{S}{N} \simeq 0.291561 k_B. \tag{4.107}$$

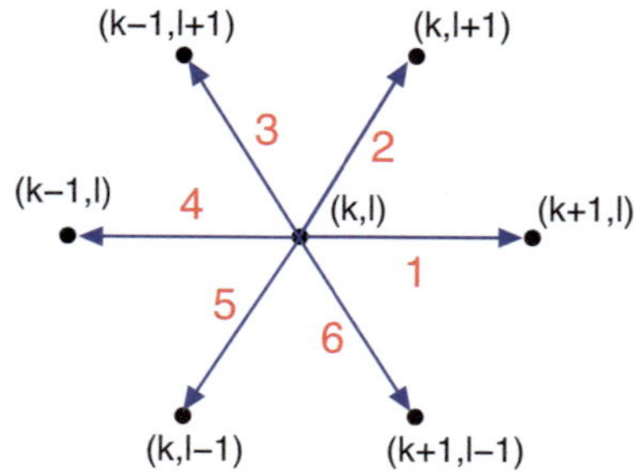


**Fig. 4.25** Triangular lattice. Sites are labeled using a tilted coordinate system. The six directions, labeled as $\nu = 1, 2, \ldots 6$ are shown

## *4.8.3 Triangular Lattice*

In this subsection, we derive the free energy of the Ising model on a triangular lattice. The triangular lattice consists of six directions, as shown in Fig. 4.25. The angle between adjacent bonds is $2\pi/6$, so the weight for an angle change is given by

$$\alpha = e^{i\pi/6}. \tag{4.108}$$

Here, we consider the case where all exchange couplings are equal to $J$, and we define:

$$t = \tanh(\beta J). \tag{4.109}$$

In this case, the weight for the path of $n+1$ steps, $W_{n+1}(i, j, \nu)$, is given as

$$\begin{cases} W_{n+1}(i,j,1) = tW_n(i-1,j,1) + t\alpha^{-1}W_n(i,j-1,2) \\ \qquad +t\alpha^{-2}W_n(i+1,j-1,3) \\ \qquad +t\alpha^{2}W_n(i,j+1,5) + t\alpha W_n(i-1,j+1,6), \\ W_{n+1}(i,j,2) = t\alpha W_n(i-1,j,1) + tW_n(i,j-1,2) + t\alpha^{-1}W_n(i+1,j-1,3) \\ \qquad +t\alpha^{-2}W_n(i+1,j,4) + t\alpha^{2}W_n(i-1,j+1,6), \\ W_{n+1}(i,j,3) = t\alpha^{2}W_n(i-1,j,1) + t\alpha W_n(i,j-1,2) + tW_n(i+1,j-1,3) \\ \qquad +t\alpha^{-1}W_n(i+1, ,4) + t\alpha^{-2}W_n(i,j+1,5), \\ W_{n+1}(i,j,4) = t\alpha^{2}W_n(i,j-1,2) + t\alpha W_n(i+1,j-1,3) + tW_n(i+1, ,4) \\ \qquad +t\alpha^{-1}W_n(i,j+1,5) + t\alpha^{-2}W_n(i-1,j+1,6), \\ W_{n+1}(i,j,5) = t\alpha^{-2}W_n(i-1,j,1) + t\alpha^{2}W_n(i+1,j-1,3) + t\alpha W_n(i+1,4) \\ \qquad +tW_n(i,j+1,5) + t\alpha^{-1}W_n(i-1,j+1,6), \\ W_{n+1}(i,j,6) = t\alpha^{-1}W_n(i-1,j,1) + t\alpha^{-2}W_n(i,j-1,2) + t\alpha^{2}W_n(i+1,4) \\ \qquad +t\alpha W_n(i,j+1,5) + tW_n(i-1,j+1,6). \end{cases} \tag{4.110}$$

Using the Fourier transformation:

$$W_n(p,q,k,\nu) = \sum_{l=1}^{L}\sum_{m=1}^{L} e^{ipk+iql}W_n(l,m,k,\nu), \tag{4.111}$$

(4.110) is expressed as

$$\boldsymbol{W}_{n+1} = \Lambda(p,q)\boldsymbol{W}_n, \tag{4.112}$$

where

$$\Lambda(p,q) = t\begin{pmatrix} e^{-ip} & e^{-iq} & \alpha^{-2}e^{i(p-q)} & 0 & \alpha^2 e^{iq} & \alpha e^{i(-p+q)} \\ \alpha e^{-ip} & e^{-iq} & \alpha^{-2}e^{i(p-q)} & \alpha^2 e^{ip} & 0 & \alpha^2 e^{i(-p+q)} \\ \alpha^2 e^{-ip} & \alpha e^{-iq} & e^{i(p-q)} & \alpha^{-1}e^{ip} & \alpha^{-2}e^{iq} & 0 \\ 0e^{-ip} & \alpha^2 e^{-iq} & \alpha e^{i(p-q)} & e^{ip} & \alpha^{-1}e^{iq} & \alpha^{-2}e^{i(-p+q)} \\ \alpha^{-2}e^{-ip} & 0 & \alpha^2 e^{i(p-q)} & \alpha e^{ip} & e^{iq} & \alpha^{-1}e^{i(-p+q)} \\ \alpha^{-1}e^{-ip} & \alpha^{-2}e^{-iq} & 0 & \alpha^2 e^{ip} & \alpha e^{iq} & e^{i(-p+q)} \end{pmatrix} \tag{4.113}$$

The free energy per spin is given by

$$\beta f = -\frac{1}{N}\ln Z = -\ln 2 + 3\ln(\cosh 2K) - \frac{1}{2N}\ln \mathrm{Det}(I - \Lambda)$$

$$= -\ln 2 + 3\ln(\cosh 2K) - \frac{1}{2N}\sum_{p'=1}^{L}\sum_{q'=1}^{L}\ln \mathrm{Det}(I - \Lambda(p,q)), \tag{4.114}$$

where

$$\det(I - \Lambda(p,q))$$

$$= (1+t)^2\left(t^4 - 2t^2 + 6t^2 - 2t + 1 - 2t(t-1)^2(\cos(p) + \cos(p-q) + \cos(q))\right). \tag{4.115}$$

In the limit $L \to \infty$, the sum is expressed by an integral:

$$\frac{1}{2N}\sum_{p'=1}^{L}\sum_{q'=1}^{L} \to \frac{1}{2\pi}\int_0^{2\pi} dp \frac{1}{2\pi}\int_0^{2\pi} dq, \tag{4.116}$$

and

$$\beta f = -\ln 2 + 3\ln(\cosh 2K) - \frac{1}{2}\frac{1}{(2\pi)^2}\int_0^{2\pi}\int_0^{2\pi} dp dq F(t,p,q), \tag{4.117}$$

where

$$F(t,p,q) = \ln[(1+t)^2(t^4 - 2t^2 + 6t^2 - 2t + 1 - 2t(t-1)^2$$
$$(\cos(p) + \cos(p-q) + \cos(q)))]. \tag{4.118}$$

#### 4.8.3.1 Ferromagnetic Model ($J > 0$)

In this case, the minimum of $\det(I - \Lambda(p, q))$ is at $p = q = 0$:

$$\det(I - \Lambda(0, 0)) = (1 + x^2)(x^2 - 4x + 1)^2. \tag{4.119}$$

This determinant becomes zero when

$$\frac{k_{\mathrm{B}} T_{\mathrm{C}}}{J} = \frac{4}{\ln 3} \simeq 3.64\ldots, \tag{4.120}$$

below which the present model has non-zero spontaneous magnetization.

#### 4.8.3.2 Antiferromagnetic Model ($J > 0$)

In this case, the minimum of $\det(I - \Lambda(p, q))$ is located at

$$(p, q) = \left(\frac{2\pi}{3}, \frac{4\pi}{3}\right). \tag{4.121}$$

At this point,

$$\det(I - \Lambda\,(2\pi/3, 4\pi/3)) = 3t(-1t)^2(1 + t)^2, \tag{4.122}$$

which becomes 0 when $t = -1$. This corresponds to

$$\frac{k_{\mathrm{B}} T_{\mathrm{C}}}{J} = 0. \tag{4.123}$$

In this case, the free energy is given by

$$\beta f = -\ln 2 - 3\ln(\cosh 2K) - \frac{1}{8\pi^2} \int_0^{2\pi}\int_0^{2\pi} dpdq$$

$$\ln((1 + t)^2(t^4 - 2t^2 + 6t^2 - 2t + 1 - 2t(t - 1)^2$$
$$\left(2\cos\left(\frac{p+q}{2}\right)\cos\left(\frac{p-q}{2}\right) + \cos(p - q))\right). \tag{4.124}$$

As $t \to -1$, we introduce the following variables:

$$\xi = \frac{p+q}{2}, \quad \eta = \frac{p-q}{2}, \tag{4.125}$$

and the free energy simplifies to

$$\beta f = \frac{J}{k_{\rm B}T} + \ln 2 - \frac{1}{2\pi^2}\int_0^{\pi}\int_0^{\pi} d\xi d\eta \ln\left(1 + 4(\cos^2\xi + \cos\xi\cos\eta)\right) - \ln 2. \tag{4.126}$$

Given that the ground-state energy per spin at $T = 0$ is $E(0)/N = J$, the entropy at $T = 0$ is given by

$$\frac{S}{N} = \frac{E - F}{NT} = -\frac{J}{NT} - \frac{f}{NT}. \tag{4.127}$$

From this, the residual entropy per spin at $T = 0$ is obtained as

$$\frac{S(0)}{Nk_{\rm B}} = k_{\rm B}\frac{1}{2\pi^2}\int_0^{\pi}\int_0^{\pi} d\xi d\eta \ln\left(1 + 4(\cos^2\xi + \cos\xi\cos\eta)\right) \simeq 0.323066\ldots. \tag{4.128}$$

### 4.8.4 Large Unit Cell

The appearance of singularities in random bond systems is an important issue in statistical physics [117]. The method discussed here can be applied to lattices with large unit cells, in which nearest neighbor interactions ($J_{ij} = \pm 1$) are arbitrarily distributed. This approach has been used to study the detailed properties of a random bond system up to a size of $701 \times 701$ [118].

### 4.8.5 Combinatorial Method to Obtain the Free Energy of the Ising Model

In the combinatorial approach, the adjacency matrix $A(i, j|i', j')$ is used. This matrix, of size $N \times N$, encodes all possible paths from a site $(i, j)$ to another site $(i', j')$. The diagonal elements of the matrix $(A^n)_{(i,j),(i,j)}$ correspond to the number of all possible paths that return to the original point in $n$ steps. However, these diagonal elements may include configurations that result in overcounting, as illustrated in Fig. 4.26.

The configuration depicted in Fig. 4.26a consists of 7 bonds and belongs to the term of $t^7$. This term does not contribute to the partition function. To form a closed loop, one bond is used more than once, as shown in Fig. 4.26b, which is not allowed in the product (4.75). This configuration also appears as a superposition of two square loops, as illustrated in Fig. 4.26c.

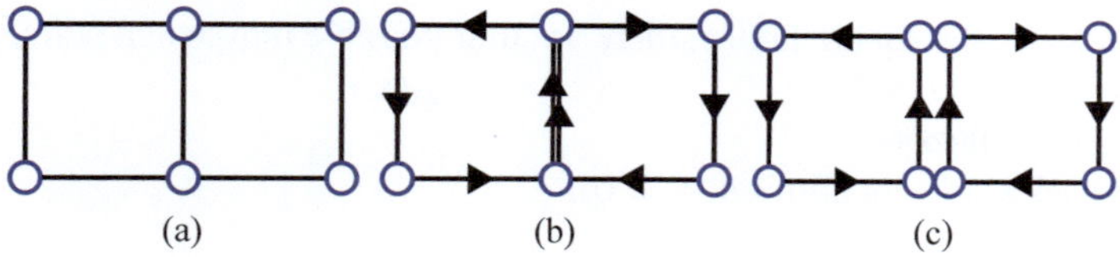


**Fig. 4.26** Loops that do not contribute to the partition function. **a** A 7-stroke loop, **b** the corresponding loop consisting of 8 strokes, and **c** a superposition of two square loops

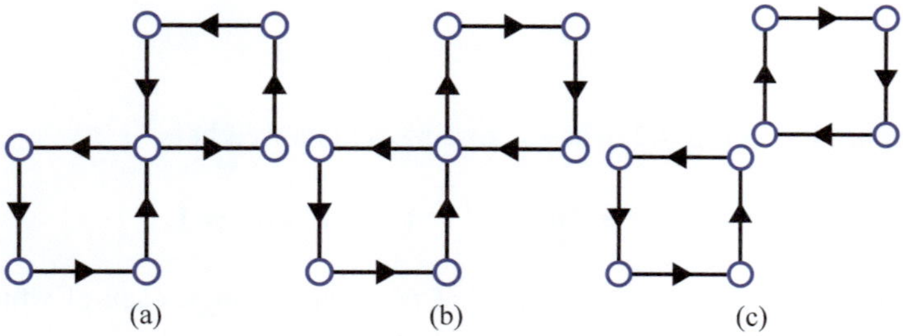


**Fig. 4.27** A loop of 8 strokes on a corner-sharing configuration. **a** 8-stroke loop without a crossing, **b** 8-stroke loop with a crossing, and **c** a superposition of two square loops

In this way, the correspondence between the product (4.75) for the partition function and the counting based on the simple adjacency matrix is not exact. Thus, a method to avoid such overcounting is required.

Overcounts can also occur in configurations where loops share a corner, as depicted in Fig. 4.27. In this case, the graph consists of an 8-stroke loop, and distinct possible paths can be formed at the corner shared by two square loops. However, the graph must be counted only once in the partition function.

To address the issue of overcounts in path configurations, a sophisticated method has been developed [102, 104–106, 108, 109]. Here, we provide a brief outline of this method. Instead of using a simple adjacency matrix $A(i, j|i', j')$, which is an $N \times N$ matrix, the specification of the path situation at a point $(i, j)$ is extended. There, new path weights are introduced, depending on the change in the path angle at a point $(i, j)$. For any closed single loop, the sum of the angle changes $\Delta\phi_j$ is $2\pi$. Based on this fact, factors on the path that depend on the angle change $\Delta\phi_j$ of the path at a site are introduced as

$$\alpha_j = e^{i\Delta\phi_j/2}. \tag{4.129}$$

Then, the product of $\alpha_j$ over a loop is $-1$, i.e.,

$$\prod_j \alpha_j = e^{-i\sum_j \Delta\phi_j} = e^{i\pi} = -1. \tag{4.130}$$

For example, on the square lattice, there are four possible moves at a point, namely:
(1) move straight ($\alpha = 1$),
(2) turn to the left ($\alpha = e^{i\pi/4}$),
(3) move backward (not allowed, $\alpha = 0$),
(4) turn to the right ($\alpha = e^{-i\pi/4}$).
Since there are four types of paths, the matrix for path propagation becomes a $4N \times 4N$ matrix.

With these factors, the weight of the configuration in Fig. 4.27a is

$$\alpha\alpha\alpha^{-1}\alpha\alpha\alpha^{-1}\alpha = \alpha^4 = -1. \tag{4.131}$$

The configuration in Fig. 4.27b has a crossing, and its weight is

$$\alpha \times 1 \times \alpha^{-1}\alpha^{-1}\alpha^{-1} \times 1 \times \alpha\alpha = \alpha^0 = 1. \tag{4.132}$$

The configuration in Fig. 4.27c consists of two square loops, each of which has a weight of $-1$. Thus, the weight of this configuration is

$$(-1) \times (-1) = 1. \tag{4.133}$$

Using these signs for the configurations, it is known that the overcounts can be avoided, and the product (4.75) is reproduced [102, 104–106, 108, 109]. Let the sum of the configurations for a single loop of $r$ strokes (including crossings) be $f_r$. The product (4.75) is then given by

$$\prod_{\langle ij\rangle}(1 + t_{ij}\sigma_i\sigma_j) = 1 + \sum_{s=1}^{\infty}(-1)^s \frac{1}{s!} \sum_{r_1,r_2,\ldots,r_s}^{\infty} f_{r_1} f_{r_2} \cdots f_{r_s}\, t^{r_1+r_2+\cdots+r_s}. \tag{4.134}$$

For each configuration, we have two possible directions. Thus, if we denote the number of loops with $r$ strokes as $v_r$, then

$$\prod_{\langle ij\rangle}(1 + t_{ij}\sigma_i\sigma_j) = 1 + \sum_{s=1}^{\infty}(-1)^s \frac{1}{s!\,2^s} \sum_{r_1+r_2+\cdots+r_s=1}^{\infty} v_{r_1} v_{r_2} \cdots v_{r_s}\, t^{r_1+r_2+\cdots+r_s}. \tag{4.135}$$

Moreover, if we count the configurations by starting at a site within the loop, the total number of loops $b_r$ is given by

$$b_r = v_r r, \tag{4.136}$$

which is obtained from the relation:

$$b_r t^r = \mathrm{Tr}\,\Lambda_{\mathrm{all}}^r. \tag{4.137}$$

Using $b_r$, we have

$$\prod_{\langle ij\rangle}(1+t_{ij}\sigma_i\sigma_j) = 1+\sum_{s=1}^{\infty}(-1)^s\frac{1}{s!\,2^s}\sum_{r_1,r_2,\ldots,r_s}^{\infty}\frac{b_{r_1}}{r_1}\frac{b_{r_2}}{r_2}\cdots\frac{b_{r_s}}{r_s}t^{r_1+r_2+\cdots+r_s} \quad (4.138)$$

$$= \sum_{s=0}^{\infty}(-1)^s\frac{1}{s!\,2^s}\left(\sum_{r=0}^{\infty}\frac{b_r t^r}{r}\right)^s = \sum_{s=0}^{\infty}(-1)^s\frac{1}{s!\,2^s}\left(\sum_{r=0}^{\infty}\frac{\mathrm{Tr}\,\Lambda_{\mathrm{all}}^r}{r}\right)^s. \quad (4.139)$$

Thus, we obtain

$$\prod_{\langle ij\rangle}(1+t_{ij}\sigma_i\sigma_j) = \exp\left(-\frac{1}{2}\sum_{r=0}^{\infty}\frac{1}{r}\,\mathrm{Tr}\,\Lambda_{\mathrm{all}}^r\right) = \exp\left(-\frac{1}{2}\sum_{r=0}^{\infty}\frac{1}{r}\sum_i\lambda_i^r\right), \quad (4.140)$$

where $\lambda_i$ are the eigenvalues of $\Lambda_{\mathrm{all}}$. Then, we have

$$\exp\left(-\frac{1}{2}\sum_i\sum_{r=0}^{\infty}\frac{1}{r}\lambda_i^r\right) = \exp\left(\frac{1}{2}\sum_i\ln(1-\lambda_i)\right) = \prod_i\sqrt{1-\lambda_i}, \quad (4.141)$$

and thus, the partition function is given by

$$Z = \sum_{\{\sigma_i=\pm1\}} e^{-\beta\mathcal{H}} = 2^N\left(\prod_{\langle ij\rangle}(1-t_{ij}^2)^{-1/2}\right)\prod_i\sqrt{1-\lambda_i}. \quad (4.142)$$

The free energy per site is then given by

$$\frac{F}{N k_{\mathrm{B}}T} = -\ln Z = -\ln 2+\frac{1}{2N}\sum_{\langle ij\rangle}\ln\left(1-t_{ij}^2\right)-\frac{1}{2N}\sum_i\ln(1-\lambda_i). \quad (4.143)$$

For lattices with a periodic structure, the matrix $\Lambda_{\mathrm{all}}$ can be block-diagonalized into smaller matrices $\Lambda(p,q)$ using a Fourier transformation.

## 4.9 Appendix B: The Vertex Model

For the exact properties of frustrated models, the so-called vertex model is very powerful, although the methods can be applied only for two-dimensional systems. Here we briefly review the method.

The vertex model is a model in which arrows are assigned to each bond. On a square lattice, there are 16 ($=2^4$) possible patterns around a vertex because the four

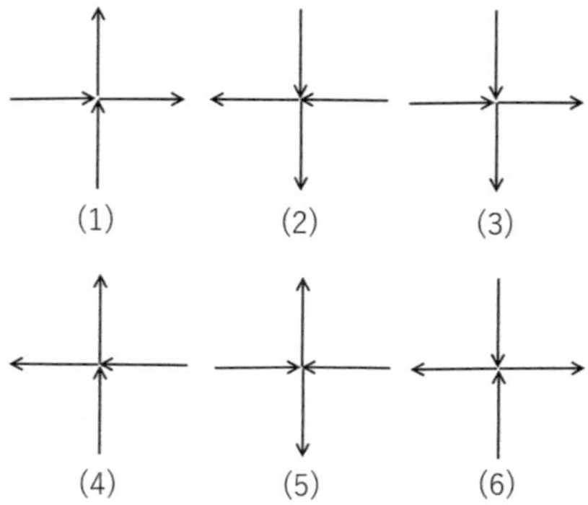


**Fig. 4.28** Six-vertex model configurations satisfying the ice rule

arrows connecting to a vertex can independently point "in" or "out". Let the energy of the $i$-th configuration be $\epsilon_i$, and assign a weight to each configuration:

$$w_i = e^{-\beta\epsilon_i}, \tag{4.144}$$

which represents the Boltzmann weight for the appearance of a given vertex pattern.

The ice configurations (Fig. 4.4b) are expressed by the six arrangements of two arrows point-in and two arrows point-out at a vertex, as depicted in Fig. 4.28.

For the ice model, all six configurations have the same weight:

$$w_1 = w_2 = w_3 = w_4 = w_5 = w_6, \tag{4.145}$$

and weight of other configuration is zero.

### 4.9.1 6-Vertex Models

The 6-vertex model is defined by the weights $w_i$ $(i = 1, \ldots, 6)$. In particular, the so-called symmetric six-vertex model, in which $w_1 = w_2 = a$, $w_3 = w_4 = b$, and $w_5 = w_6 = c$, has been studied in detail [110]. In the ice model, the weights for all six vertex configurations are the same, and thus, the model does not exhibit temperature dependence.

The so-called KDP model was introduced as a two-dimensional model to for the ferroelectric material $KH_2PO_4$ [119]. The energies of the vertex configurations in the KDP model are given by

$$\epsilon_1 = \epsilon_2 = 0, \quad \epsilon_3 = \epsilon_4 = \epsilon_5 = \epsilon_6 > 0. \tag{4.146}$$

and the corresponding weights are

$$w_1 = w_2 = 1, \quad w_3 = w_4 = w_5 = w_6 = w < 1. \tag{4.147}$$

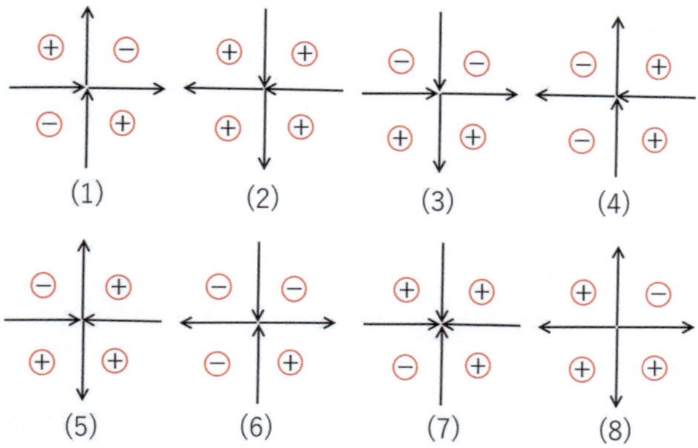


**Fig. 4.29** Eight-vertex model and the corresponding spin configurations of the Ising model

The most likely state (the lowest-energy state) has all horizontal arrows pointing in the same direction, and likewise for all vertical arrows. Such a state is a ferroelectric state. The thermodynamic properties were obtained by Lieb [120].

For the antiferroelectric case, the so-called F model was introduced [121], where

$$\epsilon_1 = \epsilon_2 = \epsilon_3 = \epsilon_4 > 0, \quad \epsilon_5 = \epsilon_6 = 0. \tag{4.148}$$

The lowest-energy state of this model is dominated by vertex configurations (5) and (6) in Fig. 4.28. In this state, adjacent horizontal bonds necessarily have arrows in opposite directions, and the same is true for vertical bonds. This state is an antiferroelectric state.

The arrows do not necessarily represent only polarization; they may also be interpreted as any two-fold variables. For example, the model is used to study the surface configuration of a BCC lattice, which exhibits the so-called roughening transition [122]. The 6-vertex model has been extensively studied and provides important exact properties of phase transitions [123].

### 4.9.2 *Eight-Vertex Model*

The six-vertex model is extended by adding two more configurations (all-in and all-out) as depicted in Fig. 4.29, which is called eight-vertex model.

The eight configurations of the model can be related to the spin configurations of the Ising model with nearest-neighbor ($J$), next-nearest-neighbor ($J_2$), and four-spin ($J_4$) interactions on the dual lattice (where a spin is placed on each square), as depicted in Fig. 4.29. Downward arrows or rightward arrows are drawn between parallel spins, while upward arrows or leftward arrows are drawn between anti-parallel spins. Let the Hamiltonian of the Ising model be

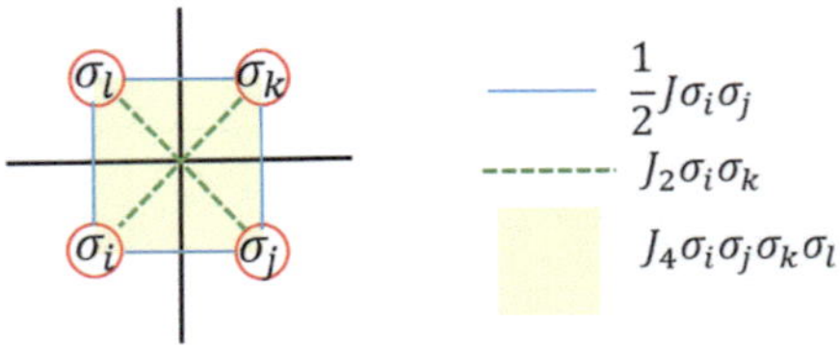


**Fig. 4.30** The interactions among the Ising model (4.149)

$$\mathcal{H} = -J \sum_{\langle ij \rangle:\text{nn}} \sigma_i \sigma_j - J_2 \sum_{\langle ik \rangle:\text{nnn}} \sigma_i \sigma_k - J_4 \sum_{\langle ijkl \rangle:\text{squares}} \sigma_i \sigma_j \sigma_k \sigma_l, \tag{4.149}$$

as depicted in Fig. 4.30, where half of the nearest-neighbor interaction $J$ is assigned to each vertex because the bond is shared by two vertices.

The direct relationship between the vertex weights and the corresponding Ising model is given by

$$\begin{aligned} &w_1 = e^{\beta(-2J+2J_2+J_4)}, \quad w_2 = e^{\beta(2J+2J_2+J_4)}, \quad w_3 = w_4 = e^{\beta(-2J_2+J_4)}, \\ &w_5 = w_6 = w_7 = w_8 = e^{-\beta J_4}. \end{aligned} \tag{4.150}$$

Generally, models with weights of the form

$$w_1 = w_2 = a, \quad w_3 = w_4 = b, \quad w_5 = w_6 = c, \quad w_7 = w_8 = d \tag{4.151}$$

have been exactly solved by Baxter [110]. However, the set of weights in (4.150) does not correspond to the form in (4.151). To apply Baxter's solution, we set $J = 0$, then $w_1 = w_2 = e^{\beta(2J_2+J_4)}$. Then, the condition (4.151) is satisfied. If we set $K_4 = 0$, the system consists of two non-interacting Ising models with the exchange coupling $J_2$. The partition function of the model with $w_1 = w_2 = e^{2\beta J_2}$, $w_3 = w_4 = e^{-2\beta J_2}$, and $w_5 = w_6 = w_7 = w_8 = 1$ gives the square of the partition function of the regular Ising model.

Moreover, Baxter obtained critical properties of the model. For the following set of weights (the order of weights does not matter):

$$w_1 = w_2 = w_3 = w_4 = a, \quad w_5 = w_6 = d, \quad w_7 = w_8 = 1, \tag{4.152}$$

the critical point is given by

$$2a + d = 1, \tag{4.153}$$

and the critical exponent $\alpha$ (for the specific heat) is given by

$$\alpha = 2 - \frac{\pi}{\mu}, \quad \cos\mu = \frac{a^2 - d^2}{a^2 + d^2}, \quad 0 < \mu < \pi. \tag{4.154}$$

In Fig. 4.6, several trajectories for models, e.g., the regular Ising model, the Villain odd model, and models discussed in Sect. 4.2, are plotted in the phase diagram in the $(a, d)$ space.

## References


1. G. Toulouse, Commun. Phys. **2**, 115 (1977)
2. *Magnetic Systems with Competing Interactions (Frustrated Spin Systems)*, ed. by H. T. Diep (World Scientific, 1994). ISBN:981-02-1715-3
3. G.M. Wannier, Phys. Rev. **79**, 357 (1950)
4. G.M. Wannier, (errata) phys. Rev. B **7**, 5017 (1973)
5. R.M.F. Houttappel, Physica **16**, 425 (1950)
6. H. Husimi, I. Syozi, Prog. Theor. Phys. **5**, 177–186 (1950), and **5**, 341–351 (1950)
7. J. Stephenson, J. Math. Phys. **5**, 1009, (1964); J. Math. Phys. **11**, 420, (1970)
8. P.W. Kasteleyn, J. Math. Phys. **4**, 287 (1963)
9. J. Villain, J. Phys. Chem. **10**, 1717 (1977)
10. G. Forgacs, Phys. Rev. B **22**, 4473 (1980)
11. K. Kano, S. Naya, Prog. Theor. Phys. **10**, 158 (1953)
12. J.H. Barry, M. Khatun, T. Tanaka, Phys. Rev. B **37**, 5193 (1988)
13. A.N. Berker, G.S. Grest, C.M. Soukoulis, D. Blankschtein, M. Ma, J. Appl. Phys. **55**, 2416 (1984)
14. F. Matsubara, S. Inawashiro, J. Phy. Soc. Jpn. **56**, 2666 (1987)
15. A. Danielian, Phys. Rev. Lett. **6**, 670 (1961); Phys. Rev. **133**, A1344 (1964); D.D. Betts, C.J. Elliott, Phys. Lett. **18** 18; M.K. Phani, J.L. Lebowiz, Phys. Rev. B **21**, 4027 (1980)
16. R. Moessner, Phys. Rev. B **57**, R5587 (1998)
17. L.D.C. Jaubert, R. Moessner, Phys. Rev. B **91**, 214422 (2015)
18. T. Sakakibara, T. Tayama, Z. Hiroi, K. Matsuhira, S. Takagi, Phys. Rev. Lett. **90**, 207205 (2003)
19. *Spin Ice, Springer Series in Solid-Sate Science*, ed. by M. Udagawa, L. Jaubert (2021), vol. 197.
20. L. Pauling, J. Am. Chem. Soc. **57**, 2680–2684 (1935)
21. E.H. Lieb, Phys. Rev. **162**, 162–172 (1967)
22. E.H. Lieb, Exact solution of the problem of the entropy of two-dimensional ice. Phys. Rev. Lett. **18**, 692–694 (1967)
23. S. Miyashita, J. Phys. Soc. Jpn. **52**, 780 (1983)
24. M. Suzuki, Prog. Theor. Phys. **51**, 1992–1993 (1974)
25. L. P. Kadanoff and A. C. Brown, Ann. Phys. (N.Y.) **121**. 318 (1979)
26. J. Villain, J. Phys. (Paris) **36**, 581 (1975)
27. I. Peschel, Z. Phys, B-Condensed Matter **45**, 399 (1882)
28. A. Luther, I. Peschel, Phys. Rev. B **12**, 3908 (1975)
29. E.H.F. Knops, in *Fundamental Problems in Statistical Mechanics*, ed. by E.G.D. Cohen, pp. 7–20 (1980); E.H.F. Knops, Ann. Phys. **128**, 448 (1980); E.H.F. Knops, L.W.J. Ouden, Ann. Phys. **138**, 155 (1982)
30. O. Nagai, S. Miyashita, T. Horiguchi, Phys. Rev. B **47**, 202 (1993)
31. J. Villian, R. Bidaux, J.P. Carton, R.J. Conte, J. Phys. (Paris) **41**, 1263 (1980)
32. O. Nagai, K. Kang, Y. Yamada, T. Horiguchi, Phys. Lett. A **195**, 263 (1994)
33. A. Lopowski, T. Horoguchi, D. Lopowska, Phys. Rev. Lett. **74**, 3888 (1995)
34. S. Miyashita, M. Suzuki, J. Phys. Soc. Jpn. **50**, 1840–1845 (1981)
35. S. Miyashita, Prog. Theor. Phys. **69**, 714–724 (1983)
36. I. Syozi, "Two-Dimensional Ferroelectric Models", in C. Domb; M. S. Green (eds.), Phase Transitions and Critical Phenomena, vol. 1, New York: Academic Press, pp. -330 (1972)

37. J. Stephenson, Phys. Rev. B **1**, 4405 (1970)
38. H. Kitatani, S. Miyashita, M. Suzuki, Phys. Lett. **108A**, 45 (1985)
39. H. Kitatani, S. Miyashita, M. Suzuki, J. Phys. Soc. Jpn. **55**, 865 (1986)
40. T. Cikyu, M. Suzuki, Prog. Thor. Phys. **81**, 927 (1989)
41. M. Mekata, J. Phys. Soc. Jpn. **42**, 76 (1977)
42. M. Mekata, K. Adachi, J. Phys. Soc. Jpn. **44**, 806 (1978)
43. S. Fujiki, K. Shutoh,Y. Abe, S. Katsura, J. Phys. Soc. Jpn. **52**, 1531 (1983); H. Takayama, K. Matsumoto,H. Kawahara, K. Wada, J. Phys. Soc. Jpn. **52**, 2888 (1983); D.P. Landau, Phys. Rev. B **27**, 5604 (1983); S. Miyashita, H. Kitatani, Y. Kanada, J. Phys. Soc. Jpn. **60**, 1523 (1991); M. Nishino, S. Miyashita, Phys. Rev. B **94**, 184434 (2016)
44. J.M. Kosterlitz, J. Phys. C7 (1974) 1046; J.M. Kosterlitz, D.J. Thouless, J. Phys. **C6**, 1181 (1973); V.L. Berezinskii, Sov. Phys. JETP **32**, 493(1971), ibid. **34**, 610 (1972)
45. J.V. Jose, L.P. Kadanoff, S. Kirkpatrick, D. Nelson, Phys. Rev. B **16**, 1217 (1977)
46. S. Elitzur, R.B. Pearson, J. Shigemitsu, Phys. Rev. D **19**, 3698 (1979)
47. J.L. Cardy, J. Phys. A **13**, 1507 (1980)
48. M. Nishino, S. Miyashita, Phys. Rev. B **92**, 184404 (2015)
49. O. Koseki, F. Matsubara, J. Phys. Soc. Jpn. **69**, 1202 (2000)
50. S. Miyashita, J. Phys. Soc. Jpn. **66**, 3411 (1997)
51. M. Oshikawa, Phys. Rev. B **61**, 3430 (2000)
52. Y. Ueno, K. Kasono, Phys. Rev. B **48**, 16471 (1993)
53. N. Todoroki, Y. Ueno, S. Miyashita, Phys. Rev. B **66**, 214405 (2002)
54. N. Todoroki, S. Miyashita, J. Phys. Soc. Jpn. **73**, 412 (2004)
55. T. Kawasaki and S. Miyashita, Prog. Theor. Phys. **89**, 985 (1993). Prog. Theor. Phys. ibid. **93**, 47 (1995)
56. M.E. Fisher, Phys. Rev. **176**, 257 (1968)
57. B.M. McCoy, T.T. Wu, Phys. Rev. Lett. **21**, 549 (1968); and Phys. Rev. **176**, 631 (1968)
58. A.B. Harris, J. Phys. C: Solid State Phys. **7**, 1671 (1974)
59. V.S. Dotsenko, S. Dotenko et al., J. Phys. C: Solid Stat Phys. **15**, 495–507 (1982)
60. M. Inoue, J. Phys. Soc. Jpn. **64**, 3699 (1995)
61. J.-S. Wang, W. Selke W., Vl. S. Dotesnko, V.B. Andreichenko, Physica A**164**, 221 (1990), Nucl. Phys. B**344**, 531-556 (1990)
62. D. Stauffer, Phys. Rep **54**, 1 (1979); J.W. Essam, Rep. Prog. Phys. **43** 833 (1980); D. Stauffer, A. Aharony, *An Introduction to Percolation Theory* (Taylor and Francis, London, 1985)
63. R.B. Griffiths, Phys. Rev. Lett. **23**, 17 (1969); B. McCoy, Phys. Rev. Lett. **23** 383 (1969); R.B. Griffiths,*Phase Transitions and Critical Phenomena*, ed. by C. Domb, M.S. Green (Academic Press, London, 1972), vol. 1, pp 7–109
64. H. Takano, S. Miyashita, J. Phys. Soc. Jpn. **58**, 3871 (1989)
65. H. Takano, H. Nakanishi, S. Miyashita, Phys. Rev. B **37**, 3716 (1988)
66. D.A. Huse, D.S. Fisher, Phys. Rev. B **35**, 6841 (1987)
67. Y. Imry, S.-K. Ma, Phys. Rev. Lett. **35**, 1399 (1975); A. Aharony, Y. Imry, S.-K. Ma, Phys. Rev. Lett. **37** 1364 (1976); S. Fishman, A. Aharony, J. Phys. C: Solid Stat Phys. **12**, L729 (1979)
68. D.A. Garanin, E.M. Chudnovsky, T. Proctor, Phys. Rev. B **88**, 224418 (2013); T.C. Proctor, D.A. Garanin, E.M. Chudnovsky, PRL **112**, 097201 (2014)
69. A.P. Young, *Spin Glass and Random Field* (World Scientic Publishing, Singapore, 1998)
70. S.F. Edwards, P.W. Anderson, J. Phys. F: Met. Phys. **5**, 965 (1975)
71. D. Sherrington, S. Kirkpartick, Phys. Rev. Lett. **35**, 1792 (1975)
72. D.J. Thouless, P.W. Anderson, R.G. Palmer, Phil. Mag. **35**, 593 (1977)
73. K. Binder and A. P. Young, Rev. Mod. Phys. **58**, 801 (1986). M. Mezard, G. Parisi and M. A. Virasoro, *Spin Glass Thoey and Beyond*, World Scientific (1987)
74. S. Chikazawa, T. Saito, T. Sato, Y. Miyako, J. Phys. Soc. Jpn. **47**, 335 (1979)
75. M. Suzuki, Prog. Thor. Phys. **58**, 1151 (1977)
76. E. Vincent, *Ageing, Rejuvenation and Memory:The Example of Spin-Glasses*, Lecture Note in Phys. **716** pp 7-60 (Springer-Verlag, Berlin (2007))

77. H. Nishimori, J. Phys. C: Solid State Phys. **13**, 4071 (1980)
78. G. Parisi, Phys. Rev. Lett. **43**, 1754 (1979)
79. G. Parisi, Phys. Rev. Lett. **50**, 1946 (1983)
80. M. Mezard, G. Parisi, G. Toulouse and M. A. Virasoro, Phys. Rev. Lett., **52**, 1156 (1984). J. Physique **45**, 843 (1984)
81. J. Phys. C**18**, L529 (1985)
82. D.C. Mattis, Phys. Lett. A **56**, 421 (1976)
83. S. Miyashita, M. Suzuki, J. Phys. Soc. Jpn. **50**, 1840 (1981)
84. D.S. Fisher, D.A. Huse, Phys. Rev. Lett. **56**, 1601 (1986)
85. H. Takano and S. Miyashita, J. Phys. Soc. Jpn. **64**, 423 (1995). H. Takano and S. Miyashita, J. Phys. Soc. Jpn. **64**, 3688 (1995)
86. H. Kawamura, Chiral ordering in heisenberg spin glasses in two and three dimensions. Phys. Rev. Lett. **68**, 3785–3788 (1992)
87. E. Vincent, *Spin glass experiments in Elsevier Encyclopedia of Condensed Matter Physics, 2nd*, edtion. (By T, Chakraborty, 2023)
88. F. Lefloch, J. Hammann, M. Ocio, E. Vincent, Europhys. Lett. **18**, 647 (1992)
89. S. Miyashita, E. Vincent, Eur. Phys. J. B **22**, 203 (2001)
90. S. Tanaka, S. Miyashita, Prog. of Theore. Phys. Supplements **157**, 34 (2005)
91. V. Dupuis, "Dynamique lente des systèmes magnétiques désordonnés. PhD Thesis France: Orsay University. https://tel.archives-ouvertes.fr/tel-00002623/document
92. A.J. Bray, M.A. Moore, Phys. Rev. Lett. **58**, 57 (1987)
93. L.F. Cugliandolo, J. Kurchan, Phys. Rev. B **60**, 922 (1999)
94. Q. Zhai, R. L. Orbach, and D. L. Schlagel, Phys. Rev. B **105**, 014434. (2022)
95. M. Sasaki, V. Dupuis, J.-P. Bouchaud and E. Vincent, Euro. Phys. J. B **29**, 469 (2002). D. Zhang, T. Chen, M. Vucelja, S.H.Lee and G. W Chern, Phys. Rev. E **104**, 064105 (2021)
96. H. Hoffmann and P. Sibani, Phys. Rev. Lett. **63**, 2853. (1989). M. Sasaki and K. Nemoto, J. Phys.l Soc. Jpn., **70**, 1099 (2001). J.-P. Bouchaud, V. Dupuis, J. Hammann, and E. Vincent, Phys. Rev. B **65**, 024439 (2001)
97. I. Paga et al., J. Stat. Mech.: Theor. Exp. **033301** (2021); M. Baity-Jesi, et al., Nature Phys. **19**, 978 (2023)
98. L.A. Kramers, G.H. Wannier, Phys. Rev. **60**, 252–263 (1941)
99. L. Onsager, Phys. Rev. **65**, 117 (1944); G.H. Wannier, Rev. Mod. Phys. **17**, 50 (1945); B. Kaufman, Phys. Rev. **76**, 1232 (1949); B. Kaufman, L. Onsager, Phys. Rev. **76**, 1244 (1949); Y. Nambu, Prog. Thor. Phys. **1**, 5 (1950); C.N. Yang, Phys. Rev. **85**, 808 (1052)
100. R.M. MaCoy, T.T. Wu, *The Tow-Dimensional Ising model* (Harvard University Press, Cambridge, 1973)
101. T.D. Schultz, D.C. Mattis, E.H. Lieb, Rev. Mod. Phys. **36**, 856 (1964)
102. M. Kac, J.C. Ward, Phys. Rev. **88**, 1332 (1952)
103. C.A. Hurst, H.S. Green, J. Chem. Phys. **33**, 1059 (1960)
104. P.W. Kastelyn, J. Math. Phys. **4**, 287 (1963)
105. E.W. Montroll, R. Potts, J.C. Ward, J. Math. Phys. **4**, 308 (1963)
106. Soviet Physics JETP, **20**, 477 (1965)
107. V.G. Vaks, A.I. Larlin, N. Ovchinikov, Sov. Phys. JETP **22**, 820 (1966)
108. R.P. Feynman, *Statistical Mecanics* (W. A. Benjamin Inc., 1972)
109. S. Sharman, J. Math. Phys. **1**,202 (1960), **4**, 1213 (1963)
110. R.J. Baxter, *Exact Solvable Models in Statistical Mechanics* (Academic Press, 1982)
111. H. Nakano, Prog. Thoer. Phys. **40**, 231 (1968); I. Syozi, Prog. Thoer. Phys. **39**, 1367 (1968); S. Miyazima, Prog. Thoer. Phys. **40**, 462 (1968)
112. J.R. Hamm, Phys. Rev. B **15**, 5391 (1977)
113. V.V. Bryksin, A.Y. Gol'tsev, E.E. Kudinov, J. Phys. C: Solid St. Phys. **13**, 5999 (1980)
114. P. Hoever, Z. Hhys. B **48**, 137 (1982); P. Hoever, W.F. Wolff, J. Zittartz, Z. Hhys. B **41**, 41 (1981)
115. M. Karder, A.N. Berker, Phys. Rev. B **26**, 219 (1981)
116. S. Miyashita, Prog. Thor. Phys. **69**, 714 (1983)

117. B.M. McCoy, T.T. Wu, Phys. Rev. **76**, 31 (1968); *The Two-Dimensional Ising Model*, (Harvard University. Press, Cambridge, 1973). VI.S. Dotsenko, S.S. Dotsenko, Adv. Phys. **32**, 19 (1983); R. Shankar, Phys. Rev. Lett. **58**, 2466 (1987), **61** 2390 (1988)
118. M. Inoue, J. Phys. Soc. Jpn. **64**, 3699 (1995)
119. J.C. Slater, J. Chem. Phys. **9**, 16–33 (1941)
120. E.H. Lieb, Phys. Rev. Lett. **19**, 108–110 (1967)
121. F. Rys, Helv. Phys. Acta **36**, 537 (1963)
122. H. van Beijeren, Phys. Rev. Lett. **38**, 993 (1977)
123. E.H. Lieb, F.Y. Wu, Two-dimensional ferroelectric models, in C. Domb, M.S. Green (eds.), *Phase Transitions and Critical Phenomena* (Academic Press, New York), vol. 1, pp. 331–490 (1972); B. Sutherland, Phys. Rev. Lett., **19**, 103-104 (1967); C.P. Yang, Phys. Rev. Lett., **19**, 586–588 (1967)

# Chapter 5
# Ordering in Frustrated Systems with Continuous Spins

**Abstract** In contrast to the case of discrete spins, in systems with continuous variables, such as XY and Heisenberg spins, frustration leads to non-collinear ground-state spin configurations that arise as a compromise between competing interactions. In this chapter, we study the ground-state configurations and their properties at finite temperatures on the antiferromagnetic triangular lattice (AFT) and the kagome lattice for continuous variables.

## 5.1 Order Parameter in Frustrated Continuous Spins

A typical example of non-collinear ground-state spin configurations in frustrated systems is the spiral structure observed in one-dimensional systems with ferromagnetic nearest-neighbor interactions and antiferromagnetic next-nearest-neighbor interactions [1]. With non-collinear ground-state spin configurations, macroscopic degeneracy due to frustration is resolved in most cases. However, energetically degenerate (or nearly degenerate) spin configurations appear. In such cases, entropy plays an important role in determining the thermodynamic equilibrium state, a phenomenon known as "order-by-disorder" [2, 3].

We consider models with continuous spins, which are represented by unit vectors:

$$\boldsymbol{S}_i = (S_i^x, S_i^y, S_i^z), \quad |\boldsymbol{S}_i| = 1. \tag{5.1}$$

The spin vector $\boldsymbol{S}_i$ can be expressed in polar coordinates as

$$\boldsymbol{S}_i = \begin{pmatrix} \sin\theta_i \cos\phi_i \\ \sin\theta_i \sin\phi_i \\ \cos\theta_i \end{pmatrix}. \tag{5.2}$$

© The Author(s), under exclusive license to Springer Nature Singapore Pte Ltd. 2025 
S. Miyashita, *Physics of Entropy*, Springer Series in Solid-State Sciences 212,
https://doi.org/10.1007/978-981-95-1758-9_5

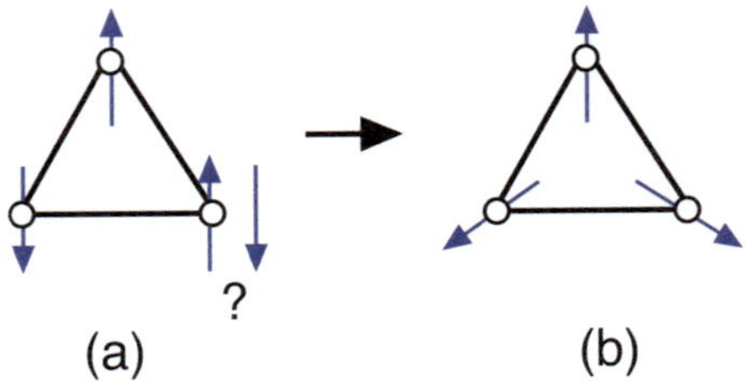


**Fig. 5.1** **a** Degenerate ground states of Ising antiferromagnets on a single triangle lattice. **b** 120° structure of antiferromagnets with continuous spins on a single triangle lattice

## 5.2 Antiferromagnets on the Triangular Lattice of Ising-Like Heisenberg Model (AFT-IH)

We first consider a classical Ising-like Heisenberg model with an anisotropy parameter $A \geq 1$ on the triangular lattice, which includes the Heisenberg model ($A = 1$) as a special case. The Hamiltonian of the system is given by

$$\mathcal{H} = J \sum_{\langle ij \rangle} \left( S_i^x S_j^x + S_i^y S_j^y + A S_i^z S_j^z \right) - H \sum_i S_i^z, \tag{5.3}$$

where $\langle ij \rangle$ denotes the sum over all nearest-neighbor pairs on the lattice.

The triangular lattice can be divided into three sublattices, denoted as A, B, and C. The ground-state spin configuration of the entire system is obtained by repeating a spin configuration that minimizes the energy of the three spins forming a triangle. Thus, we first investigate the ground-state spin configuration of three spins, $\boldsymbol{S}_{\mathrm{A}}, \boldsymbol{S}_{\mathrm{B}}, \boldsymbol{S}_{\mathrm{C}}$, on the triangular plaquette.

The antiferromagnetic interactions on the three bonds of the triangle are not compatible to each other. As a result, the system forms a spin structure where the spins compromise with each other, in contrast to the degenerate ground states observed in the Ising spin systems on a triangular plaquette as depicted in Fig. 5.1a. For $A = 1$, the spins form the so-called "120° structure", as shown in Fig. 5.1b.[1]

The symmetry of the order parameter for the 120° structure differs from that of the spin itself. The order parameter of the 120° structure is specified by two spin vectors, e.g., the spin vectors of sublattice A and sublattice B, denoted as $\boldsymbol{S}_{\mathrm{A}}$ and $\boldsymbol{S}_{\mathrm{B}}$, respectively. To specify the 120° structure more conveniently, a vector perpendicular to the spin plane of the 120° structure is introduced. This vector is defined as

$$\boldsymbol{\kappa} = \frac{1}{3\sqrt{3}} \left( \boldsymbol{S}_{\mathrm{A}} \times \boldsymbol{S}_{\mathrm{B}} + \boldsymbol{S}_{\mathrm{B}} \times \boldsymbol{S}_{\mathrm{C}} + \boldsymbol{S}_{\mathrm{C}} \times \boldsymbol{S}_{\mathrm{A}} \right), \tag{5.4}$$

[1] The $-120°$ structure also represents a ground state.

which is known as "vector chirality".[2] Typically, the 120° structure is specified by the vector chirality $\boldsymbol{\kappa}$ and one of the spin vectors, e.g., $\boldsymbol{S}_\mathrm{A}$. This order parameter causes interesting phase transitions as will be explained in Sect. 5.7.

When the system exhibits an easy-axis anisotropy ($A > 1$), the 120° structure is modified, but the specification of the order parameter using $\boldsymbol{\kappa}$ and $\boldsymbol{S}_\mathrm{A}$ remains valid. Hereafter, we denote the sublattices A, B, and C as 1, 2, and 3, respectively.

### 5.2.1 Conditions for Minimum (Extreme) Energy

We obtain the ground-state configuration in the presence of a magnetic field $H$ applied in the $z$-direction. The energy of the system is explicitly given by

$$\begin{aligned} E &= \sin\theta_1 \sin\theta_2 \cos(\phi_1 - \phi_2) + A\cos\theta_1 \cos\theta_2 - H\cos\theta_1 \\ &+ \sin\theta_2 \sin\theta_3 \cos(\phi_2 - \phi_3) + A\cos\theta_2 \cos\theta_3 - H\cos\theta_2 \\ &+ \sin\theta_3 \sin\theta_1 \cos(\phi_3 - \phi_1) + A\cos\theta_3 \cos\theta_1 - H\cos\theta_3. \end{aligned} \tag{5.5}$$

To obtain the minimum energy configuration, we calculate the derivatives of the energy with respect to the angles $\theta_i$ and $\phi_i$. Hereafter, we use the notation:

$$c_i = \cos\theta_i, \quad s_i = \sin\theta_i, \quad i = 1, 2, 3. \tag{5.6}$$

Derivatives are explicitly given by

$$\frac{\partial E}{\partial \theta_1} = 0 \Longrightarrow c_1\left(s_2\cos(\phi_1 - \phi_2) + s_3\cos(\phi_3 - \phi_1)\right) - As_1\left(c_2 + c_3\right) + Hs_1 = 0, \tag{5.7}$$

$$\frac{\partial E}{\partial \theta_2} = 0 \Longrightarrow c_2\left(s_1\cos(\phi_1 - \phi_2) + s_3\cos(\phi_2 - \phi_3)\right) - As_2\left(c_1 + c_3\right) + Hs_2 = 0, \tag{5.8}$$

$$\frac{\partial E}{\partial \theta_3} = 0 \Longrightarrow c_3\left(s_1\cos(\phi_3 - \phi_1) + s_2\cos(\phi_2 - \phi_3)\right) - As_3\left(c_1 + c_2\right) + Hs_3 = 0, \tag{5.9}$$

$$\frac{\partial E}{\partial \phi_1} = 0 \Longrightarrow -s_1\left(s_2\sin(\phi_1 - \phi_2) + s_3\sin(\phi_3 - \phi_1)\right) = 0, \tag{5.10}$$

$$\frac{\partial E}{\partial \phi_2} = 0 \Longrightarrow s_2\left(s_1\sin(\phi_1 - \phi_2) - s_3\sin(\phi_2 - \phi_3)\right) = 0, \tag{5.11}$$

$$\frac{\partial E}{\partial \phi_3} = 0 \Longrightarrow s_3\left(-s_1\sin(\phi_3 - \phi_1) + s_2\sin(\phi_2 - \phi_3)\right) = 0. \tag{5.12}$$

[2] The term "chirality" ($\boldsymbol{\chi}$) would be more appropriate, but the symbol $\boldsymbol{\kappa}$ is used here to follow the notation introduced in the first paper that studied this problem [4].

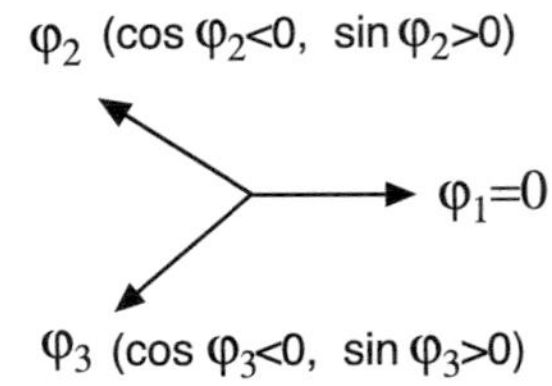


**Fig. 5.2** Top view (from the $z$ axis) of the configuration with $\phi_1 = 0$, $\phi_2$, and $\phi_3$

Due to the rotational symmetry around the $z$-axis, we set $\phi_1 = 0$ without loss of generality. In the following, we obtain possible stable configurations and compare their energies.

### 5.2.2 *The Azimuthal Angles ($\phi_1 = 0, \phi_2, \phi_3$) Expressed by Variables $s_1$, $s_2$, and $s_3$*

First, we consider the case $s_1 s_2 s_3 \neq 0$, and we obtain $\phi_2$ and $\phi_3$ as functions of $s_1$, $s_2$, and $s_3$. Here, we assume the configuration depicted in Fig. 5.2, where $(\cos\phi_1 = 1, \sin\phi_1 = 0)$, $(\cos\phi_2 < 0, \sin\phi_2 > 0)$, and $(\cos\phi_3 < 0, \sin\phi_3 < 0)$.

From (5.10), (5.11) and (5.12), we have the following relation:

$$s_1(s_2\sin(\phi_2) + s_3\sin(\phi_3)) = 0 \implies \sin(\phi_3) = -\frac{s_2}{s_3}\sin(\phi_2). \tag{5.13}$$

Denoting

$$x = \sin(\phi_2) > 0, \tag{5.14}$$

other quantities are expressed in terms of $x$ as

$$\cos(\phi_2) = \pm\sqrt{1 - \sin^2(\phi_2)} = -\sqrt{1 - x^2}, \tag{5.15}$$

and

$$\sin(\phi_3) = -x\frac{s_2}{s_3}, \quad \cos(\phi_3) = -\sqrt{1 - x^2\frac{s_2^2}{s_3^2}}, \tag{5.16}$$

where we have chosen the negative sign since we consider the case where $\cos(\phi_1)\cos(\phi_2) < 0$ and $\cos(\phi_1)\cos(\phi_3) < 0$, as illustrated in Fig. 5.2.

#### 5.2.2.1 Solution for the Case ($\sin\phi_2 \neq 0$ and $\sin\phi_3 \neq 0$)

Using $x$, (5.11) can be rewritten as

$$s_1 x - s_3\left(-x\sqrt{1 - x^2\frac{s_2^2}{s_3^2}} - x\frac{s_2}{s_3}\sqrt{1-x^2}\right) = s_1 x + x\left(\sqrt{s_3^2 - x^2 s_2^2} + s_2\sqrt{1-x^2}\right) = 0$$

$$\Longrightarrow -s_1 = \sqrt{s_3^2 - x^2 s_2^2} + s_2\sqrt{1-x^2}. \tag{5.17}$$

The solution of this equation determines $\phi_2$ and $\phi_3$ as functions of $s_1$, $s_2$ and $s_3$ for the case of $\phi_1 = 0$. After some calculations,[3] $x$ is obtained as

$$x = \pm\frac{\sqrt{4s_2^2 s_3^2 - (s_1^2 - s_2^2 - s_3^2)^2}}{2s_1 s_2}$$

$$= \pm\frac{\sqrt{(s_1 - s_2 + s_3)(s_1 + s_2 - s_3)(s_2 + s_3 + s_1)(s_2 + s_3 - s_1)}}{2s_1 s_2}. \tag{5.18}$$

Here, we take positive sign for $x(> 0)$ as depicted in Fig. 5.2. The choice of $x < 0$ simply corresponds to an exchange of $\phi_2$ and $\phi_3$.

The azimuthal angles $\phi_2$ and $\phi_3$ are given by[4]

$$\sin\phi_2 = x, \quad \cos\phi_2 = -\frac{s_2^2 + s_1^2 - s_3^2}{2s_1 s_2}, \tag{5.20}$$

and

---

3

$$s_1^2 = s_3^2 - x^2 s_2^2 + 2s_2\sqrt{s_3^2 - x^2 s_2^2}\sqrt{1-x^2} + s_2^2(1-x^2)$$

$$\left(s_1^2 - s_3^2 + x^2 s_2^2 - s_2^2(1-x^2)\right)^2 = 4s_2^2(s_3^2 - x^2 s_2^2)(1-x^2)$$

$$(s_1^2 - s_2^2 - s_3^2)^2 + 4s_2^2 x^2(s_1^2 - s_2^2 - s_3^2) + 4s_2^4 x^4 = 4s_2^4 x^4 - 4s_2^2 x^2(s_2^2 + s_3^2) + 4s_2^2 s_3^2$$

$$4s_2^2 x^2(s_1^2 - s_2^2 - s_3^2) + 4s_2^2 x^2(s_2^2 + s_3^2) = 4s_2^2 s_3^2 - (s_1^2 - s_2^2 - s_3^2)^2.$$

.

4 Here we use the following relation:

$$1 - x^2 = 1 - \frac{4s_2^2 s_3^2 - (s_1^2 - s_2^2 - s_3^2)^2}{4s_1^2 s_2^2} = \frac{4s_1^2 s_2^2 - 4s_2^2 s_3^2 + (s_1^2 - s_2^2 - s_3^2)^2}{4s_1^2 s_2^2}$$

$$= \frac{s_1^4 + s_2^4 + s_3^4 + 2s_1^2 s_2^2 - 2s_2^2 s_3^2 - 2s_1^2 s_3^2}{4s_1^2 s_2^2} = \frac{(s_1^2 + s_2^2 - s_3^2)^2}{4s_1^2 s_2^2}. \tag{5.19}$$

.

$$\sin \phi_3 = -\frac{s_2}{s_3}x, \quad \cos \phi_3 = -\frac{s_3^2 + s_1^2 - s_2^2}{2s_1 s_3}, \tag{5.21}$$

where we have taken the negative sign for $\cos \phi_2$ and $\cos \phi_3$ to match the configuration shown in Fig. 5.2. From them,

$$\sin(\phi_2 - \phi_3) = -\frac{s_1}{s_3}x, \quad \cos(\phi_2 - \phi_3) = \frac{s_1^2 - s_2^2 - s_3^2}{2s_2 s_3}. \tag{5.22}$$

Substituting (5.20), (5.21), and (5.22) into (5.7):

$$-c_1 \left( s_2 \frac{s_2^2 + s_1^2 - s_3^2}{2s_1 s_2} + s_3 \frac{s_3^2 + s_1^2 - s_2^2}{2s_1 s_3} \right) = As_1 (c_2 + c_3) - Hs_1.$$

Similarly, by substituting into (5.8) and (5.9), we obtain

$$\begin{aligned} c_1 + A(c_2 + c_3) &= H, \\ c_2 + A(c_3 + c_1) &= H, \\ c_3 + A(c_1 + c_2) &= H. \end{aligned} \tag{5.23}$$

These equations are expressed as

$$\begin{pmatrix} 1 & A & A \\ A & 1 & A \\ A & A & 1 \end{pmatrix} \begin{pmatrix} c_1 \\ c_2 \\ c_3 \end{pmatrix} = \begin{pmatrix} H \\ H \\ H \end{pmatrix}. \tag{5.24}$$

The determinant is calculated as

$$\begin{vmatrix} 1 & A & A \\ A & 1 & A \\ A & A & 1 \end{vmatrix} = 1 + 2A^3 - 3A^2 = (A - 1)^2(1 + 2A). \tag{5.25}$$

For $A \neq 1$ and $A = -1/2$, we have the following solution:

$$c_1 = c_2 = c_3 = \frac{H}{2A + 1} \equiv c \tag{5.26}$$

for the case $|H/(2A + 1)| \leq 1$. Otherwise, no extreme configuration exists. For $H > 2A + 1$, the polarized state,

$$c_1 = c_2 = c_3 = 1, \tag{5.27}$$

yields the minimum energy. As a natural solution, $\theta_1 = \theta_2 = \theta_3$ holds, and then

$$s_1 = s_2 = s_3 \equiv s. \tag{5.28}$$

Using (5.20) and (5.21), we obtain

$$\sin \phi_2 = \frac{\sqrt{3}}{2}, \quad \cos \phi_2 = -\frac{1}{2} \Longrightarrow \phi_2 = \frac{2\pi}{3}, \tag{5.29}$$

and

$$\sin \phi_3 = -\frac{\sqrt{3}}{2}, \quad \cos \phi_3 = -\frac{1}{2} \Longrightarrow \phi_3 = \frac{4\pi}{3}. \tag{5.30}$$

This configuration is known as the "umbrella configuration", which is depicted in Fig. 5.3a. Using (5.26), the energy of the umbrella configuration for a triangular unit is given by

$$\begin{aligned} E_{\text{umbrella}} &= -\frac{3}{2}s^2 + 3Ac^2 - 3Hc = -\frac{3}{2} + \left(\frac{H}{2A+1}\right)^2 \left(\frac{3}{2} + 3A\right) - \frac{3H^2}{2A+1} \\ &= -\frac{3}{2}\frac{2A+1+H^2}{2A+1}. \end{aligned} \tag{5.31}$$

As mentioned above, for the case $|H/(2A+1)| \geq 1$, the polarized state (5.27) yields the minimum energy, and the energy is

$$E_{\text{polarized}} = 3A - 3H. \tag{5.32}$$

#### 5.2.2.2 Case of the Heisenberg Model ($A = 1$)

When the determinant (5.25) becomes zero ($(A-1)^2(1+2A) = 0$), we need to analyze the ground-state configuration separately. For $A = 1$, all the equations in (5.23) coincide as

$$c_1 + c_2 + c_3 = H, \tag{5.33}$$

which implies that any combination of $c_1$, $c_2$, and $c_3$ satisfying this relation gives a ground state. Thus, the model exhibits non-trivial degeneracy for $A = 1$. The magnetization along the $z$-axis is

$$S_1^z + S_2^z + S_3^z = c_1 + c_2 + c_3 = H. \tag{5.34}$$

From (5.7), $c_2 + c_3 = H - c_3$ and substituting into (5.27), we have

$$c_1 (s_2 \cos(\phi_1 - \phi_2) + s_3 \cos(\phi_3 - \phi_1)) = s_1 (H - c_1) - H s_1$$

$$\Longrightarrow c_1 (s_1 + s_2 \cos(\phi_1 - \phi_2) + s_3 \cos(\phi_3 - \phi_1)) = 0. \tag{5.35}$$

For $c_1 \neq 0$, we have the relation:

$$s_1 + s_2 \cos(\phi_1 - \phi_2) + s_3 \cos(\phi_3 - \phi_1) = 0, \tag{5.36}$$

which implies $S_1^x + S_2^x + S_3^x = 0$. From (5.20) and (5.21), we obtain

$$s_2 \sin \phi_2 + s_3 \sin \phi_3 = 0 \Longrightarrow S_1^y + S_2^y + S_3^y = 0. \tag{5.37}$$

With these relations, we arrive at the well-known result for $A = 1$, where the condition for the ground state is given by

$$\boldsymbol{S}_1 + \boldsymbol{S}_2 + \boldsymbol{S}_3 = \boldsymbol{H}. \tag{5.38}$$

This result holds in the limit $c_1 \to 0$, and thus (5.38) is valid for all cases. It indicates the nontrivial degeneracy of the present model. This degeneracy is lifted by entropy effects at finite temperatures, which will be explained later [5].

For $A = -1/2$, no extremum configuration exists, and the polarized state given by (5.27) represents the minimum energy state. We do not analyze this case in detail.

### 5.2.3 *Planar Configuration Including the z-Axis* ($\sin \phi_1 = 0, \sin \phi_2 = 0, \sin \phi_3 = 0$)

Next, we study the case where the three spins lie in a plane that includes the $z$-axis. In this case, all the spins are in the same plane. Thus, without loss of generality, we may set $\phi_1 = \phi_2 = \phi_3 = 0$, allowing $c_i$ to take values in the range $-1 \leq c_i \leq 1$.

Under this condition, the extremum conditions (5.7)–(5.9) are given by

$$c_1 (s_2 + s_3) = A s_1 (c_2 + c_3) - H s_1, \tag{5.39}$$

$$c_2 (s_3 + s_1) = A s_2 (c_1 + c_3) - H s_2, \tag{5.40}$$

$$c_3 (s_1 + s_2) = A s_3 (c_1 + c_2) - H s_3. \tag{5.41}$$

Hereafter, we use the following notations:

$$c = \cos\left(\frac{\theta_2 + \theta_3}{2}\right), \quad s = \sin\left(\frac{\theta_2 + \theta_3}{2}\right), \tag{5.42}$$

and

$$c' = \cos\left(\frac{\theta_2 - \theta_3}{2}\right), \quad s' = \sin\left(\frac{\theta_2 - \theta_3}{2}\right). \tag{5.43}$$

Using sum formulas for trigonometric functions, e.g.,

$$\sin\theta_2 + \sin\theta_3 = 2\sin\left(\frac{\theta_2 + \theta_3}{2}\right)\cos\left(\frac{\theta_2 - \theta_3}{2}\right),$$

the following relations hold:

$$s_2 + s_3 = 2\,sc', \quad c_2 + c_3 = 2cc', \quad s_2 - s_3 = 2cs', \quad c_2 - c_3 = -2\,ss', \tag{5.44}$$

which are used in the following calculations.

First, we study the case of $H = 0$. From (5.39), we have

$$c_1(s_2 + s_3) = As_1(c_2 + c_3) \implies c_1\,2\,sc' = As_1\,2cc'. \tag{5.45}$$

If $c' = \cos\left((\theta_2 - \theta_3)/2\right) \neq 0$, then

$$c_1 s = As_1 c. \tag{5.46}$$

From (5.40) $\pm$ (5.41), we have

$$s_1(c_2 + c_3) + c_2 s_3 + c_3 s_2 = A(c_1(s_2 + s_3) + s_2 c_2 + s_3 c_2), \tag{5.47}$$

$$s_1(c_2 - c_3) + c_2 s_3 - c_3 s_2 = A(c_1(s_2 - s_3) + s_2 c_3 - s_3 c_2). \tag{5.48}$$

These equations can be rewritten as

$$s_1\,2cc' + 2cs = A(c_1\,2\,sc' + 2cs) \implies cs(1 - A) = Ac_1 sc' - s_1 cc' = (A^2 - 1)s_1 cc', \tag{5.49}$$

$$s_1(-2\,s's) - 2\,s'c' = A(c_1\,2\,s'c + 2\,s'c') \implies -(A + 1)s'c' = Ac_1 s'c + s_1 ss'. \tag{5.50}$$

For the case $c \neq 0$ and $s' \neq 0$, the following relations hold:

$$s = -(1 + A)s_1 c', \tag{5.51}$$

$$-\,s_1 s = (A + 1)c' + Ac_1 c. \tag{5.52}$$

Using (5.51), we have

$$-\,s_1 s = -\frac{s}{s_1} + Ac_1 c \implies -s_1^2\, s = -s + As_1 c_1 c$$

$$\implies s(1 - s_1^2) = As_1 c_1 c \implies sc_1 = As_1 c, \tag{5.53}$$

which is the same as (5.46).

Thus, we find that the extremum conditions (5.46), (5.51), and (5.52) are not independent. Therefore, for any value of $\theta_1$, we can find values of $\theta_2$ and $\theta_3$ that satisfy the conditions for an extremum. This property can be also verified as follows. From (5.46), we have

$$s^2 = A^2 \left(\frac{s_1}{c_1}\right)^2 c^2 = A^2 \frac{s_1^2}{c_1^2}\left(1 - s^2\right) \Longrightarrow s^2 = \frac{A^2 s_1^2}{c_1^2 + A^2 s_1^2}, \tag{5.54}$$

$$c^2 = 1 - s^2 = \frac{c_1^2}{c_1^2 + A^2 s_1^2}. \tag{5.55}$$

From (5.51), we have

$$c'^2 = \frac{s^2}{(1+A)^2 s_1^2} = \frac{A^2}{(A+1)^2} \frac{1}{c_1^2 + A^2 s_1^2}, \quad s'^2 = 1 - \frac{A^2}{(A+1)^2} \frac{1}{c_1^2 + A^2 s_1^2}. \tag{5.56}$$

Then, from (5.52), multiplying by $(s_1 s)$, and using (5.46), we have

$$-s_1^2 s^2 = (A+1)c' s_1 s + A c_1 c s_1 s = (A+1)c' s_1 s + c_1^2 s^2$$
$$\Longrightarrow -s^2 = (A+1)c' s_1 s \Longrightarrow s = -(A+1)c' s_1. \tag{5.57}$$

The values given by (5.54)–(5.56) satisfy this relation, and thus, $\theta_2$ and $\theta_3$ are functions of $\theta_1$.

The energy $E$ in the present configuration, as defined in (5.6), is

$$E = s_1(s_2 + s_3) + s_2 s_3 + A c_1 (c_2 + c_3) + A c_2 c_3$$

$$= 2 s_1 s c' + (s c' + c s')(s c' - c s') + 2\, A c_1 c c' + A(c c' - s s')(c c' + s s')$$

$$= 2 s_1 s c' + s^2 c'^2 - c^2 s'^2 + 2\, A c_1 c c' + A(c^2 c'^2 - s^2 s'^2). \tag{5.58}$$

The energy is given by an explicit function of $A$ as

$$E = -\frac{A^2 + A + 1}{A + 1}, \tag{5.59}$$

after some calculations.[5] This result is indeed independent of $\theta_1$. Thus, the ground state is continuously degenerate with respect to the choice of $\theta_1$. This non-trivial degeneracy is a special property of the present model.

---

[5] Since

$$sc' = -\sqrt{\frac{A^2 s_1^2}{c_1^2 + A^2 s_1^2} \frac{A^2}{(A+1)^2} \frac{1}{c_1^2 + A^2 s_1^2}} = -\frac{s_1 A^2}{c_1^2 + A^2 s_1^2} \frac{1}{A+1},$$

and

$$cc' = -\sqrt{\frac{c_1^2}{c_1^2 + A^2 s_1^2} \frac{A^2}{(A+1)^2} \frac{1}{c_1^2 + A^2 s_1^2}} = -\frac{c_1 A}{c_1^2 + A^2 s_1^2} \frac{1}{A+1},$$

$$2 s_1 sc' + 2\,A c_1 cc' = -2\left(\frac{s_1^2\,A^2}{c_1^2 + A^2 s_1^2} \frac{1}{A+1} + \frac{c_1^2\,A^2}{c_1^2 + A^2 s_1^2} \frac{1}{A+1}\right) = -\frac{2\,A^2}{c_1^2 + A^2 s_1^2} \frac{1}{A+1},$$

where we chose the signs to satisfy (5.46), assuming $s_1 > 0$ and $c_1 > 0$, which does not lose generality. And,

$$s^2 c'^2 - c^2 s'^2 + A(c^2 c'^2 - s^2 s'^2)$$

$$= \frac{A^2 s_1^2}{c_1^2 + A^2 s_1^2} \frac{A^2}{(A+1)^2} \frac{1}{c_1^2 + A^2 s_1^2} - \frac{c_1^2}{c_1^2 + A^2 s_1^2}\left(1 - \frac{A^2}{(A+1)^2} \frac{1}{c_1^2 + A^2 s_1^2}\right)$$

$$+ A\left(\frac{c_1^2}{c_1^2 + A^2 s_1^2} \frac{A^2}{(A+1)^2} \frac{1}{c_1^2 + A^2 s_1^2} - \frac{A^2 s_1^2}{c_1^2 + A^2 s_1^2}\left(1 - \frac{A^2}{(A+1)^2} \frac{1}{c_1^2 + A^2 s_1^2}\right)\right)$$

$$= \left(\frac{1}{c_1^2 + A^2 s_1^2}\right)^2 \left[\frac{A^3}{(A+1)^2}(A s_1^2 + c_1^2) - (c_1^2 + A^3 s_1^2)\frac{(c_1^2 + A^2 s_1^2)(A+1)^2 - A^2}{(A+1)^2}\right]$$

$$= \left(\frac{1}{c_1^2 + A^2 s_1^2}\right)^2 \frac{1}{(A+1)^2}\left[A^3(A s_1^2 + c_1^2) - (c_1^2 + A^3 s_1^2)((c_1^2 + A^2 s_1^2)(A+1)^2 - A^2)\right]$$

$$= -\frac{c_1^2 + A^3 s_1^2}{c_1^2 + A^2 s_1^2} + \left(\frac{1}{c_1^2 + A^2 s_1^2}\right)^2 \frac{A^2}{(A+1)^2}\left(A(A s_1^2 + c_1^2) + c_1^2 + A^3 s_1^2\right)$$

$$= -\frac{c_1^2 + A^3 s_1^2}{c_1^2 + A^2 s_1^2} + \left(\frac{1}{c_1^2 + A^2 s_1^2}\right) \frac{A^2}{(A+1)}. \tag{5.60}$$

### 5.2.4 Ground State Configurations of the Coplanar Case in a Field ($H > 0$)

For the case of $H = 0$, we observed a non-trivial degeneracy, meaning that for any value of $\theta_i$, there exists a ground-state configuration with the same energy. In contrast, when $H > 0$, this degeneracy is lifted. Here, we examine possible stable configurations and their corresponding energies.

#### 5.2.4.1 Fan-Shape Configuration: $c_1 = 1, c_2 = c_3, s_2 = -s_3$

This configuration is depicted in Fig. 5.3b'. Here, we denote $s = s_2 = -s_3$ and $c = c_2 = c_3$ (fan-shape configuration). It should be noted that $s$ and $c$ here differ from the $s$ and $c$ in the previous section. The energy is given by

$$E_{\text{Fan}} = -s^2 + A(2c + c^2) - H - 2Hc = c^2 - 1 + A(2c + c^2) - H - 2Hc. \tag{5.62}$$

The first derivative with respect to $c$ is 0, from which $c$ is obtained as

$$\frac{dE_{\text{Fan}}}{dc} = 2c + A(2 + 2c) - 2H = 0 \implies c = \frac{H - A}{A + 1}, \quad H \leq 2A + 1. \tag{5.63}$$

The corresponding energy is obtained as

$$\begin{aligned} E_{\text{Fan}} &= (A + 1)\left(\frac{H - A}{A + 1}\right)^2 + 2(A - H)\frac{H - A}{A + 1} - 1 - H \\ &= -\frac{A^2 + A + 1 + H + H^2 - AH}{A + 1}. \end{aligned} \tag{5.64}$$

---

Thus, the energy is given by

$$\begin{aligned} E &= -\frac{2A^2}{c_1^2 + A^2 s_1^2}\frac{1}{A + 1} - \frac{c_1^2 + A^3 s_1^2}{c_1^2 + A^2 s_1^2} + \left(\frac{1}{c_1^2 + A^2 s_1^2}\right)\frac{A^2}{A + 1} \\ &= -\frac{2A^2 + (c_1^2 + A^3 s_1^2)(A + 1) - A^2}{(c_1^2 + A^2 s_1^2)(A + 1)} = -\frac{A^2 + A + 1 + (A^2 - 1)(A^2 + A + 1)s_1^2}{(1 + (A^2 - 1)s_1^2)(A + 1)} \\ &= -\frac{(A^2 + A + 1)(1 + (A^2 - 1)s_1^2)}{(1 + (A^2 - 1)s_1^2)(A + 1)} = -\frac{A^2 + A + 1}{A + 1}. \end{aligned} \tag{5.61}$$

#### 5.2.4.2 Y-Shape Configuration: $c_1 = -1, c_2 = c_3, s_2 = -s_3$

This configuration is depicted in Fig. 5.3b. By noting $s = s_2 = -s_3$ and $c = c_2 = c_3$ (Y-shape configuration), the energy is given by

$$E_{\text{Y}} = -s^2 + A(2c + c^2) - H - 2Hc = c^2 - 1 + A(-2c + c^2) + H - 2Hc. \tag{5.65}$$

The first derivative with respect to $c$ is

$$\frac{dE_{\text{Y}}}{dc} = 2c + A(-2 + 2c) - 2H = 0 \implies c = \frac{H + A}{A + 1}, \quad H \leq 1. \tag{5.66}$$

The corresponding energy is

$$\begin{aligned} E_{\text{Y}} &= (A + 1)\left(\frac{H + A}{A + 1}\right)^2 + 2(A + H)\frac{H - A}{A + 1} - 1 - H \\ &= -\frac{A^2 + A + 1 - H + H^2 + AH}{A + 1}. \end{aligned} \tag{5.67}$$

For $H > 0$, we have $E_{\text{Y}} < E_{\text{Fan}}$ for $0 \leq H \leq 1$, and thus, the Fan-shape configuration is not considered in this range.

#### 5.2.4.3 Up-Up-Down (uud) Configuration: $c_1 = -1, c_2 = c_3 = 1, s_1 = s_2 = s_3 = 0$

This configuration (up-up-down (uud) configuration) is depicted in Fig. 5.3c. For $H \geq 1$, this configuration becomes the ground state, and its energy is given by

$$E_{\text{uud}} = -A - H. \tag{5.68}$$

For large $H$, the ground state changes to a polarized state (up-up-up (uuu) configuration), and thus, the uud-state must become unstable at some critical value of $H$.

To study the stability of the uud-state, we perform a harmonic analysis. Up to the second order of the deviations $(\delta\theta_1, \delta\theta_2, \delta\theta_3)$ from the uud-state, the energy $E_{\text{ha-uud}}$ is given by

$$E_{\text{ha-uud}} = -H\left(1 + \frac{1}{2}\delta\theta_1^2 - \frac{1}{2}\delta\theta_2^2 - \frac{1}{2}\delta\theta_3^2\right) + \delta\theta_1\delta\theta_2 - A\left(1 - \frac{1}{2}\delta\theta_1^2\right)\left(1 - \frac{1}{2}\delta\theta_2^2\right)$$

$$+\,\delta\theta_2\delta\theta_3 + A\left(1-\frac{1}{2}\delta\theta_2^2\right)\left(1-\frac{1}{2}\delta\theta_3^2\right) + \delta\theta_3\delta\theta_1 - A\left(1-\frac{1}{2}\delta\theta_3^2\right)\left(1-\frac{1}{2}\delta\theta_1^2\right)$$

$$\simeq \delta\theta_1\delta\theta_2 + \delta\theta_2\delta\theta_3 + \delta\theta_3\delta\theta_1 + A\delta\theta_1^2 - \frac{H}{2}(\delta\theta_1^2 - \delta\theta_2^2 - \delta\theta_3^2) + E_0, \tag{5.69}$$

where $E_0 = -A - H$. The energy is expressed as

$$E_{\text{ha-uud}} = (\delta\theta_1, \delta\theta_2, \delta\theta_3)\begin{pmatrix} A - H/2 & 1/2 & 1/2 \\ 1/2 & H/2 & 1/2 \\ 1/2 & 1/2 & H/2 \end{pmatrix}\begin{pmatrix} \delta\theta_1 \\ \delta\theta_2 \\ \delta\theta_3 \end{pmatrix} + E_0. \tag{5.70}$$

Instability occurs when the determinant of the Hessian matrix becomes zero:

$$\begin{vmatrix} 2A - H & 1 & 1 \\ 1 & H & 1 \\ 1 & 1 & H \end{vmatrix} = (H-1)\left((2A-H)(H+1) - 2\right) = 0. \tag{5.71}$$

The critical values of $H$ are given by

$$H_{c1} = 1, \quad H_{c2} = \frac{2A - 1 \pm \sqrt{(2A-1)^2 + 4(2A-2)}}{2}. \tag{5.72}$$

Since the solutions are real for $A > 1$, the Hessian matrix is positive in the interval

$$H_{c1} = 1 \le H \le H_{c2} = \frac{2A - 1 + \sqrt{4A^2 + 4A - 7}}{2}. \tag{5.73}$$

At $H_{c1} = 1$, the eigenvector for the zero eigenvalue is given by

$$\begin{pmatrix} 2A-1 & 1 & 1 \\ 1 & 1 & 1 \\ 1 & 1 & 1 \end{pmatrix}\begin{pmatrix} \delta\theta_1 \\ \delta\theta_2 \\ \delta\theta_3 \end{pmatrix} = \begin{pmatrix} 0 \\ 0 \\ 0 \end{pmatrix}. \tag{5.74}$$

When $A \neq 1$,

$$\begin{pmatrix} \delta\theta_1 \\ \delta\theta_2 \\ \delta\theta_3 \end{pmatrix} \propto \begin{pmatrix} 0 \\ 1 \\ -1 \end{pmatrix}, \tag{5.75}$$

which gives the Y shape. When $A = 1$, any set $\delta_1 + \delta_2 + \delta_3 = 0$ gives the solution as (5.38).

At $H_{c2}$, the eigenvector for the zero eigenvalue is given by

$$\begin{pmatrix} 2A - H_{c2} & 1 & 1 \\ 1 & H_{c2} & 1 \\ 1 & 1 & H_{c2} \end{pmatrix} \begin{pmatrix} \delta\theta_1 \\ \delta\theta_2 \\ \delta\theta_3 \end{pmatrix} = \begin{pmatrix} 0 \\ 0 \\ 0 \end{pmatrix}, \tag{5.76}$$

which gives

$$\begin{pmatrix} \delta\theta_1 \\ \delta\theta_2 \\ \delta\theta_3 \end{pmatrix} \propto \begin{pmatrix} 1 \\ -1/(H_{c2}+1) \\ -1/(H_{c2}+1) \end{pmatrix}. \tag{5.77}$$

Thus, for $H > H_{c2}$, the ground-state configuration transitions from the uud structure to a configuration in which the two up spins tilt downward and the down spin tilts upward, yielding

$$c_1 < 0, c_2 = c_3 > 0, s_1 > 0, s_2 = s_3 < 0, \tag{5.78}$$

which gives the V shape which will be studied in the next section.

#### 5.2.4.4 V-Shape: $c_1, c_2 = c_3 > 0, s_1 > 0, s_2 = s_3 < 0$

This configuration is depicted in Fig. 5.3d. Noting $s = s_2 = s_3$ and $c = c_2 = c_3$ (V-shape configuration), the extremum condition is given by

$$2c_1 s = 2As_1 c - s_1 H, \quad c(s + s_1) = As(c + c_1) - sH, \tag{5.79}$$

from which we can determine $c^2$ as functions of $A$ and $H$ as follows. From the first equation in (5.79),

$$\frac{s_1}{c_1} = \frac{2s}{2Ac - H} \implies s_1^2 = \frac{4(1-c^2)}{(2Ac-H)^2} \frac{1}{1 + \frac{4(1-c^2)}{(2Ac-H)^2}}$$

$$\implies s_1^2 = \frac{4(1-c^2)}{(2Ac-H)^2 + 4(1-c^2)}. \tag{5.80}$$

From the second equation in (5.79),

$$cs_1 - Asc_1 = (A-1)cs - sH \implies cs_1 - A\frac{2Ac - H}{2}s_1 = ((A-1)c - H)s. \tag{5.81}$$

which gives the relation

$$s_1 = s\frac{(A-1)c - H}{c - A(2Ac - H)/2} = 2\sqrt{1-c^2}\frac{(A-1)c - H}{2c(1-A^2) + AH}, \tag{5.82}$$

where we assume $s_1 > 0$. From these relations for $s_1$, we have the relation:

$$\frac{((A-1)c - H)^2}{(2c(1-A^2) + AH)^2} = \frac{1}{(2Ac - H)^2 + 4(1-c^2)}. \tag{5.83}$$

We solve this relation numerically to obtain the corresponding configuration. The energy of this configuration is calculated using the obtained value of $c$.

For $H = 0$, $c^2$ is explicitly given by

$$\frac{1}{(1+A)^2} = \frac{1}{A^2c^2 + 1 - c^2} \Longrightarrow c^2 = \frac{(1+A)^2 - 1}{A^2 - 1} = \frac{A^2 + 2A}{A^2 - 1}. \tag{5.84}$$

Since $|c| \leq 1$, the above relation is valid only when

$$(A^2 + 2A)^2 < (A^2 - 1)^2 \Longrightarrow A^4 + 4A^3 + 4A^2 < A^4 - 2A^2 + 1$$

$$\Longrightarrow 4A^3 + 6A^2 - 1 = (2A+1)(2A^2 + 2A + 1) < 0, \tag{5.85}$$

i.e.,

$$2A + 1 < 0 \Longrightarrow A < -\frac{1}{2}. \tag{5.86}$$

This implies that the present solution appears only for $A < -0.5$ when $H = 0$. Thus, for $H = 0$, the V-shape does not appear when $A > 1$.

#### 5.2.4.5 Polarized State (Up-Up-Up (UUU) Configuration: $c_1 = c_2 = c_3 = 1$, $s_1 = s_2 = s_3 = 0$)

This configuration is depicted in Fig. 5.3e. It represents a trivially polarized state in a high magnetic field, where the energy is given by

$$E_{\text{uuu}} = 3A - 3H. \tag{5.87}$$

We study the instability of the up-up-up (uuu) configuration in a magnetic field. The energy in the harmonic analysis with small deviations from the uuu-state, denoted as $(\delta\theta_1, \delta\theta_2, \delta\theta_3)$, is given by

$$E_{\text{ha-uuu}} = -H\left(3 - \frac{1}{2}\delta\theta_1^2 - \frac{1}{2}\delta\theta_2^2 - \frac{1}{2}\delta\theta_3^2\right)$$

$$+ \delta\theta_1\delta\theta_2 + A\left(1 - \frac{1}{2}\delta\theta_1^2\right)\left(1 - \frac{1}{2}\delta\theta_2^2\right) + \delta\theta_2\delta\theta_3 + A\left(1 - \frac{1}{2}\delta\theta_2^2\right)\left(1 - \frac{1}{2}\delta\theta_3^2\right)$$

$$+ \delta\theta_3\delta\theta_1 + A\left(1 - \frac{1}{2}\delta\theta_3^2\right)\left(1 - \frac{1}{2}\delta\theta_1^2\right)$$

$$\begin{aligned}&\simeq \delta\theta_1\delta\theta_2 + \delta\theta_2\delta\theta_3 + \delta\theta_3\delta\theta_1 - A(\delta\theta_1^2 + \delta\theta_2^2 + \delta\theta_3^2)\\ &+\frac{H}{2}(\delta\theta_1^2 + \delta\theta_2^2 + \delta\theta_3^2) + E_0,\end{aligned} \tag{5.88}$$

where $E_0 = 3A - 3H$. This energy can be expressed in matrix form as

$$E_{\text{ha-uuu}} = (\delta\theta_1, \delta\theta_2, \delta\theta_3)\begin{pmatrix} -A + H/2 & 1/2 & 1/2 \\ 1/2 & -A + H/2 & 1/2 \\ 1/2 & 1/2 & -A + H/2 \end{pmatrix}\begin{pmatrix}\delta\theta_1\\ \delta\theta_2\\ \delta\theta_3\end{pmatrix} + E_0. \tag{5.89}$$

Instability occurs when the determinant of the Hessian matrix becomes zero:

$$\begin{vmatrix} -2A + H & 1 & 1 \\ 1 & -2A + H & 1 \\ 1 & 1 & -2A + H \end{vmatrix} = (H - 2A + 2)(H - 2A - 1)^2 = 0. \tag{5.90}$$

The critical values of $H$ are then given by

$$H_{c_0} = 2A - 2, \quad H_{c_3} = 2A + 1. \tag{5.91}$$

Thus, the uuu-state is stable for $H > 2A + 1$ (it is metastable for $H < 2A - 2$.).

The eigenvector corresponding to $H_{c_0}$ is

$$\begin{pmatrix}\delta\theta_1\\ \delta\theta_2\\ \delta\theta_3\end{pmatrix} = \begin{pmatrix}1\\1\\1\end{pmatrix}, \tag{5.92}$$

which represents a uniform rotation. However, this deviation results in a higher energy than other stable solutions.

The eigenvectors corresponding to $H_{c_3}$ are

$$\begin{pmatrix}\delta\theta_1\\ \delta\theta_2\\ \delta\theta_3\end{pmatrix}, \qquad \delta\theta_1 + \delta\theta_2 + \delta\theta_3 = 0 \Longrightarrow \delta\theta_3 = -\delta\theta_1 - \delta\theta_2, \tag{5.93}$$

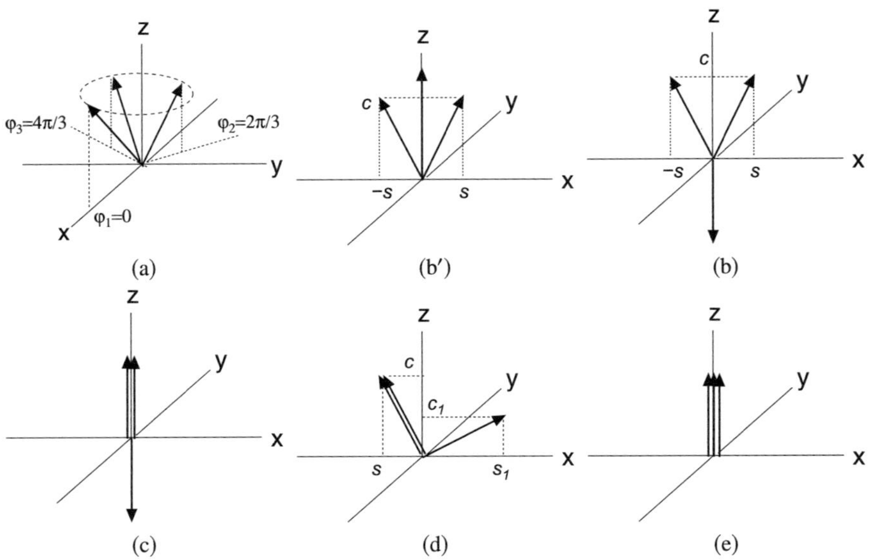


**Fig. 5.3** Configurations in a field. **a** Umbrella structure, **b'** Fan-shape structure, **b** Y-shape structure, **c** uud structure, **d** V-shape structure, and **e** Polarized (uuu) structure

for arbitrary values of $\delta\theta_1$ and $\delta\theta_2$. In fact, the V-shaped configuration becomes the ground state below $H_{c_3}$.

### 5.2.5 *Magnetic Field (H)-Dependence of Energies of the Ground State (Summary)*

The ground-state configurations in the magnetic field are depicted in Fig. 5.3. Their energies are listed as follows:

(a) umbrella (5.31): $c_1 = c_2 = c_3, s_1 = s_2 = s_2$

$$E_{\text{umbrella}} = -\frac{3}{2}\frac{2A + 1 + H^2}{2A + 1}.$$

(b) Y-shape (5.67): $c_1 = -1, s_2 = -s_3, c_2 = c_3$

$$E_{\text{Y}} = -\frac{A^2 + A + 1 - H + H^2 + AH}{A + 1}, \quad H \leq 1.$$

(b') Fan-shape (5.64): $c_1 = 1, s_2 = -s_3, c_2 = c_3$

$$E_{\text{Fan}} = -\frac{A^2 + A + 1 + H + H^2 - AH}{A + 1}.$$

As we pointed out, $E_{\text{Fan}} > E_{\text{Y}}$ for $H > 2A + 1$. For $A > 1$, $E_{\text{Fan}} < E_{\text{umbrella}}$, because

$$2\frac{A^2 + A + 1 + H + H^2 - AH}{A + 1} - 3\frac{2A + 1 + H^2}{2A + 1}$$

$$= \frac{(A - 1)(2A + 1 + H)^2}{(A + 1)(2A + 1)} > 0. \tag{5.94}$$

(c) uud (5.68): $c_1 = -1, c_2 = c_3 = 1$

$$E_{\text{uud}} = -A - H, \quad 1 \le H \le 1 < \frac{2A - 1 + \sqrt{4A^2 + 4A - 7}}{2}.$$

(d) V-shape: $c_1, c_2 = c_3, s_1, s_2 = s_3$

$$E_{\text{V}} = -2s_1 s + s^2 + A(2c_1 c + c^2) - H(c_1 + 2c).$$

(e) uuu (5.87): $c_1 = c_2 = c_3 = 1$

$$E_{\text{uuu}} = 3A - 3H, \quad H > 2A + 1.$$

The ground state for $0 < H \le H_{\text{c1}} = 1$ is the Y-shape configuration, while it is the polarized state (up, up, up) for $H \ge H_{\text{c3}} = 2A + 1$. In the interval of $H_{\text{c1}} < H < H_{\text{c3}}$, the system undergoes a transition from the uud configuration to the V-shape configuration. To determine the critical field $H_{\text{c2}}$, we compare the energies. Their energies for $A = 2$ are depicted in Fig. 5.4.

Here, we find that the energies of the Y-shape, uud-shape, V-shape, and uuu-shape are smoothly connected. The corresponding spin configurations and the magnetization $m = c_1 + c_2 + c_3$ are depicted in Fig. 5.5.

For $A = 1$, the region of the uud configuration shrinks to a point. The energies of the configurations for $A = 1$ are depicted in Fig. 5.6a, and the configurations are shown in Fig. 5.6b.

For $A < 1$, the umbrella-shape gives the ground state in the magnetic field until the saturated ferromagnetic state $H_{\text{c3}}$.

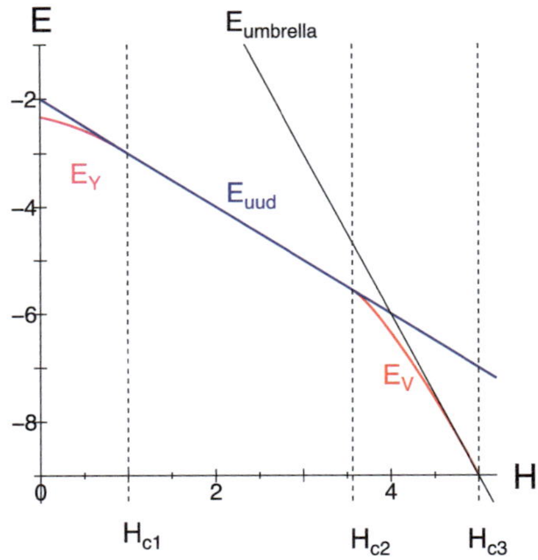


**Fig. 5.4** Magnetic field ($H$) dependence of the ground-state energies for the configurations with $A = 2$. The Y-shape, uud, and V-shape configurations appear before the fully polarized state

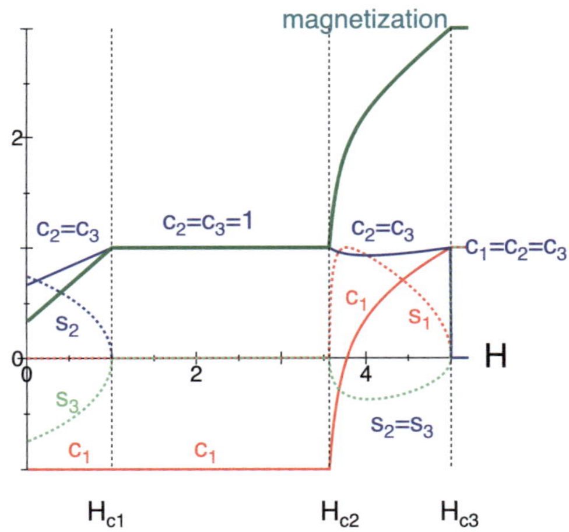


**Fig. 5.5** Magnetic field ($H$) dependence of ground-state configurations for $A = 2$. $c_i$ (solid) and $s_i$ (dotted) represent $\cos\theta_i$ and $\sin\theta_i$, respectively. $(c_1, s_1)$ are plotted with red (solid and dotted) lines, $(c_2, s_2)$ with blue lines, and $(c_3, s_3)$ with green lines. The magnetization $m = S_1^z + S_2^z + S_3^z = c_1 + c_2 + c_3$ is represented by a bold dark green line

## 5.3 Entropy in the Harmonic Analysis Around a Given Ground State Configuration

Because of frustration, some configurations are energetically degenerate at $T = 0$ or have nearly equal energies. In these situations, entropy plays an important role in determining the thermodynamically most stable state. Thus, in this section,

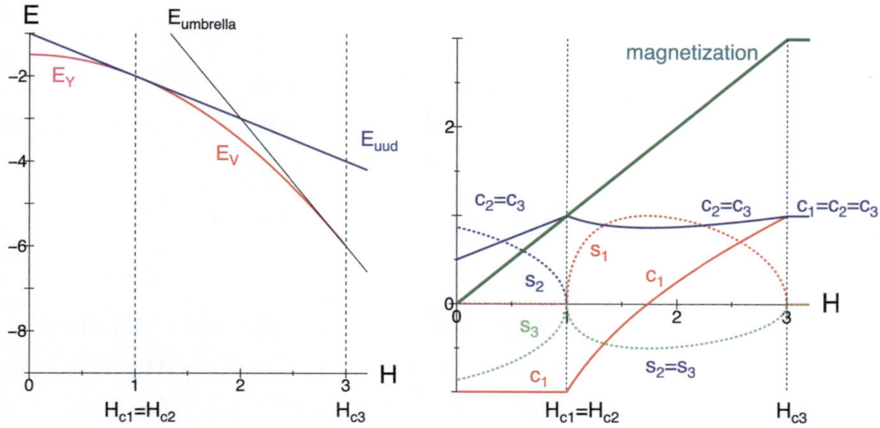


(a) (b)

**Fig. 5.6** **a** Magnetic field ($H$) dependence of the ground-state energies of configurations for $A = 1$. The Y-shape and V-shape configurations appear before the fully polarized state. The uud configuration appears only at $H = H_{c1}$. **b** $H$ dependence of the ground-state configurations for $A = 1$. $c_i$ (solid) and $s_i$ (dotted) represent $\cos\theta_i$ and $\sin\theta_i$. $(c_1, s_1)$ are plotted with red (solid and dotted) lines, respectively. $(c_2, s_2)$ with blue lines and $(c_3, s_3)$ with green lines. The magnetization $m = S_1^z + S_2^z + S_3^z = c_1 + c_2 + c_3$ is represented by a bold dark green line

we evaluate the entropy in harmonic analysis, i.e., the entropy arising from small fluctuations around the stable configurations depicted in Fig. 5.3.

### 5.3.1 Harmonic Analysis: Energy with Small Fluctuations

The energy of the system was given in (5.6). At low temperatures, spins fluctuate around the ground state, and thermal effects can be analyzed using a harmonic approximation. Since the model has three sublattices, we express a ground-state configuration as

$$(\theta_i, \phi_i) = (\theta_i^0, \phi_i^0), \quad (i = 1, 2, 3). \tag{5.95}$$

We denote the small spin fluctuations around the ground state by

$$\theta_i = \theta_i^0 + \delta\theta_i, \quad \phi_i = \phi_i^0 + \delta\phi_i, \quad (i = 1, 2, 3). \tag{5.96}$$

In the following analysis, we denote

$$\sin\theta_i^0 = s_i, \quad \cos\theta_i^0 = c_i, \quad (i = 1, 2, 3). \tag{5.97}$$

The ground-state energy $O(\delta_i^0)$ is given by

$$\begin{aligned}E_0 &= s_1 s_2 \cos(\phi_1^0 - \phi_2^0) + A c_1 c_2 - H c_1 \\ &+ s_2 s_3 \cos(\phi_2^0 - \phi_3^0) + A c_2 c_3 - H c_2 \\ &+ s_3 s_1 \cos(\phi_3^0 - \phi_1^0) + A c_3 c_1 - H c_3.\end{aligned} \tag{5.98}$$

The first order of deviations ($O(\delta_i^1)$) is given by

$$\begin{aligned}\delta E_1 &= (c_1 s_2 \cos(\phi_1^0 - \phi_2^0) - A s_1 c_2 + H s_1 + s_3 c_1 \cos(\phi_3^0 - \phi_1^0) - A c_3 s_1)\delta\theta_1 \\ &+ (s_1 c_2 \cos(\phi_1^0 - \phi_2^0) - A c_1 s_2 + H s_2 + c_2 s_3 \cos(\phi_2^0 - \phi_3^0) - A s_2 c_3)\delta\theta_2 \\ &+ (s_1 c_3 \cos(\phi_3^0 - \phi_1^0) - A c_1 s_3 + H s_3 + c_3 s_2 \cos(\phi_2^0 - \phi_3^0) - A s_3 c_2)\delta\theta_3 \\ &- s_1 s_2 \sin(\phi_1^0 - \phi_2^0)(\delta\phi_1 - \delta\phi_2) - s_2 s_3 \sin(\phi_2^0 - \phi_3^0)(\delta\phi_2 - \delta\phi_3) \\ &- s_3 s_1 \sin(\phi_3^0 - \phi_1^0)(\delta\phi_3 - \delta\phi_1).\end{aligned} \tag{5.99}$$

The condition for the extremum of the energy is given by $\delta E_1 = 0$, which was used to study the ground-state configuration, (5.7)–(5.12).

In the second order of deviations ($O(\delta_i^1)$):

$$\begin{aligned}\delta E_2 &= \frac{1}{2}\frac{\partial^2 E}{\partial\theta_1^2}\delta\theta_1^2 + \frac{1}{2}\frac{\partial^2 E}{\partial\theta_1\partial\theta_2}(\delta\theta_1\delta\theta_2 + \delta\theta_2\delta\theta_1) + \cdots \\ &+\frac{1}{2}\frac{\partial^2 E}{\partial\phi_1^2}\delta\phi_1^2 + \frac{1}{2}\frac{\partial^2 E}{\partial\phi_1\partial\phi_2}(\delta\phi_1\delta\phi_2 + \delta\phi_2\delta\phi_1) + \cdots + \\ &\frac{1}{2}\frac{\partial^2 E}{\partial\theta_1\partial\phi_1}(\delta\theta_1\delta\phi_1 + \delta\phi_1\delta\theta_1) + \cdots \\ &= A_{11}\delta\theta_1^2 + (A_{12}\delta\theta_1\delta\theta_2 + A_{21}\delta\theta_2\delta\theta_1) + \cdots + B_{11}\delta\phi_1^2 + (B_{12}\delta\phi_1\delta\phi_2 + B_{21}\delta\phi_2\delta\phi_1) \\ &+\cdots(D_{14}\delta\theta_1\delta\phi_1 + D_{41}\delta\phi_1\delta\theta_1) + \cdots.\end{aligned} \tag{5.100}$$

Because $\delta\theta_1\delta\theta_2$ and $\delta\theta_2\delta\theta_1$ give the same contribution, the definition of $A_{12}$ and $A_{21}$ has an ambiguity, but here we use the symmetric expression, i.e., $A_{12} = A_{21}$. For other pairs of $A_{ij}$ and $B_{ij}$, we use the same expression. However, for $D_{ij}$, $D_{15}$ and $D_{24}$ are coefficients for $\delta\theta_1\delta\phi_2$ and $\delta\theta_2\delta\phi_1$, respectively, and thus, they are independent quantities. For $D_{ij}$, $D_{14}$ and $D_{41}$ are coefficients of $\delta\theta_1\delta\phi_1$ and $\delta\phi_1\delta\theta_1$, and we can assign $D_{14} = D_{41}$. This quadratic form is expressed in matrix form with symmetric $3 \times 3$ matrices $\hat{A}$ and $\hat{B}$. For the part involving $D_{ij}$, we introduce a $3 \times 3$ matrix $\hat{D}$, renaming as

$$D_{14} = D_{41} \rightarrow D_{11}, \quad \text{etc.,} \tag{5.101}$$

resulting in

$$\hat{D} = \begin{pmatrix} D_{11} & D_{12} & D_{13} \\ D_{21} & D_{22} & D_{23} \\ D_{31} & D_{32} & D_{33} \end{pmatrix}. \tag{5.102}$$

Then, $\delta E_2$ is given in the form:

$$\delta E_2(\delta\theta_1, \delta\theta_2, \delta\theta_3, \delta\phi_1, \delta\phi_2, \delta\phi_3)$$

$$= (\delta\theta_1, \delta\theta_2, \delta\theta_3, \delta\phi_1, \delta\phi_2, \delta\phi_3) \begin{pmatrix} A_{11} & A_{12} & A_{13} & D_{11} & D_{12} & D_{13} \\ A_{21} & A_{22} & A_{23} & D_{21} & D_{22} & D_{23} \\ A_{31} & A_{32} & A_{33} & D_{31} & D_{32} & D_{33} \\ D_{11} & D_{21} & D_{31} & B_{11} & B_{12} & B_{13} \\ D_{12} & D_{22} & D_{32} & B_{21} & B_{22} & B_{23} \\ D_{13} & D_{23} & D_{33} & B_{31} & B_{32} & B_{33} \end{pmatrix} \begin{pmatrix} \delta\theta_1 \\ \delta\theta_2 \\ \delta\theta_3 \\ \delta\phi_1 \\ \delta\phi_2 \\ \delta\phi_3 \end{pmatrix}, \tag{5.103}$$

where

$$A_{11} = (-s_1 s_2 \cos(\phi_1^0 - \phi_2^0) - A c_1 c_2 + H c_1 - s_3 s_1 \cos(\phi_1^0 - \phi_3^0) - A c_1 c_3)/2, \tag{5.104}$$

$$A_{22} = (-s_1 s_2 \cos(\phi_1^0 - \phi_2^0) - A c_1 c_2 + H c_2 - s_3 s_2 \cos(\phi_2^0 - \phi_3^0) - A c_2 c_3)/2, \tag{5.105}$$

$$A_{33} = (-s_1 s_3 \cos(\phi_1^0 - \phi_3^0) - A c_1 c_3 + H c_3 - s_3 s_2 \cos(\phi_2^0 - \phi_3^0) - A c_2 c_3)/2, \tag{5.106}$$

$$A_{12} = A_{21} = (c_1 c_2 \cos(\phi_1^0 - \phi_2^0) + A s_1 s_2)/2, \tag{5.107}$$

$$A_{13} = A_{31} = (c_1 c_3 \cos(\phi_3^0 - \phi_1^0) + A s_1 s_3)/2, \tag{5.108}$$

$$A_{23} = A_{32} = (c_2 c_3 \cos(\phi_2^0 - \phi_3^0) + A s_2 s_3)/2, \tag{5.109}$$

$$B_{11} = (-s_1 s_2 \cos(\phi_1^0 - \phi_2^0) - s_1 s_3 \cos(\phi_3^0 - \phi_1^0))/2, \tag{5.110}$$

$$B_{22} = (-s_1 s_2 \cos(\phi_1^0 - \phi_2^0) - s_2 s_3 \cos(\phi_2^0 - \phi_3^0))/2, \tag{5.111}$$

$$B_{33} = (-s_2 s_3 \cos(\phi_2^0 - \phi_3^0) - s_1 s_3 \cos(\phi_3^0 - \phi_1^0))/2, \tag{5.112}$$

$$B_{12} = B_{21} = s_1 s_2 \cos(\phi_1^0 - \phi_2^0)/2, \tag{5.113}$$

$$B_{13} = B_{31} = s_1 s_3 \cos(\phi_3^0 - \phi_1^0)/2, \tag{5.114}$$

$$B_{23} = B_{32} = s_2 s_3 \cos(\phi_2^0 - \phi_3^0)/2. \tag{5.115}$$

The diagonal elements, $D_{ii}$ for $\delta\theta_i \delta\phi_i$ $(i = 1, 2, 3)$ are given by

$$2D_{11} = \frac{\partial^2 E}{\partial\theta_1 \partial\phi_1} = -c_1 s_2 \sin(\phi_1 - \phi_2) + s_3 c_1 \sin(\phi_3 - \phi_1), \quad \text{etc.,} \tag{5.116}$$

where the factor 2 comes from the fact that there are two terms corresponding to this term, i.e., $D_{ij}(\delta\theta_i\delta\phi_j + \delta\phi_j\delta\theta_j)$. The off-diagonal elements $D_{ij}(i \neq j)$ $i \neq j)$ are

$$\delta\theta_1\delta\phi_2 : \quad D_{12} = c_1 s_2 \sin(\phi_1^0 - \phi_2^0)/2, \tag{5.117}$$

$$\delta\theta_2\delta\phi_1 : \quad D_{21} = -s_1 c_2 \sin(\phi_1^0 - \phi_2^0)/2, \tag{5.118}$$

$$\delta\theta_1\delta\phi_3 : \quad D_{13} = -c_1 s_3 \sin(\phi_3^0 - \phi_1^0)/2, \tag{5.119}$$

$$\delta\theta_3\delta\phi_1 : \quad D_{31} = c_3 s_1 \sin(\phi_3^0 - \phi_1^0)/2, \tag{5.120}$$

$$\delta\theta_2\delta\phi_3 : \quad D_{23} = c_2 s_3 \sin(\phi_2^0 - \phi_3^0)/2, \tag{5.121}$$

$$\delta\theta_3\delta\phi_2 : \quad D_{32} = -c_3 s_2 \sin(\phi_2^0 - \phi_3^0)/2. \tag{5.122}$$

In the coplanar structure including the $z$-axis ($\phi_1 = \phi_2 = \phi_3 = 0$) where $\sin(\phi_i - \phi_j) = 0$, and thus all the $D_{ij} = 0$. Therefore, the matrix is decomposed into two $3 \times 3$ matrices ($\hat{A}$ and $\hat{B}$). On the other hand, in the umbrella structure, $c_1 = c_2 = c_3 = H/(2A + 1)$, $\quad s_1 = s_2 = s_3 = \sqrt{1 - c^2}$, and $\sin(\phi_2^0 - \phi_1^0) = \sin(\phi_3^0 - \phi_2^0) = \sin(\phi_3^0 - \phi_1^0) = \sqrt{3}/2$. Thus, $D_{ij}$ $(i \neq j)$ remains nonzero, although

$$D_{11} = D_{22} = D_{33} = 0. \tag{5.123}$$

### 5.3.2 Free Energy in the Harmonic Approximation

We first study the case of a single triangle lattice composed of three spins. Generally, the partition function is given by

$$Z = \int_0^{2\pi} d\phi_1 \int_0^{\pi} \sin\theta_1 d\theta_1 \int_0^{2\pi} d\phi_2 \int_0^{\pi} \sin\theta_2 d\theta_2 \int_0^{2\pi} d\phi_3 \int_0^{\pi} \sin\theta_3 d\theta_3 e^{-\beta\mathcal{H}}. \tag{5.124}$$

The Hamiltonian in the harmonic approximation is given by

$$\mathcal{H}_{\rm HA} = E_0 + \delta E_2(\delta\theta_1, \delta\theta_2, \delta\theta_3, \delta\phi_1, \delta\phi_2, \delta\phi_3), \tag{5.125}$$

where $E_0$ is the ground-state energy of a single triangle of three spins (5.99), and $\delta E_2$ is given by (5.103). The case for the triangular lattice will be studied later.

For convenience in performing the integration, we introduce the following variables:

$$x_i = d\theta_i, \quad y_i = \sin\theta_i d\phi_i, \tag{5.126}$$

and extend the range of integration to $(-\infty, \infty)$ assuming the relevant fluctuations are small. Then, we have

$$\delta E_2 = (x_1, x_2, x_3, y_1, y_2, y_3) \begin{pmatrix} A_{11} & A_{12} & A_{13} & F_{11} & F_{12} & F_{13} \\ A_{21} & A_{22} & A_{23} & F_{21} & F_{22} & F_{23} \\ A_{31} & A_{32} & A_{33} & F_{31} & F_{32} & F_{33} \\ F_{11} & F_{21} & F_{31} & C_{11} & C_{12} & C_{13} \\ F_{12} & F_{22} & F_{32} & C_{21} & C_{22} & C_{23} \\ F_{13} & F_{23} & F_{33} & C_{31} & C_{32} & C_{33} \end{pmatrix} \begin{pmatrix} x_1 \\ x_2 \\ x_3 \\ y_1 \\ y_2 \\ y_3 \end{pmatrix} \equiv^t \boldsymbol{v} E_2 \boldsymbol{v}, \tag{5.127}$$

where

$$C_{ij} = \frac{B_{ij}}{s_i^0 s_j^0}, \quad F_{ij} = \frac{D_{ij}}{s_j^0}. \tag{5.128}$$

In the cases where $s_j = 0$, we use the value of the limit as $s_j \to 0$, which is not singular.

The partition function is given by

$$Z = \int_{-\infty}^{\infty} dx_1 \int_{-\infty}^{\infty} dy_1 \int_{-\infty}^{\infty} dx_2 \int_{-\infty}^{\infty} dy_2 \int_{-\infty}^{\infty} dx_3 \int_{-\infty}^{\infty} dy_3 e^{-\beta E_0 - \beta\left(\sum_{i=1}^{6}\sum_{j=1}^{6}(E_2)_{ij} v_i v_j\right)}. \tag{5.129}$$

Using the property of the bilinear form, the integrand is expressed in terms of the eigenvalues of $E_2$ ($\lambda_i, i = 1, \cdots, 6$) as

$$e^{-\beta E_0 - \beta\left(\sum_{i=1}^{6}\sum_{j=1}^{6}(E_2)_{ij} v_i v_j\right)} = e^{-\beta E_0 - \beta\left(\sum_{i=1}^{6}\lambda_i \xi_i^2\right)}, \tag{5.130}$$

where $-\infty < \xi_i < \infty$. Noting the relation:

$$\int_{-\infty}^{\infty} e^{-\beta a x^2} dx = \sqrt{\frac{\pi}{\beta a}}, \tag{5.131}$$

the partition function is given by

$$Z = e^{-\beta E_0} \sqrt{\frac{\pi}{\beta\lambda_1}} \sqrt{\frac{\pi}{\beta\lambda_2}} \sqrt{\frac{\pi}{\beta\lambda_3}} \sqrt{\frac{\pi}{\beta\lambda_4}} \sqrt{\frac{\pi}{\beta\lambda_5}} \sqrt{\frac{\pi}{\beta\lambda_6}} = e^{-\beta E_0} \frac{\pi^3}{\beta^3} \frac{1}{\sqrt{\det(E_2)}}, \tag{5.132}$$

where we used the property that the determinant of the matrix is given by the product of eigenvalues:

$$\det(E_2) = \lambda_1\lambda_2\lambda_3\lambda_4\lambda_5\lambda_6. \tag{5.133}$$

The free energy is given by

$$F = -k_{\mathrm{B}}T \ln Z = E_0 + k_{\mathrm{B}}T \left(3\ln(\beta/\pi) + \frac{1}{2}\ln \det E_2\right). \tag{5.134}$$

The entropy is

$$\begin{aligned} S = -\frac{\partial F}{\partial T} &= -k_{\mathrm{B}} \left(3\ln(\beta/\pi) + \frac{1}{2}\ln \det E_2\right) + 3k_{\mathrm{B}} \\ &= -\frac{1}{2}k_{\mathrm{B}} \ln \det E_2 + S_0, \quad S_0 = 3k_{\mathrm{B}}(1 - \ln(\beta/\pi)). \end{aligned} \tag{5.135}$$

If the ground-state energy $E_0$ is the same for different ground-state configurations, the configuration with the larger entropy $S$ is thermodynamically more favorable and represents the equilibrium ground state. The term of $S_0$ is a constant for a given temperature. Thus, the configuration giving the maximum of the part $-\frac{1}{2}k_{\mathrm{B}} \ln \det E_2$ is the most favorable. It should be noted that the entropy diverges to $-\infty$ when the temperature approaches zero in systems of classical harmonic oscillators. Here we consider only the difference between entropies of configurations. At finite temperatures, even if a state has an energy higher than $E_0$, it can be thermodynamically stable when it has a larger entropy, resulting in lower free energy $F = E - TS$.

### 5.3.3 Coplanar Configuration Including z-Axis

For the coplanar configuration including the $z$-axis studied in Sect. 5.2.3:

$$\phi_1 = \phi_2 = \phi_3 = 0, \quad c_1 = -1, c_2 = c_3 = \frac{A+H}{A+1}, \quad s_1 = 0, s_2 = s_3 = \sqrt{1-c^2} = -s_3, \tag{5.136}$$

the matrix element:

$$C_{11} = -\frac{1}{2}\frac{s_1 s_2 + s_1 s_3}{s_1^2} = -\frac{1}{2}\frac{s_2 + s_3}{s_1}, \tag{5.137}$$

is indeterminate (zero over zero), so we need to take the limit. From the relation

$$c_1 s_2 - A s_1 c_2 + s_3 c_1 - A c_3 s_1 + H_1 s_1 = 0, \tag{5.138}$$

we can express the ratio as

$$\frac{s_2 + s_3}{s_1} = \frac{A(c_3 + c_2) - H}{c_1}, \tag{5.139}$$

and similarly, for $C_{22}$ and $C_{33}$. Thus, we can express them as

$$C_{11} = -\frac{1}{2}\frac{s_2 + s_3}{s_1} = -\frac{1}{2}\frac{A(c_3 + c_2) - H}{c_1}, \tag{5.140}$$

$$C_{22} = -\frac{1}{2}\frac{s_1 + s_3}{s_2} = \frac{1}{2}, \quad C_{33} = -\frac{1}{2}\frac{s_1 + s_2}{s_3} = \frac{1}{2}, \tag{5.141}$$

$$C_{12} = C_{21} = \frac{1}{2}, \quad C_{13} = C_{31} = \frac{1}{2}, \quad C_{23} = C_{32} = \frac{1}{2}. \tag{5.142}$$

As we pointed out in the previous section, for the coplanar structure including the $z$-axis ($\phi_1 = \phi_2 = \phi_3$), $D_{ij} = 0$, and the Hamiltonian is given in a block-diagonal form:

$$\mathcal{H}_{\text{HA}} = E_0 + \sum_{i=1}^{3}\sum_{j=1}^{3} A_{ij} x_i x_j + \sum_{i=1}^{3}\sum_{j=1}^{3} C_{ij} y_i y_j. \tag{5.143}$$

The partition function (5.132) is given by

$$Z = e^{-\beta E_0}\frac{\pi^3}{\beta^3}\frac{1}{\sqrt{\det(E_2)}} = e^{-\beta E_0}\frac{\pi^3}{\beta^3}\frac{1}{\sqrt{\det(A)\det(C)}}, \tag{5.144}$$

and the free energy is given by

$$F = -k_B T \ln Z = E_0 + k_B T\left(3\ln\left(\frac{\beta}{\pi}\right) + \frac{1}{2}\ln \det A + \frac{1}{2}\ln \det C\right). \tag{5.145}$$

The entropy is

$$S = -\frac{dF}{dT} = -k_B\left(3\ln\left(\frac{\beta}{\pi}\right) + \frac{1}{2}\ln \det A + \frac{1}{2}\ln \det C\right) - 3k_B. \tag{5.146}$$

Here, $\det A$ and $\det C$ are given by

$$\det A = A_{11}A_{22}A_{33} + 2A_{12}A_{23}A_{31} - \left(A_{11}A_{23}^2 + A_{22}A_{13}^2 + A_{33}A_{12}^2\right), \tag{5.147}$$

and

$$\det C = C_{11}C_{22}C_{33} + 2C_{12}C_{23}C_{31} - \left(C_{11}C_{23}^2 + C_{22}C_{13}^2 + C_{33}C_{12}^2\right), \quad (5.148)$$

respectively.

#### 5.3.3.1 Case of the Zero Field ($H = 0$)

When $H = 0$, after some calculations, we obtain the following properties (derivations of them are given in the next subsection Sect. 5.3.3.2):

$$A1 = A_{11}A_{22}A_{33} = \frac{A^2}{4(A+1)}, \quad (5.149)$$

$$A2 = A_{11}A_{23}^2 + A_{22}A_{13}^2 + A_{33}A_{12}^2 = \frac{A^2(A^2+A+1)}{4(A+1)^3}, \quad (5.150)$$

$$A3 = A_{12}A_{23}A_{31} = -\frac{A^3}{8(A+1)^3}, \quad (5.151)$$

and

$$C_{11}C_{22}C_{33} = \frac{A^2}{4(A+1)}, \quad (5.152)$$

$$C_{11}C_{23}^2 + C_{22}C_{13}^2 + C_{33}C_{12}^2 = \frac{A^2+A+1}{4(A+1)}, \quad (5.153)$$

$$C_{12}C_{23}C_{31} = \frac{1}{8}. \quad (5.154)$$

It should be noted that, simple substitution of these results leads to

$$\det A = 0, \quad \det C = 0. \quad (5.155)$$

This result comes from the zero eigenvalues. The eigenvalue of eigenstate corresponding to uniform rotation is zero ($\lambda = 0$) because there is no energy loss due to such a motion. Moreover, in the coplanar configurations including the $z$-axis, the system exhibits non-trivial degeneracy in the choice of $\theta_1$. Therefore, to compare the entropies of different configurations, we need to remove these modes. Thus, the analysis of the model on a triangle (with 3 spins) is somewhat complicated.

However, as we will see in Sect. 5.4, on a triangular lattice consisting of many triangles, the zero eigenvalues corresponding to the global uniform motions over the lattice appear only at special points in the wave vector space $\boldsymbol{k}$, and they do not affect the integrals over $\boldsymbol{k}$.

#### 5.3.3.2 Derivation of the Relations (5.149)–(5.154)

Because concrete derivations of the relations (5.149)–(5.154) is not very simple, we present their details. The matrix elements are given as follows:

$$A_{11} = \frac{-s_1s_2 - Ac_1c_2 - s_3s_1 - Ac_1c_3}{2} = \frac{-s_1(s_2+s_3) - Ac_1(c_2+c_3)}{2}$$

$$= \frac{s_2+s_3}{2}\left(-s_1 - Ac_1\frac{c_2+c_3}{s_2+s_3}\right). \tag{5.156}$$

Using the relation (5.39): $c_2 + c_3 = c_1(s_2+s_3)/As_1$,

$$A_{11} = -\frac{s_2+s_3}{2}\left(s_1 + \frac{c_1^2}{s_1}\right) = -\frac{s_2+s_3}{2s_1}. \tag{5.157}$$

Similarly,

$$A_{22} = \frac{-s_1s_2 - Ac_1c_2 - s_3s_2 - Ac_2c_3}{2} = -\frac{s_1+s_3}{2s_2}. \tag{5.158}$$

and

$$A_{33} = \frac{-s_1s_3 - Ac_1c_3 - s_3s_2 - Ac_2c_3}{2} = -\frac{s_1+s_2}{2s_3}. \tag{5.159}$$

For

$$A_{12} = A_{21} = \frac{c_1c_2 + As_1s_2}{2} = \frac{c_1}{2}\left(c_2 + A\frac{s_1}{c_1}s_2\right),$$

using the relation (5.39), and taking the negative sign of $c'$, it is expressed as

$$A_{12} = \frac{c_1}{2}\left(c_2 + \frac{s_2+s_3}{c_2+c_3}s_2\right) = \frac{c_1}{2}\frac{c_2^2 + c_2c_3 + s_2^2 + s_2s_3}{c_2+c_3}$$

$$= \frac{c_1}{2}\frac{1+\cos(\theta_2-\theta_3)}{2cc'} = \frac{c_1}{2}\frac{2c'^2}{2cc'} = c_1\frac{c'}{c} = -\frac{A}{2(A+1)}. \tag{5.160}$$

Similarly,

$$A_{13} = A_{31} = c_1c_3 + As_1s_3 = -\frac{A}{2(A+1)}, \tag{5.161}$$

and

$$A_{23} = A_{32} = c_2 c_3 + A s_2 s_3 = -\frac{A}{2(A+1)}. \tag{5.162}$$

Now, we calculate $A_{11} + A_{22} + A_{33}$ by substituting the above expressions:

$$A_{11} + A_{22} + A_{33} = -\frac{s_2 + s_3}{2s_1} - \frac{s_1 + s_3}{2s_2} - \frac{s_1 + s_2}{2s_3} = -\frac{2sc'}{2s_1} - \frac{s_1 s_3 + s_3^2 + s_1 s_2 + s_2^2}{2s_2 s_3} \tag{5.163}$$

$$= -c'\frac{Ac}{c_1} - \frac{s_1(s_2 + s_3) + (s_2 + s_3)^2 - 2s_2 s_3}{2s_2 s_3} = \frac{Ac'}{\sqrt{c_1^2 + A^2 s_1^2}} - \frac{s_1 s c' + (2sc')^2}{s_2 s_3} + 1,$$

where we took the negative sign of $c'$ ($c' < 0$).

Using the following relation:

$$s_2 s_3 = \sin\left(\frac{\theta_2 + \theta_3}{2} + \frac{\theta_2 - \theta_3}{2}\right) \sin\left(\frac{\theta_2 + \theta_3}{2} - \frac{\theta_2 - \theta_3}{2}\right) = (sc' + cs')(sc' - cs')$$

$$= s^2 c'^2 - c^2 s'^2 = (1 - c^2)c'^2 - c^2(1 - c'^2) = c'^2 - c^2 c'^2 - c^2 + c^2 c'^2 = c'^2 - c^2, \tag{5.164}$$

this form can be reformed as

$$A_{11} + A_{22} + A_{33} = \frac{A^2}{A+1}\frac{1}{c_1^2 + A^2 s_1^2} - \frac{s_1 A^2 s_1/(A+1) - 2A^4 s_1^2/(A+1)^2}{(c^2 - c'^2)(c_1^2 + A^2 s_1^2)} + 1. \tag{5.165}$$

Furthermore, using the relations obtained from (5.55) and (5.56):

$$(c^2 - c'^2)(c_1^2 + A^2 s_1^2) = c_1^2 - \frac{A^2}{(A+1)^2},$$

$A_{11} + A_{22} + A_{33}$ is written as

$$A_{11} + A_{22} + A_{33} = \frac{A^2}{A+1}\frac{1}{c_1^2 + A^2 s_1^2} - \frac{s_1 A^2 s_1/(A+1) - 2A^4 s_1^2/(A+1)^2}{c_1^2 - \frac{A^2}{(A+1)^2}} + 1$$

$$= \frac{A^2}{A+1}\frac{1}{c_1^2 + A^2 s_1^2} - s_1^2 A^2 \frac{(A+1)(c_1^2 + A^2 s_1^2) - 2A^2}{(c_1^2(A+1)^2 - A^2)(c_1^2 + A^2 s_1^2)} + 1. \tag{5.166}$$

Using the following relation

$$(A+1)(c_1^2+A^2s_1^2)-2A^2=(A+1)(c_1^2+A^2(1-c_1^2))-2A^2=c_1^2(1-A^2)(A+1)+A^2(A+1)-2A^2$$

$$=(1-A)(c_1^2(A+1)^2-A^2),$$

we obtain

$$A_{11}+A_{22}+A_{33}=\frac{A^2}{A+1}\frac{1}{c_1^2+A^2s_1^2}-s_1^2A^2\frac{(1-A)}{c_1^2+A^2s_1^2}+1=\frac{A^2-A^2(1-c_1^2)(1-A^2)}{(A+1)(c_1^2+A^2s_1^2)}+1.$$

Noticing the following relation

$$2A^2-2A^2(1-c_1^2)(1-A^2)=2A^2(1-(1-A^2-c_1^2+c_1^2A^2))=2A^2(A^2+c_1^2(1-A^2)),$$

we finally obtain the explicit formula without referring to the angles such as $c_1$:

$$A_{11}+A_{22}+A_{33}=\frac{A^2}{A+1}+1=\frac{A^2+A+1}{A+1}.$$

For the relation (5.149),

$$A_{11}A_{22}A_{33}=-\frac{(s_2+s_3)(s_1+s_3)(s_1+s_2)}{8s_1s_2s_3}=-\frac{(s_2+s_3)(s_1^2+s_1s_2+s_1s_3+s_2s_3)}{8s_1s_2s_3}$$

$$=-\frac{s_2s_1^2+s_1s_2^2+s_1s_2s_3+s_2^2s_3+s_3s_1^2+s_1s_2s_3+s_1s_3^2+s_2s_3^2}{8s_1s_2s_3}$$

$$=-\frac{1}{8}\left(\frac{s_1}{s_3}+\frac{s_2}{s_3}+\frac{s_2}{s_1}+\frac{s_1}{s_2}+\frac{s_3}{s_2}+\frac{s_3}{s_1-2}\right)=\frac{A_{11}+A_{22}+A_{33}}{4}-\frac{1}{4}$$

and thus

$$A1(=A_{11}A_{22}A_{33})=\frac{A^2+A+1}{4(A+1)}-\frac{1}{4}=\frac{A^2}{4(A+1)}.$$

For the relation (5.150), using (5.160)–(5.162),

$$A2(= A_{11}A_{23}^2 + A_{22}A_{13}^2 + A_{33}A_{12}^2) = A_{11}A_{23}^2 + A_{22}A_{13}^2 + A_{33}A_{12}^2$$
$$= \left(\frac{A}{2(A+1)}\right)^2 (A_{11} + A_{22} + A_{33}) = \left(\frac{A}{A+1}\right)^2 \frac{A^2+A+1}{4(A+1)} = \frac{A^2(A^2+A+1)}{4(A+1)^3}. \quad (5.167)$$

For the relation (5.151),

$$A3(= A_{23}A_{31}A_{12}) = -\left(\frac{A}{8(A+1)}\right)^3. \quad (5.168)$$

For the matrix $C$,

$$C_{11} = \frac{-s_1s_2 - s_1s_3}{2s_1^2} = \frac{s_2+s_3}{2s_1} = A_{11}, \quad (5.169)$$

$$C_{22} = \frac{-s_1s_2 - s_2s_3}{2s_2^2} = \frac{s_1-s_3}{2s_2} = A_{22}, \quad (5.170)$$

$$C_{33} = \frac{-s_2s_3 - s_1s_3}{2s_3^2} = \frac{-s_2+s_1}{2s_2} = A_{33}, \quad (5.171)$$

$$C_{12} = \frac{s_1s_2}{2s_1s_2} = \frac{1}{2}, \quad C_{13} = \frac{s_1s_3}{2s_1s_3} = \frac{1}{2}, \quad C_{23} = \frac{s_2s_3}{2s_2s_3} = \frac{1}{2}. \quad (5.172)$$

## 5.4 Free Energy of the Model on the Triangular Lattice

Now, we study the properties of the model on the triangular lattice with $N$ spins. The triangular lattice consists of three sublattices, denoted by open-circles, squares and closed-circles in Fig. 5.7. The number of spins in each sublattice is $N/3$. The unit cell of the lattice contains three spins, each of which belongs to one of the sublattices. Here, we denote the sites on sublattices 1, 2, and 3 by $i$, $j$, and $k$, respectively.

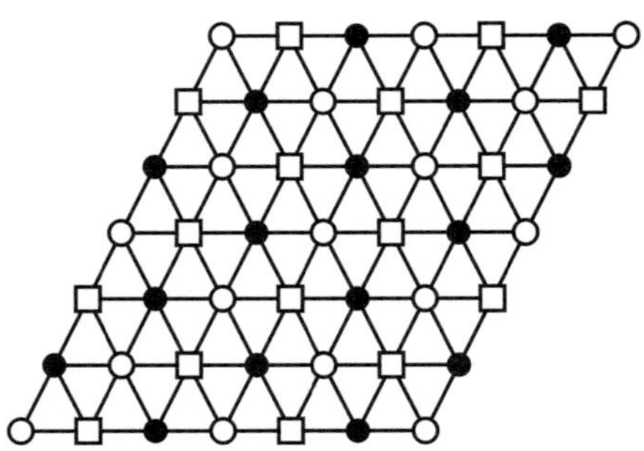

**Fig. 5.7** Triangular lattice. The three sublattices (1, 2, 3) are represented by open circles, squares, and closed circles, respectively

The energy is given by

$$\begin{aligned} E = &\sum_{\langle ij\rangle} \left(\sin\theta_i \sin\theta_j \cos(\phi_i - \phi_j) + A\cos\theta_i \cos\theta_j\right) \\ &+ \sum_{\langle jk\rangle} \left(\sin\theta_j \sin\theta_k \cos(\phi_j - \phi_k) + A\cos\theta_j \cos\theta_k\right) \\ &+ \sum_{\langle ki\rangle} (\sin\theta_k \sin\theta_i \cos(\phi_k - \phi_i) + A\cos\theta_k \cos\theta_i) \\ &- h\sum_{\ell=1}^{N} \cos\theta_\ell, \end{aligned} \tag{5.173}$$

where $\langle ij\rangle$, $\langle jk\rangle$, and $\langle ki\rangle$ denote the nearest neighbor pairs in the triangular lattice. The amplitude of the magnetic field is modified as $h = H_{\text{triangular lattice}} = H_{\text{triangle lattice}}/3$, because a site on the triangular lattice is sheared by three triangles.

The lattice points are located in the positions:

$$\boldsymbol{r} = n_x \begin{pmatrix} 1 \\ 0 \end{pmatrix} + n_y \begin{pmatrix} 1/2 \\ \sqrt{3}/2 \end{pmatrix}, \tag{5.174}$$

where $n_x = 0, 1, \ldots, N_x$ and $n_y = 0, 1, \ldots, N_y$. Here, $N_x$ and $N_y$ are multiples of 3, and $N = N_x N_y$. We use periodic boundary conditions in both directions, i.e.,

$$\theta_{N_x,n_x} = \theta_{0,n_x}, \quad \theta_{n_y,N_y} = \theta_{n_y,0}, \quad n_x = 0, 1, \ldots, N_x - 1, \quad n_y = 0, 1, \ldots, N_y - 1. \tag{5.175}$$

### 5.4.1 Ground State Energy

The ground state on the lattice is given by the repetition of the ground state of the three spins on a triangle (Fig. 5.7), and the energy for the bond interaction is given by

$$E^{\text{bond}}_{\text{triangular lattice}} = E^{\text{bond}}_{\text{triangle}} \times N/3, \tag{5.176}$$

where $N$ is the total number of spins.

#### 5.4.1.1 Harmonic Analysis

To study the entropy in the harmonic approximation, we need to calculate the contributions up to the second-order derivatives. The contribution from the first-order terms is zero for a stationary state. Because each site belongs to three neighboring triangles, the coefficient of $\delta\theta_i^2$ is $3 \times A_{ii}$. The Hamiltonian for the second order is given by

$$\mathcal{H}_2 = \sum_{i \in \text{sublattice1}}^{N/3} \frac{3}{2} A_{11} \delta\theta_i^2 + \sum_{j \in \text{sublattice2}}^{N/3} \frac{3}{2} A_{22} \delta\theta_j^2 + \sum_{k \in \text{sublattice3}}^{N/3} \frac{3}{2} A_{33} \delta\theta_k^2$$

$$+ \sum_{i \in \text{sublattice1}}^{N/3} A_{12} \delta\theta_i (\delta\theta_{j_1} + \delta\theta_{j_2} + \delta\theta_{j_3}) + \sum_{j \in \text{sublattice2}}^{N/3} A_{23} \delta\theta_j (\delta\theta_{k_1} + \delta\theta_{k_2} + \delta\theta_{k_3})$$

$$+ \sum_{k \in \text{sublattice3}}^{N/3} A_{31} \delta\theta_k (\delta\theta_{i_1} + \delta\theta_{i_2} + \delta\theta_{i_3})$$

$$+ \sum_{i \in \text{sublattice1}}^{N/3} \frac{3}{2} C_{11} \delta y_i^2 + \sum_{j \in \text{sublattice2}}^{N/3} \frac{3}{2} C_{22} \delta y_j^2 + \sum_{k \in \text{sublattice3}}^{N/3} \frac{3}{2} C_{33} \delta y_k^2$$

$$+ \sum_{i \in \text{sublattice1}}^{N/3} C_{12} y_i (y_{j_1} + y_{j_2} + y_{j_3}), \quad + \sum_{j \in \text{sublattice2}}^{N/3} C_{23} y_j (y_{k_1} + y_{k_2} + y_{k_3})$$

$$+ \sum_{k \in \text{sublattice3}}^{N/3} C_{31} y_k (y_{i_1} + y_{i_2} + y_{i_3})$$

$$+2 \sum_{i \in \text{sublattice1}}^{N/3} F_{12} \delta\theta_i (y_{j_1} + y_{j_2} + y_{j_3}), \quad +2 \sum_{j \in \text{sublattice2}}^{N/3} F_{23} \delta\theta_j (y_{k_1} + y_{k_2} + y_{k_3})$$

$$+2 \sum_{k \in \text{sublattice3}}^{N/3} F_{31} \delta\theta_k (y_{i_1} + y_{i_2} + y_{i_3}), \tag{5.177}$$

where

$$y_i = \sin\theta_i^0 d\phi_i, \quad y_j = \sin\theta_j^0 d\phi_j, \quad y_k = \sin\theta_k^0 d\phi_k. \tag{5.178}$$

Here, $j_1$, $j_2$, and $j_3$ denote the sites of sublattice-2 neighboring the site $i$ of sublattice-1 as depicted by solid (red), dotted (blue), and dashed (green) arrows in Fig. 5.8a, respectively. Similarly, $k_1$, $k_2$, and $k_3$ denote the sites of sublattice-3 neighboring the site $j$ of sublattice-2 (Fig. 5.8b), and $i_1$, $i_2$, and $i_3$ denote the sites of sublattice-1 neighboring the site $k$ of sublattice-3 (Fig. 5.8c).

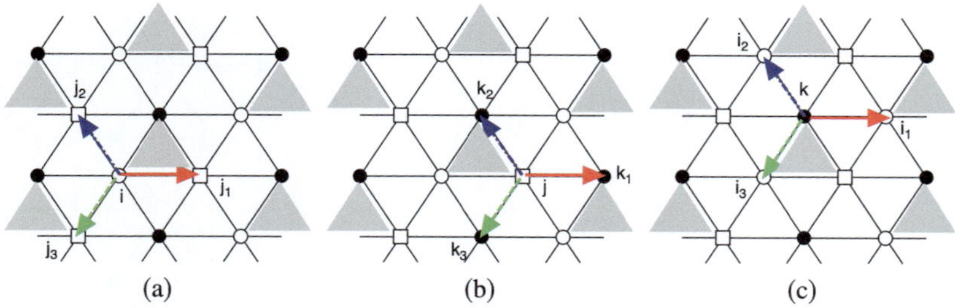


**Fig. 5.8** **a** Transition vectors from a site of sublattice-1 to sites of sublattice-2. **b** Transition vectors from a site of sublattice-2 to sites of sublattice-3. **c** Transition vectors from a site of sublattice-1 to sites of sublattice-3

## 5.5 Fourier Transformation of the Hamiltonian for Second Order Variations ($\mathcal{H}_2$)

Utilizing the translational symmetry, we introduce Fourier-transformed variables as follows:

$$x_m^{(n)} = \sum_{\boldsymbol{k}} x_{\boldsymbol{k}}^{(n)} e^{i\boldsymbol{k}\boldsymbol{r}_m^{(n)}}, \quad x_{\boldsymbol{k}}^{(n)} = \frac{1}{N/3} \sum_{\boldsymbol{r}_m^{(n)}} x_m^{(n)} e^{-i\boldsymbol{k}\boldsymbol{r}_m^{(n)}}, \tag{5.179}$$

$$y_m^{(n)} = \sum_{\boldsymbol{k}} y_{\boldsymbol{k}}^{(n)} e^{i\boldsymbol{k}\boldsymbol{r}_m^{(n)}}, \quad y_{\boldsymbol{k}}^{(n)} = \frac{1}{N/3} \sum_{\boldsymbol{r}_m^{(n)}} y_m^{(n)} e^{-i\boldsymbol{k}\boldsymbol{r}_m^{(n)}}, \tag{5.180}$$

where $n(= 1, 2, 3)$ denotes the sublattice, and $m$ demotes the location of $m$-th shaded triangle representing the sublattice structure. The position of the $m$-th site in sublattice $n$, $\boldsymbol{r}_m^{(n)}$, is

$$\boldsymbol{r}_{m=(m_1,m_2)}^{(1)} = m_1\boldsymbol{a}_1 + m_2\boldsymbol{a}_2, \tag{5.181}$$

$$\boldsymbol{r}_{m=(m_1,m_2)}^{(2)} = \begin{pmatrix} 1 \\ 0 \end{pmatrix} + m_1\boldsymbol{a}_1 + m_2\boldsymbol{a}_2, \tag{5.182}$$

$$\boldsymbol{r}_{m=(m_1,m_2)}^{(3)} = \frac{1}{2}\begin{pmatrix} 1 \\ \sqrt{3} \end{pmatrix} + m_1\boldsymbol{a}_1 + m_2\boldsymbol{a}_2. \tag{5.183}$$

Here the basis vectors of the sublattice are

$$\boldsymbol{a}_1 = \begin{pmatrix} 3/2 \\ \sqrt{3}/2 \end{pmatrix}, \quad \boldsymbol{a}_2 = \begin{pmatrix} 0 \\ \sqrt{3} \end{pmatrix}, \tag{5.184}$$

which are illustrated in Fig. 5.9.

The sum $\sum_{\boldsymbol{r}_m^{(n)}}$ denotes the summation over the sites of sublattice $n$ ($n = 1, 2,$ or 3), and $\sum_{\boldsymbol{k}}$ is taken over the first Brillouin zone of the triangular lattice's

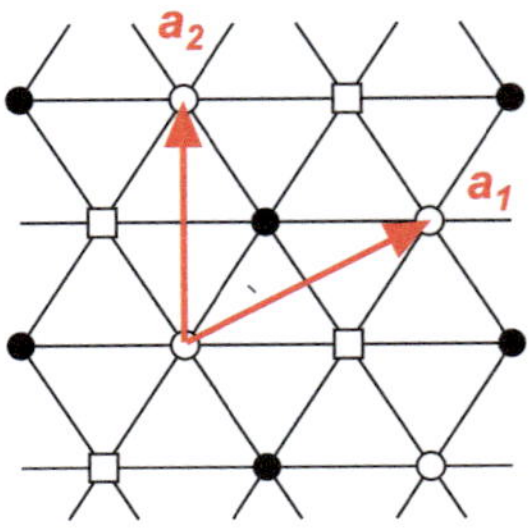


**Fig. 5.9** Basis vectors of the sublattice of the triangular lattice

sublattice. The length of the bonds of the sublattice is $\sqrt{3}$. The Brillouin zone will be explained in the next subsection.

Substituting these transformed variables, the energy is expressed in a diagonal form:

$$\sum_{m\in \text{sublatticen}}^{N/3} \frac{3}{2}A_{nn}\delta\theta_i^2 = \sum_{i\in \text{sublatticen}}^{N/3} \frac{3}{2}A_{nn}\sum_{\boldsymbol{k}} x_{\boldsymbol{k}}^{(n)} e^{i\boldsymbol{k}\boldsymbol{r}_m^{(n)}} \sum_{\boldsymbol{k}'} x_{\boldsymbol{k}'}^{(n)} e^{i\boldsymbol{k}'\boldsymbol{r}_m^{(n)}}$$

$$= \frac{3}{2}A_{nn}\sum_{\boldsymbol{k}}\sum_{\boldsymbol{k}'}\sum_{i\in \text{sublattice1}}^{N/3} e^{i(\boldsymbol{k}+\boldsymbol{k}')\boldsymbol{r}_m^{(n)}} x_{\boldsymbol{k}}^{(n)} x_{\boldsymbol{k}'}^{(n)}$$

$$= \frac{3}{2}A_{nn}\sum_{\boldsymbol{k}}\sum_{\boldsymbol{k}'}\frac{N}{3}\delta_{\boldsymbol{k}+\boldsymbol{k}',0} x_{\boldsymbol{k}}^{(n)} x_{\boldsymbol{k}'}^{(n)} = \frac{3}{2}A_{nn}\frac{N}{3}\sum_{\boldsymbol{k}} x_{\boldsymbol{k}}^{(n)} x_{-\boldsymbol{k}}^{(n)}. \tag{5.185}$$

For the term of $A_{12}$,

$$\sum_{m\in \text{sublattice1}}^{N/3} A_{12}\delta\theta_i(\delta\theta_{j_1}+\delta\theta_{j_2}+\delta\theta_{j_3}) = \sum_{m\in \text{sublattice1}}^{N/3} \frac{3}{2}A_{12}\sum_{\boldsymbol{k}} x_{\boldsymbol{k}}^{(1)} e^{i\boldsymbol{k}\boldsymbol{r}_m^{(n)}} \times$$

$$\left(\sum_{\boldsymbol{k}'} x_{\boldsymbol{k}'}^{(2)} e^{i\boldsymbol{k}'(\boldsymbol{r}_m^{(1)}+\boldsymbol{\delta}_1)} + \sum_{\boldsymbol{k}'} x_{\boldsymbol{k}'}^{(2)} e^{i\boldsymbol{k}'(\boldsymbol{r}_m^{(1)}+\boldsymbol{\delta}_2)} + \sum_{\boldsymbol{k}'} x_{\boldsymbol{k}'}^{(2)} e^{i\boldsymbol{k}'(\boldsymbol{r}_m^{(1)}+\boldsymbol{\delta}_3)}\right)$$

$$= \sum_{m\in \text{sublattice1}}^{N/3} \frac{3}{2}A_{12}\sum_{\boldsymbol{k}} x_{\boldsymbol{k}}^{(2)} e^{i\boldsymbol{k}\boldsymbol{r}_m^{(1)}} \sum_{\boldsymbol{k}'} x_{\boldsymbol{k}'}^{(2)} e^{i\boldsymbol{k}'\boldsymbol{r}_m^{(1)}} \left(e^{i\boldsymbol{k}'\boldsymbol{\delta}_1} + e^{i\boldsymbol{k}'\boldsymbol{\delta}_2} + e^{i\boldsymbol{k}'\boldsymbol{\delta}_3}\right)$$

$$= \frac{3}{2} A_{12} \sum_{\boldsymbol{k}} \sum_{\boldsymbol{k}'} x^{(1)}_{\boldsymbol{k}} x^{(2)}_{\boldsymbol{k}'} \sum_{m \in \text{sublattice}1}^{N/3} e^{i(\boldsymbol{k}+\boldsymbol{k}')\boldsymbol{r}^{(1)}_m} \left( e^{i\boldsymbol{k}'\boldsymbol{\delta}_1} + e^{i\boldsymbol{k}'\boldsymbol{\delta}_2} + e^{i\boldsymbol{k}'\boldsymbol{\delta}_3} \right)$$

$$= \frac{3}{2} A_{12} \frac{N}{3} \sum_{\boldsymbol{k}} x^{(1)}_{\boldsymbol{k}} x^{(2)}_{-\boldsymbol{k}} \left( e^{-i\boldsymbol{k}\boldsymbol{\delta}_1} + e^{-i\boldsymbol{k}\boldsymbol{\delta}_2} + e^{-i\boldsymbol{k}\boldsymbol{\delta}_3} \right), \tag{5.186}$$

where

$$\boldsymbol{\delta}_1 = \begin{pmatrix} 1 \\ 0 \end{pmatrix}, \quad \boldsymbol{\delta}_2 = \begin{pmatrix} -1/2 \\ \sqrt{3}/2 \end{pmatrix}, \quad \boldsymbol{\delta}_3 = \begin{pmatrix} -1/2 \\ -\sqrt{3}/2 \end{pmatrix}.$$

Here, we define

$$\varepsilon_{\boldsymbol{k}} = e^{-i\boldsymbol{k}\boldsymbol{\delta}_1} + e^{-i\boldsymbol{k}\boldsymbol{\delta}_2} + e^{-i\boldsymbol{k}\boldsymbol{\delta}_3} = e^{-ik_x} + e^{ik_x/2 + i\sqrt{3}k_y/2} + e^{ik_x/2 - i\sqrt{3}k_y/2}. \tag{5.187}$$

Similarly, for the term of $A_{23}$,

$$\sum_{m \in \text{sublattice}1}^{N/3} A_{23} \delta\theta_j (\delta\theta_{k_1} + \delta\theta_{k_2} + \delta\theta_{k_3}) = \frac{3}{2} A_{23} \frac{N}{3} \sum_{\boldsymbol{k}} x^{(2)}_{\boldsymbol{k}} x^{(3)}_{-\boldsymbol{k}} \varepsilon_{\boldsymbol{k}}, \tag{5.188}$$

and

$$\sum_{m \in \text{sublattice}3}^{N/3} A_{31} \delta\theta_k (\delta\theta_{i_1} + \delta\theta_{i_2} + \delta\theta_{i_3}) = \frac{3}{2} A_{31} \frac{N}{3} \sum_{\boldsymbol{k}} x^{(3)}_{\boldsymbol{k}} x^{(1)}_{-\boldsymbol{k}} \varepsilon_{\boldsymbol{k}}. \tag{5.189}$$

For the term of $A_{21}$,

$$\sum_{i \in \text{sublattice}1}^{N/3} A_{21}\, \delta\theta_j (\delta\theta_{i'_1} + \delta\theta_{i'_2} + \delta\theta_{i'_3}),$$

where the sites $i'_1$, $i'_2$, and $i'_3$ are the sites of sublattice-1 neighboring the site $j$. In this case, the shifts of the positions are given by $-\boldsymbol{\delta}_1$, $-\boldsymbol{\delta}_2$, and $-\boldsymbol{\delta}_3$, and thus the coefficient becomes $\varepsilon^*_{\boldsymbol{k}}$. The same argument holds for the terms of $A_{32}$ and $A_{13}$. For the terms involving $C_{ij}$ and $F_{ij}$, we have similar relations.

With these variables, the Hamiltonian can be written in the form:

$$\mathcal{H}_{\text{HA}} = E_0 + \sum_{\boldsymbol{k}} {}^t\boldsymbol{x}_{\boldsymbol{k}} A_{\boldsymbol{k}} \boldsymbol{x}_{-\boldsymbol{k}} + \sum_{\boldsymbol{k}} \boldsymbol{y}_{\boldsymbol{k}} C_{\boldsymbol{k}} \boldsymbol{y}_{-\boldsymbol{k}} + \sum_{\boldsymbol{k}} \boldsymbol{x}_{\boldsymbol{k}} F_{\boldsymbol{k}} \boldsymbol{y}_{-\boldsymbol{k}} + \sum_{\boldsymbol{k}} \boldsymbol{y}^t_{\boldsymbol{k}} F^*_{\boldsymbol{k}} \boldsymbol{x}_{-\boldsymbol{k}}, \tag{5.190}$$

where

$$\boldsymbol{x}_{\boldsymbol{k}} = \begin{pmatrix} x_{\boldsymbol{k}}^{(1)} \\ x_{\boldsymbol{k}}^{(2)} \\ x_{\boldsymbol{k}}^{(3)} \end{pmatrix}, \quad \boldsymbol{y}_{\boldsymbol{k}} = \begin{pmatrix} y_{\boldsymbol{k}}^{(1)} \\ y_{\boldsymbol{k}}^{(2)} \\ y_{\boldsymbol{k}}^{(3)} \end{pmatrix}, \tag{5.191}$$

and the sum of $\boldsymbol{k}$ is taken over the first Brillouin zone, which will be given explicitly in the following section. Each part for component of $\boldsymbol{k}$ is independent of the others, and the Hamiltonian is block-diagonalized. The part for $\boldsymbol{k}$ can be represented by a $6 \times 6$ matrix $\Lambda_{\boldsymbol{k}}$ as

$$\boldsymbol{x}_{\boldsymbol{k}} A_{\boldsymbol{k}} \boldsymbol{x}_{-\boldsymbol{k}} + \boldsymbol{y}_{\boldsymbol{k}} C_{\boldsymbol{k}} \boldsymbol{y}_{-\boldsymbol{k}} + \boldsymbol{x}_{\boldsymbol{k}} F_{\boldsymbol{k}} \boldsymbol{y}_{-\boldsymbol{k}} + \boldsymbol{y}_{\boldsymbol{k}} F_{\boldsymbol{k}} \boldsymbol{x}_{-\boldsymbol{k}} = (\boldsymbol{x}_{\boldsymbol{k}}, \boldsymbol{y}_{\boldsymbol{k}}) \, \Lambda_{\boldsymbol{k}} \begin{pmatrix} \boldsymbol{x}_{\boldsymbol{k}} \\ \boldsymbol{y}_{\boldsymbol{k}} \end{pmatrix}, \tag{5.192}$$

where

$$\Lambda_{\boldsymbol{k}} = \begin{pmatrix} A_{\boldsymbol{k}} & F_{\boldsymbol{k}} \\ {}^{t}F_{\boldsymbol{k}}^{*} & C_{\boldsymbol{k}} \end{pmatrix}, \tag{5.193}$$

and

$$A_{\boldsymbol{k}} = \begin{pmatrix} A_{11} & A_{12}\,\varepsilon_{\boldsymbol{k}} & A_{13}\,\varepsilon_{\boldsymbol{k}}^{*} \\ A_{21}\,\varepsilon_{\boldsymbol{k}}^{*} & A_{22} & A_{23}\,\varepsilon_{\boldsymbol{k}} \\ A_{31}\,\varepsilon_{\boldsymbol{k}} & A_{32}\,\varepsilon_{\boldsymbol{k}}^{*} & A_{33} \end{pmatrix}, \quad C_{\boldsymbol{k}} = \begin{pmatrix} C_{11} & C_{12}\,\varepsilon_{\boldsymbol{k}} & C_{13}\,\varepsilon_{\boldsymbol{k}}^{*} \\ C_{21}\,\varepsilon_{\boldsymbol{k}}^{*} & C_{22} & C_{23}\,\varepsilon_{\boldsymbol{k}} \\ C_{31}\,\varepsilon_{\boldsymbol{k}} & C_{32}\,\varepsilon_{\boldsymbol{k}}^{*} & C_{33} \end{pmatrix},$$

$$F_{\boldsymbol{k}} = \begin{pmatrix} F_{11} & F_{12}\,\varepsilon_{\boldsymbol{k}} & F_{13}\,\varepsilon_{\boldsymbol{k}}^{*} \\ F_{21}\,\varepsilon_{\boldsymbol{k}}^{*} & F_{22} & F_{23}\,\varepsilon_{\boldsymbol{k}} \\ F_{31}\,\varepsilon_{\boldsymbol{k}} & F_{32}\,\varepsilon_{\boldsymbol{k}}^{*} & F_{33} \end{pmatrix}. \tag{5.194}$$

### 5.5.1 *The Value of the Wave Numver (k) of the Triangular Lattice with Periodic Boundary Conditions*

For the $L_x \times L_y$ lattice with periodic boundary conditions (Fig. 5.7 illustrates the case of $M_x = M_y = 2$), the following conditions hold:

$$e^{i\boldsymbol{k}\cdot(\boldsymbol{a}_1+\boldsymbol{a}_2)M_y} = 1, \quad e^{i\boldsymbol{k}\cdot(2\boldsymbol{a}_1-\boldsymbol{a}_2)M_x} = 1. \tag{5.195}$$

Form these conditions, values of $\boldsymbol{k} = (k_x, k_y)$ satisfy the following relations with integers $m_x$ and $m_y$:

$$\boldsymbol{k} \cdot (\boldsymbol{a_1} + \boldsymbol{a_2}) M_y = \left( \frac{3}{2} k_x + 3 \frac{\sqrt{3}}{2} k_y \right) M_y = 2\pi m_y, \tag{5.196}$$

leading to

$$(2\boldsymbol{a_1} - \boldsymbol{a_2}) M_x = (3k_x) M_x = 2\pi m_x, \tag{5.197}$$

from which, the values of $(k_x, k_y)$ are determined as

$$k_x = \frac{2\pi}{3M_x} m_x, \quad \frac{3}{2} k_x + 3 \frac{\sqrt{3}}{2} k_y = \frac{2\pi}{M_y} m_y. \tag{5.198}$$

The values of $k_y$ is given by

$$k_y = \frac{4\pi}{3\sqrt{3}M_y} m_y - \frac{1}{\sqrt{3}} k_x \implies k_y = \frac{4\pi}{3\sqrt{3}M_y} m_y - \frac{2\pi}{3\sqrt{3}M_x} m_x. \tag{5.199}$$

The condition:

$$e^{i\boldsymbol{k} \cdot (2\boldsymbol{a_2} - \boldsymbol{a_1}) M_y} = 1, \tag{5.200}$$

is not independent of (5.195), because

$$\boldsymbol{a_1} + \boldsymbol{a_2} - (2\boldsymbol{a_1} - \boldsymbol{a_2}) = 2\boldsymbol{a_2} - \boldsymbol{a_1}. \tag{5.201}$$

#### 5.5.1.1 Periodic Boundary Conditions in a Rectangular Lattice

Instead of the tilted lattice shown in Fig. 5.7, we can also take a rectangular shape. In this case, the number of layers in the $y$-direction is $3M_y/2$, and the PBCs are expressed as

$$e^{i\boldsymbol{k} \cdot \boldsymbol{a}_2 3M_y/2} = 1, \quad e^{i\boldsymbol{k} \cdot (2\boldsymbol{a}_1 - \boldsymbol{a}_2) M_x} = 1. \tag{5.202}$$

These conditions yield

$$\boldsymbol{k} \cdot \boldsymbol{a}_2 M_y = \frac{3\sqrt{3}}{2} k_y M_y = 2\pi m_y, \tag{5.203}$$

and

$$\boldsymbol{k} \cdot (2\boldsymbol{a}_1 - \boldsymbol{a}_2) M_x = 3k_x M_x = 2\pi m_x. \tag{5.204}$$

From these relations, the values of $(k_x, k_y)$ are given by

$$k_x = \frac{2\pi}{3M_x} m_x, \quad 3k_y = \frac{4\pi}{\sqrt{3}M_y} m_y, \tag{5.205}$$

and hence

$$k_y = \frac{4\pi}{3\sqrt{3}M_y} m_y. \tag{5.206}$$

Thus, the density of $\boldsymbol{k}$ points is the same as that of the previous choice of PBC.

### 5.5.2 *The First Brillouin Zone*

The basis vectors of the sublattice are given by (5.184):

$$\boldsymbol{a}_1 = \begin{pmatrix} 3/2 \\ \sqrt{3}/2 \end{pmatrix}, \quad \boldsymbol{a}_2 = \begin{pmatrix} 0 \\ \sqrt{3} \end{pmatrix}.$$

The reciprocal lattice vectors $\boldsymbol{b}_1$ and $\boldsymbol{b}_2$ are defined by

$$\boldsymbol{b}_1 \cdot \boldsymbol{a}_1 = 2\pi, \quad \boldsymbol{b}_1 \cdot \boldsymbol{a}_2 = 0, \tag{5.207}$$

$$\boldsymbol{b}_2 \cdot \boldsymbol{a}_2 = 2\pi, \quad \boldsymbol{b}_2 \cdot \boldsymbol{a}_1 = 0. \tag{5.208}$$

From these definitions, we obtain

$$\frac{3}{2}b_x^{(1)} + \frac{\sqrt{3}}{2}b_y^{(1)} = 2\pi, \quad \sqrt{3}b_y^{(1)} = 0 \implies b_x^{(1)} = \frac{4\pi}{3}, \quad b_y^{(1)} = 0, \tag{5.209}$$

and

$$\frac{3}{2}b_x^{(2)} + \frac{\sqrt{3}}{2}b_y^{(2)} = 0, \quad \sqrt{3}b_y^{(2)} = 2\pi. \tag{5.210}$$

Therefore, we have

$$b_x^{(1)} = -\frac{1}{\sqrt{3}}b_y^{(2)}, \quad b_y^{(2)} = \frac{2\pi}{\sqrt{3}} \implies b_x^{(1)} = -\frac{2\pi}{3}, \quad b_y^{(2)} = \frac{2\pi}{\sqrt{3}}. \tag{5.211}$$

Thus, the reciprocal lattice is composed of lattice points given by

$$\nu_1 \boldsymbol{b}_1 + \nu_2 \boldsymbol{b}_2, \tag{5.212}$$

where $\nu_1$ and $\nu_2$ are integers. These lattice points are represented by small open squares in Fig. 5.10.

The first Brillouin zone is the area enclosed by the dashed lines, which are perpendicular bisectors of the red vectors connecting the origin $(0, 0)$ to the nearest reciprocal lattice points. We take $\boldsymbol{k}$ points within the first Brillouin zone.

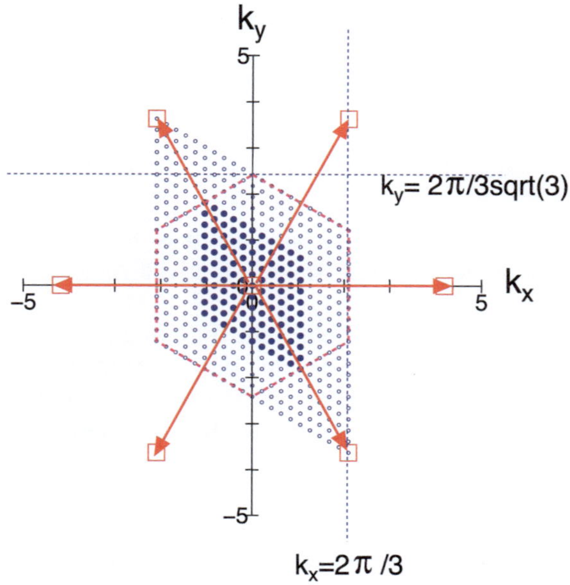


**Fig. 5.10** The reciprocal lattice of the sublattice of a triangular lattice and the Brillouin zone. The red vectors represent the path to the nearest neighbor reciprocal lattice points from (0, 0). The first Brillouin zone is the hexagonal region enclosed by the dashed lines

### *5.5.3 Non-Trivial Degeneracy at Zero Field ($H = 0$) in Entropy in the Harmonic Analysis*

For the ground-state configuration in a plane including the $z$ axis, we have obtained (5.149)–(5.154) on a triangle. On the triangular lattice where six neighboring sites exist, the quantities (5.149)–(5.154) are modified as

$$A1 = A_{11}A_{22}A_{33} = \frac{27A^2}{4(A+1)}, \tag{5.213}$$

$$A2 = A_{11}A_{23}^2 + A_{22}A_{13}^2 + A_{33}A_{12}^2 = \frac{3A^2(A^2+A+1)}{4(A+1)^3}, \tag{5.214}$$

$$A3 = A_{12}A_{23}A_{31} = -\frac{A^3}{8(A+1)^3} \tag{5.215}$$

and

$$C1 = C_{11}C_{22}C_{33} = \frac{27A^2}{8(A+1)}, \tag{5.216}$$

$$C2 = C_{11}C_{23}^2 + C_{22}C_{13}^2 + C_{33}C_{12}^2 = 3\frac{A^2+A+1}{4(A+1)}, \tag{5.217}$$

$$C3 = C_{12}C_{23}C_{31} = \frac{1}{8}. \tag{5.218}$$

The contribution to the entropy from each mode $\boldsymbol{k}$ is given by

$$\ln \det A_{\boldsymbol{k}} = \ln \left( A1 + \varepsilon_{\boldsymbol{k}} \varepsilon^*_{\boldsymbol{k}} A2 + \left( \varepsilon^3_{\boldsymbol{k}} + (\varepsilon^*_{\boldsymbol{k}})^3 \right) A3 \right), \tag{5.219}$$

$$\ln \det C_{\boldsymbol{k}} = \ln \left( C1 + \varepsilon_{\boldsymbol{k}} \varepsilon^*_{\boldsymbol{k}} C2 + \left( \varepsilon^3_{\boldsymbol{k}} + (\varepsilon^*_{\boldsymbol{k}})^3 \right) C3 \right), \tag{5.220}$$

which does not depend on $\theta_i$ and thus the entropy effect in the harmonic approximation does not resolve the non-trivial degeneracy with respect to the choice of $\theta_1$ [6].

Here, it should be noted the following points. For the part $\phi$, if $\sin\theta_i = 0$, $C_{ij}$ diverges. However, in the limit $\sin\theta_i \to 0$, the limiting value of $C_{ij}$ provides a finite value, and the above relations (5.216)–(5.217) hold. Moreover, uniform changes over the lattice give zero eigenvalue. But, its contribution to the integral over $\boldsymbol{k}$ is zero and we can ignore them.

This non-trivial degeneracy with respect to the choice of $\theta_1$ causes a peculiar dependence of the susceptibility at low temperatures, which will be presented in Sect. 5.7.1. At finite temperatures, contributions from higher-order deviations resolve the non-trivial degeneracy, where the Y-shaped structure is chosen in which the magnetization in the $xy$-plane is 0, and the fluctuation of the $xy$ components of the magnetization, $\beta(\langle M_x^2\rangle + \langle M_y^2\rangle)/N$, is observed to be finite. However, as the temperature decreases, reflecting the non-trivial degeneracy, not only the Y shape but also other degenerate configurations having nonzero $xy$ components of the magnetization begin to contribute. Therefore, $\beta(\langle M_x^2\rangle + \langle M_y^2\rangle)/N$ becomes large.[6]

## 5.6 Dependence of the Entropy on the Magnetic Field ($H$)

Now, we obtain the entropy for stable configurations in the field numerically ( $H > 0$): the umbrella configuration, the Y-shaped configuration, and the V-shaped configuration. The matrix elements:

$$A_{11} = (-3s_1s_2\cos(\phi_1^0 - \phi_2^0) - 3Ac_1c_2 + Hc_1 - 3s_3s_1\cos(\phi_1^0 - \phi_3^0) - 3Ac_1c_3)/2, \tag{5.221}$$

$$A_{22} = (-3s_1s_2\cos(\phi_1^0 - \phi_2^0) - 3Ac_1c_2 + Hc_2 - 3s_3s_2\cos(\phi_2^0 - \phi_3^0) - 3Ac_2c_3)/2, \tag{5.222}$$

$$A_{33} = (-3s_1s_3\cos(\phi_1^0 - \phi_3^0) - 3Ac_1c_3 + Hc_3 - 3s_3s_2\cos(\phi_2^0 - \phi_3^0) - 3Ac_2c_3)/2, \tag{5.223}$$

depend on $H$ while other elements of $A_{ij} (i \neq j)$, $C$, $D$ do not depend on $H$.

---

[6] This non-trivial degeneracy does not exist in the model with single-ion anisotropy $D$ instead of the anisotropic exchange coupling $A$ ($\mathcal{H} = \sum_{\langle ij\rangle} \boldsymbol{S}_i \cdot \boldsymbol{S}_j - D\sum_i (S_i^z)^2$). In this model, $\beta(\langle M_x^2\rangle + \langle M_y^2\rangle)/N$ goes to zero when $T \to 0$.

For the umbrella configuration, the matrix elements $F_{ij}$ are nonzero, and we need to calculate the eigenvalues of the $6 \times 6$ matrix $\Lambda_{\boldsymbol{k}}$. For the collinear structure including the $z$ axis, $F_{ij} = 0$, and we calculate the eigenvalues of the $3 \times 3$ matrices $\hat{A}$ and $\hat{C}$. We sum up $\ln \det \Lambda_{\boldsymbol{k}}$ or $\ln \det(\hat{A})_{\boldsymbol{k}} + \ln \det(\hat{C})_{\boldsymbol{k}}$ over $\boldsymbol{k}$ in the first Brillouin zone (Fig. 5.10).

The zero eigenvalue ($\lambda_i = 0$) indicates motion without energy loss, e.g., uniform rotation. For such a cases, $\det \Lambda_{\boldsymbol{k}=0} = 0$, but its contribution $\ln \det \Lambda_{\boldsymbol{k}}\, dk_x\, dk_y$ does not affect the integration, and thus we exclude such contributions.

To compare the entropies of different ground-state configurations, the configuration-dependent entropy (5.135):

$$\Delta S = S - N(1 - \ln \frac{\pi}{\beta}) = -\frac{1}{2} \sum_{\boldsymbol{k}} \ln \det \Lambda_{\boldsymbol{k}} \tag{5.224}$$

is calculated.

### 5.6.1 Dependence of the Entropy on the Magnetic Field ($H$) for the Isotropic Heisemberg Model ($A = 1$)

In the case of $A = 1$, the ground state remains degenerate even for $H \neq 0$, i.e., all configurations satisfying

$$c_1 + c_2 + c_3 = 3H, \quad s_1 + s_2 + s_3 = 0 \tag{5.225}$$

have the same energy. To determine which configuration has the maximum entropy, we compare the entropies of the umbrella configuration, the Y-shaped configuration, and the V-shaped configuration ($c_1, c_2 = c_3 = (H - c_1)/2, \phi_1 = \phi_2 = \phi_3 = 0$). The part of entropy $\Delta S$ is shown in Fig. 5.11. Here, we find that the umbrella configuration is less favorable than the Y and V shapes. Thus, we expect that the degeneracy in the field is lifted [5].

### 5.6.2 Dependence of the Entropy on the Magnetic Field ($H$) for the Ising-Like Heisenberg Model ($A = 2$)

Next, we study the case of $A = 2$. In this case, the coplanar configurations including the $z$ axis have lower energy than that of the umbrella configuration, and energetically, the ground-state configuration adopts a coplanar structure. Thus, the Hessian matrix is decomposed into $A_{\boldsymbol{k}}$ and $C_{\boldsymbol{k}}$. Additionally, the non-trivial degeneracy discussed in the previous section is lifted due to the Zeeman energy with $H > 0$. Therefore,

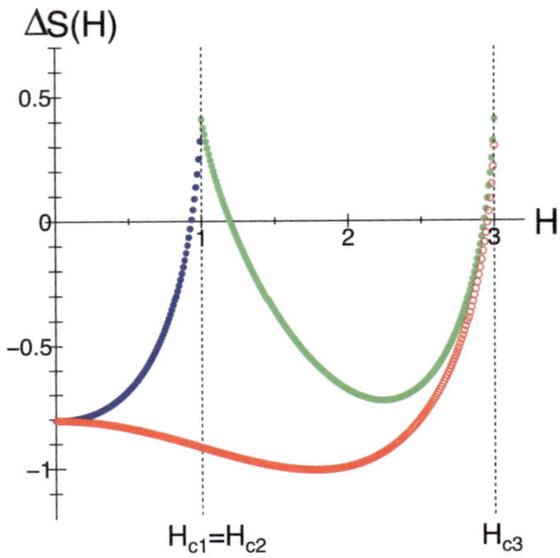


**Fig. 5.11** Magnetic field $(H)$ dependence of the entropy $\Delta S$ obtained by harmonic analysis for $A = 1$. The red, blue and green dots denote the entropy of the umbrella structure, the Y shape $(0 < H < 1)$ and the V shape $(1 < H < 3)$, respectively. For $A = 1$, $H_{c1} = H_{c2} = (2A - 1 + \sqrt{4A^2 + 4A - 7})/2 = 1$ (5.72), and $H_{c3} = 2A + 1 = 3$ (5.91)

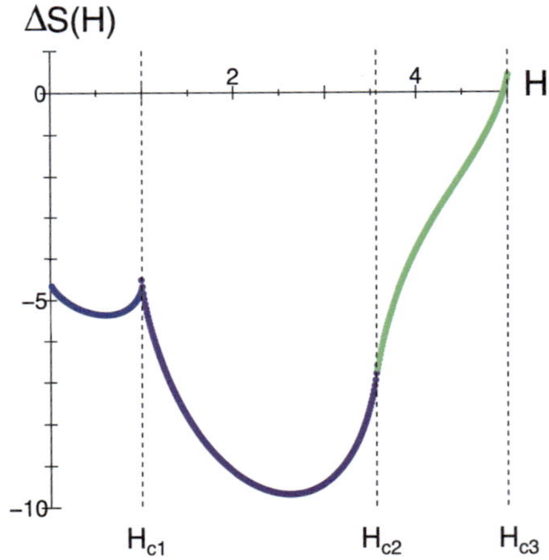


**Fig. 5.12** Magnetic field $(H)$ dependence of the entropy $\Delta S$ for $A = 2$. The red, blue, violet and green dots denote the entropies of the umbrella structure, the Y shape $(0 < H < H_{C1} = 1)$, the uud shape$(H_{c1} = 1 < H < H_{c2})$, and the V shape $(H_{c2} = 1 < H < H_{c3} = 3)$, respectively. For $A = 2$, $H_{c1} = 1$, $H_{c2} = (2A - 1 + \sqrt{4A^2 + 4A - 7})/2 = 3.56\cdots$ (5.72), and $H_{c3} = 2A + 1 = 5$ (5.91)

the ground-state configuration is determined by energy itself. In this scenario, the Y shape, uud, and V-shaped configurations appear successively at low temperatures.

The $H$ dependence of $\Delta S$ is depicted in Fig. 5.12. In the present case the umbrella configuration is unstable.

### 5.6.3 *Dependence of the Entropy on the Magnetic Field ($H$) for the XY-Like Heisenberg Model ($A = 0.98$)*

Now, we study the case that has weak planar ($xy$-like) anisotropy $A = 0.98$. Here, the ground state is uniquely determined as the umbrella configuration. However, for this value of $A$, the entropy favors configurations in the plane including the $z$ axis. The $H$ dependence of $\Delta S$ is depicted in Fig. 5.13. Thus, energy and entropy favor

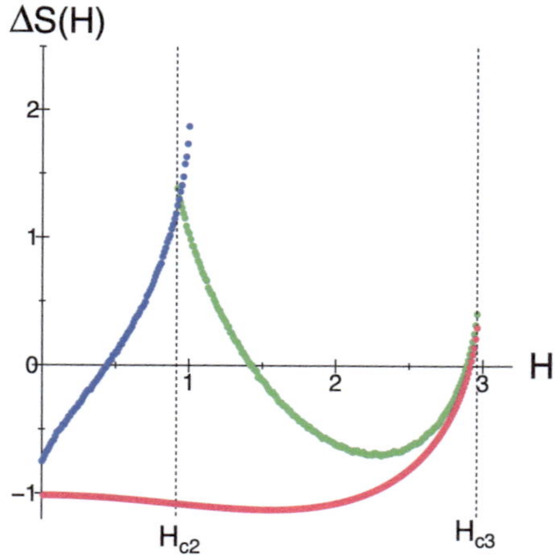


**Fig. 5.13** Magnetic field ($H$) dependence of $\Delta S$ for $A = 0.98$. The red, blue, and green dots denote the entropies of the umbrella structure, the Y shape ($0 < H < H_{c2}$), the uud shape($H_{c1} = 1 < H < H_{c3}$), and the V shape ($H_{c2} = 1 < H < H_{c3} = 3$), respectively. For $A = 0.98$, $H_{c1} = 1$, $H_{c2} = (2A - 1 + \sqrt{4A^2 + 4A - 7})/2 = 0.916\cdots$ (5.72), and $H_{c3} = 2A + 1 = 2.96$ (5.91)

different configurations. This competition causes an interesting phase diagram in the $(H, T)$ plane. When the anisotropy is strong, the umbrella structure persists at finite temperatures for weak field, but at high field, the V shape structure appears. When the anisotropy is weak, the equilibrium state at finite temperature favors the coplanar structure. There, the umbrella structure disappears, and the Y-shape structure emerges, and the sequence of phases with $H$ in the case of easy-axis anisotropy is realized. Dependence of phase diagram in the space of $(H, T)$ on the anisotropy will be discussed in Sect. 5.9.2.

## 5.7 Ordering Processes at Finite Temperatures

So far, we have studied how the ground state of antiferromagnets on a triangular lattice depends on various parameters. In this section, we briefly review the system's properties at finite temperatures. At very high temperatures, the system is in a disordered (paramagnetic) phase. As the temperature decreases, the system tends to be ordered, approaching the ground-state configuration. In the case of the ferromagnetic Ising model (non-frustrated), the ordering process is relatively simple, exhibiting a phase transition between the paramagnetic and ferromagnetic phases. For frustrated models, however, the order parameter has a more complex structure composed of several spins compromising competing interactions. Depending on the symmetry of the order parameter, the system exhibits a peculiar type of phase transition or successive phase transitions. Furthermore, in the magnetization process with an applied magnetic field $H$, the competition between energy and entropy among nearly degenerate configurations can lead to complicated phase transitions.

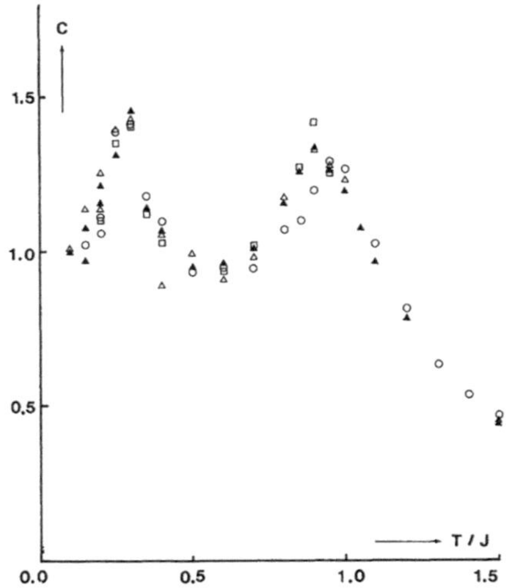


**Fig. 5.14** Temperature and size dependence of the specific heat for $A = 2$. Closed triangles, open triangles, open squares, represent system sizes $L = 12, 15, 18$, and 24, respectively. (This figure is reproduced from Ref. [6]. ©1985 The Physical Society of Japan)

### 5.7.1 *Successive Phase Transitions in a Ising-Like Heisenberg Model* ($A = 2$)

First, we study the ordering process of the antiferromagnetic Ising-like Heisenberg model on a triangular lattice (AFTIH) with $A = 2$. The temperature dependence of the specific heat is depicted in Fig. 5.14. Here, we observe two peaks corresponding to the ordering of the $z$ component and that of the $xy$ components.

In two dimensions, fluctuations are strong, and the singularity at phase transitions is weak. In the present case, the ordering of the $z$ component follows the type of the six-state clock model [7], while the ordering of the $xy$ components is described by the ferromagnetic XY model. It is known that both models exhibit phase transitions of the Kosterlitz-Thouless type [8, 9]. As a result, the transition points are not clearly observed [6].

The order parameter is composed of three spins belonging to sublattices. To detect the sublattice ordering, we observe the sublattice magnetizations:

$$S_A^x = \sum_{i \in A} S_i^x, \quad S_A^y = \sum_{i \in A} S_i^y, \quad S_A^z = \sum_{i \in A} S_i^z, \tag{5.226}$$

and similarly for sublattices B and C. The degree of sublattice ordering is detected using the quantities:

$$A^{xy} = \frac{3}{N}\left(\langle (S_A^x)^2\rangle + \langle (S_B^x)^2\rangle + \langle (S_C^x)^2\rangle + \langle (S_A^y)^2\rangle + \langle (S_B^y)^2\rangle + \langle (S_C^y)^2\rangle\right), \tag{5.227}$$

and

$$A^z = \frac{3}{N}\left(\langle (S_A^z)^2\rangle + \langle (S_B^z)^2\rangle + \langle (S_C^z)^2\rangle\right). \tag{5.228}$$

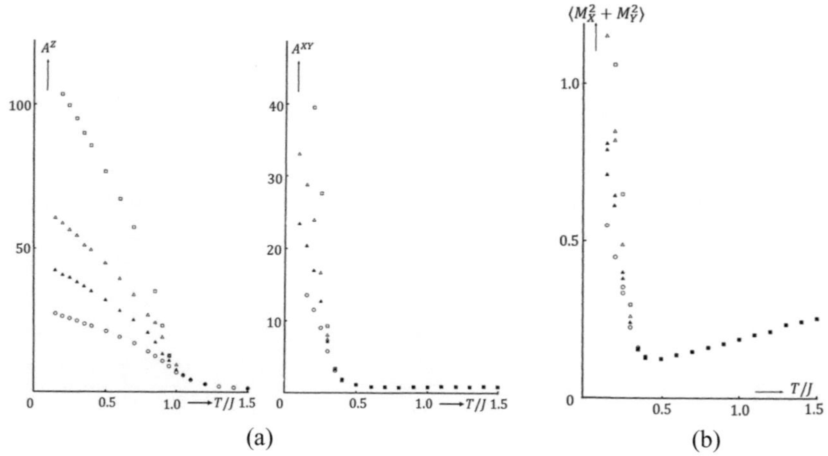


**Fig. 5.15** **a** Temperature and size dependence of the order parameter $A^z$ for the $z$ component (left) and that of the $xy$ components $A^{xy}$ for $A = 2$ (right). **b** Temperature and size dependence of the order parameter $M_{XY}$ for $A = 2$. Symbols: open circles, closed triangles, open triangles, and open squares, denote the system sizes $L = 12$, 15, 18, and 24, respectively. (These figures are reproduced from Ref. [6].©1985 The Physical Society of Japan)

We also study the ordering of ferrimagnetism using the quantity:

$$F_z = \frac{3}{N}\left(\langle (F_{\rm A}^z)^2\rangle + \langle (F_{\rm B}^z)^2\rangle + \langle (F_{\rm C}^z)^2\rangle\right), \tag{5.229}$$

where

$$F_{\rm A}^z = S_{\rm A}^z - \frac{1}{2}(S_{\rm B}^z + S_{\rm C}^z), \tag{5.230}$$

and $F_{\rm B}^z$ and $F_{\rm C}^z$ are defined similarly.

The degree of ordering of the $z$ component can be inferred from the size dependence of $A^z$. If the system only has short-range order for the $z$ component, $A^z$ is independent of $N$. If the sublattice exhibits long-range order in the $z$ component, $A^z$ is proportional to $N$. If $A^z$ exhibits quasi-long-range order, where the spin-correlation function decays as a power law, $\langle S_i^z S_j^z\rangle \propto r^{2-d-\eta}$, then $A^z$ scales as $N^{(2-\eta)/2}$. For the $xy$ components, we use the quantity $A^{xy}$.

The temperature and size dependencies of $A^z$ and $A^{xy}$ are depicted in Fig. 5.15a, b, respectively. Here, we observe successive orderings of the $z$ component at $T_1 \simeq 1.0$ and of the $xy$ components at $T_2 \simeq 0.3$. This sequential occurrence of phase transitions provides a possible scenario for the successive phase transitions of AFT systems with uniaxial anisotropy. This scenario differs from the sequential phase transitions caused by the PD (partially disordered) and Ferri (uud) phases discussed in Sect. 4.5.

To detect the appearance of spontaneous magnetization:

$$M_z = \sum_{i=1}^{N} S_i^z, \tag{5.231}$$

we examine the quantity $M_Z = \langle M_z^2 \rangle / N$ for the $z$ component. Similarly, for the $xy$ components, we use $M_{XY} = (\langle M_x^2 + M_y^2 \rangle)/N$. Magnetization $M_z$ appears in the ordered phases of the Y-shaped configuration (non-collinear ferrimagnetic state) (not shown). However, reflecting the nontrivial degeneracy caused by fluctuations in the $xy$ components, the fluctuation of the $xy$ components, $M_{XY}$, diverges at low temperatures, although the Y-shape and uud shape have no transverse magnetization, as shown in Fig. 5.15c.

## 5.7.2 AFTIH3D: Layered Triangular Lattice (Hexagonal Lattice)

In two dimensions, the divergence of fluctuations at long wavelengths (infrared divergence) prevents the existence of long-range order in systems with continuous symmetry, such as the XY model and Heisenberg model. This property has been proved by the Hohenberg-Mermin-Wagner theorem [10]. In the ferromagnetic XY system, a peculiar phase transition known as the Berezinskii-Kosterlitz-Thouless (BKT) transition occurs [8]. For systems described by the $p$-state clock model, double (dual) BKT transitions have been observed [9]. In such cases, it is difficult to identify the critical point, and the nature of the order is not immediately clear.

On the other hand, in three dimensions, long-range order can exist [11], making it easier to detect the phases. Indeed, most materials have three-dimensional structures. Therefore, we study a layered triangular lattice known as hexagonal lattice. In three dimensions, anisotropy introduced through the anisotropic coupling constant $A$ becomes singular because the interactions between chains perpendicular to the triangular planes are not frustrated. Even with small $A$, the system exhibits strong anisotropy, which grows with the chain length.

For this reason, we study a model with single-ion anisotropy $D$, rather than anisotropy in the coupling constant. The Hamiltonian of the system is given by

$$\mathcal{H} = \sum_{\langle ij \rangle} \mathbf{S}_i \cdot \mathbf{S}_j - \sum_{i=1}^{N} D(S_i^z)^2. \tag{5.232}$$

If $D > 0$, the model exhibits easy-axis anisotropy, which corresponds to the case $A > 1$, and successive phase transitions occur due to the ordering of the $z$-component (similar to a 6-state clock model) and the ordering of the $xy$ components (similar to a 3D XY model). Both of these transitions belong to the universality class of the 3D XY model [12].

## 5.8 Symmetries of Order Parameters and Topological Defects

The order parameter of an antiferromagnet on a triangular lattice composed of three spins. Depending on the types of interactions, the order parameter can exhibit various symmetries, leading to distinct phase transitions. Here, we briefly review the properties of phase transitions for the following cases.

### 5.8.1 XY-Like Model ($A < 1$)

For the easy-plane type anisotropy, which corresponds to $A < 1$ or $D < 0$, the sublattice magnetization forms the 120° structure in the $xy$ plane. This structure has two symmetries: two-fold symmetry due to the ±120° structure ( $Z_2$) and rotation symmetry around the $z$-axis ($S_1$). The ordering process reflects these symmetries, leading to two distinct phase transitions [4].

In a numerical study, the specific heat shows a sharp singularity resembling a two-dimensional Ising transition, and the spin stiffness constant exhibits a temperature dependence similar to that of a BKT transition. This observation was interpreted by scenario that the $Z_2$ symmetry induces an Ising-type phase transition at a temperature $T_C$, followed by a BKT phase transition for the $S_1$ symmetry at a lower temperature $T_{BKT} < T_C$ [4]. However, an alternative scenario, in which $T_{BKT} > T_C$, was also proposed from the following consideration. If a next-nearest ferromagnetic interaction $J_2$ is introduced, it enhances the ordering of each sublattice, thereby increasing $T_{BKT}$. For sufficiently large $J_2$, it is expected that $T_{BKT} > T_C$. If we assume $T_{BKT} < T_C$ at $J_2 = 0$, there should be a critical value of $J_2$ where the relationship between the two temperatures changes, which is not plausible. Although this issue has been debated for a long time, the case $T_{BKT} < T_C$ is currently the most widely supported scenario [13]. Nonetheless, the possibility of $T_{BKT} = T_C$ cannot be completely ruled out.

In three dimensions, the system exhibits long-range order, and the $Z_2$ symmetry breaking and spin ordering transitions occur simultaneously. This phase transition may be the so-called chiral phase transition or possibly a first-order transition, as explained below. As the typical topological fluctuation, recently, the vortex structure in $K_2CuF_4$ was observed directly in real space [14].

### 5.8.2 Isotropic Heisenberg Model ($A = 1$)

In the case of $A = 1$, the 120° structure is no longer necessarily confined to the $xy$ plane, and the structure may lie in any plane. Thus, one might naively assume that the 120° structure possesses the same symmetry as the Heisenberg model, i.e., $S_2$. However, the 120° structure is defined by assigning two vectors, such as the vectors of two spins of a triangle, or the vector chirality (5.4) and one of the vectors of sublattices. Thus, this case shows a peculiar ordering process [15].

This symmetry is equivalent to that of rigid body rotation, SO(3), and the symmetry of the order parameter is given by $P_3$ (three-dimensional projective space), while the symmetry of a single three-dimensional spin (Heisenberg spin) is $S_2$ (the surface of a three-dimensional sphere). The symmetry of the rigid body rotation corresponds to the chiral universality class [16, 17].

It is known that, in two dimensions, the model with this symmetry possesses a point defect (the first homotopy group of the order parameter space is $V = P_3$, $\Pi_1(P_3) = Z_2$). This defect is different from the usual vortex in the ferromagnetic XY model ($\Pi_1(S_1) = Z$), in which the vortex is characterized by an integer vortex number $n$:

$$\oint d\theta = 2\pi n, \quad n \in Z \text{ integer}. \tag{5.233}$$

In contrast, in the Heisenberg model, the so-called "$Z_2$ vortex" appears, where the vortex number $n$ can take only the values 0 or 1. In this case, a configuration with $n = 2$ can be continuously deformed into one with $n = 0$, whereas a configuration with $n = 1$ cannot.

The existence of this type of vortex is a unique property of the AFT Heisenberg model, and it has been experimentally detected [18]. The existence of a phase transition due to the $Z_2$ vortex has been studied extensively both theoretically and experimentally [19–25].

In three dimensions, we expect a new phase transition due to this structure of the order parameter, which is called the "chiral order parameter" [16, 17]. This type of order parameter is characterized by two vectors ($\boldsymbol{\phi}_a$, $\boldsymbol{\phi}_b$), and the corresponding phenomenological Hamiltonian is given in the form:

$$\mathcal{H} = \left(\nabla\boldsymbol{\phi}_a\right)^2 + \left(\nabla\boldsymbol{\phi}_b\right)^2 + r\left(\boldsymbol{\phi}_a^2 + \boldsymbol{\phi}_b^2\right) + u\left(\boldsymbol{\phi}_a^2 + \boldsymbol{\phi}_b^2\right)^2 + v\left((\boldsymbol{\phi}_a\boldsymbol{\phi}_b)^2 - \boldsymbol{\phi}_a^2\boldsymbol{\phi}_b^2\right). \tag{5.234}$$

This model is applicable also to the case of the XY model. The last term, involving $v$, introduces a new type of interaction that is not present in the usual $\phi^4$-model. Thus, this term causes a new type of critical point known as the "chiral fixed point", and numerical calculations proposed a new universality class with the following critical exponents [16, 17]:

$$\alpha \simeq 0.4, \quad \beta \simeq 0.25, \quad \gamma \simeq 1.1, \quad \text{and} \quad \nu \simeq 0.53, \tag{5.235}$$

which were also observed experimentally in $CsMnBr_3$ [26]. On the other hand, an examination of the renormalization-group flow chart suggests a first-order phase transition [27]. Higher-order renormalization group analysis [28] and further large Monte Carlo simulations have been performed, which indicate second-order phase transitions [29]. Studies on this problem are currently ongoing.

### 5.8.3 Ising-Like Heisenberg Model ($A > 1$)

In the case of $A > 1$, as mentioned above, the $z$-component orders first. There are six possible arrangements of the Y shape configuration on the three sublattices. At the critical point, one of them is selected. Thus, the phase transition is expected to be of the $Z_6$ type. At a lower temperature, ordering of the $xy$-components occurs, which corresponds to the ordering of the $S_1$ model. Both transitions are expected to be of the Kosterlitz-Thouless type in two dimensions. In three dimensions, both transitions are expected to belong to the universality class of the three-dimensional ferromagnetic XY model [12].

## 5.9 Magnetization Process ($M(H)$) at Finite Temperatures

The magnetization process at $T = 0$ was studied in Sect. 5.2.5. Here, we examine the magnetization process $M(H)$ at finite temperatures.

### 5.9.1 Magnetization Process ($M(H)$) for the Ising-Like Heisenberg Model ($A > 1$)

First, we study the case of AFTIH with $A = 2$ [30]. In this case, we observe the full sequence (Y-uud-V-uuu) at low temperatures. The magnetic field dependence of quantities for the two-dimensional triangular lattice is shown in Fig. 5.16a.

A typical phase diagram is shown in Fig. 5.16b. The region of uud phase denoted by 'b' increases as $H$ increases. In the three-dimensional case, a similar structure is expected. This type of sequence of phase transition was observed in experiments on $RbFe(MoO_4)_2$ and $CsFe(SO_4)_2$ [31], $GaPd_2Al_3$ [32] and $PbFe(MoO-4)_2)$ [33]. The magnetization process in quantum spin ($S = 1/2$) gives the same sequence [34].

It should be noted that $\langle M_x^2 + M_y^2\rangle$ appears in the phase of V configurations due to the uncompensated $xy$ components in its non-collinear structure (not shown). The appearance of $\langle M_x^2 + M_y^2\rangle$ in the V-shape indicates that the $xy$ components are induced by the field in the $z$ direction, which suggests that the cross susceptibility $\chi_{zx}$ diverges at the transition point.

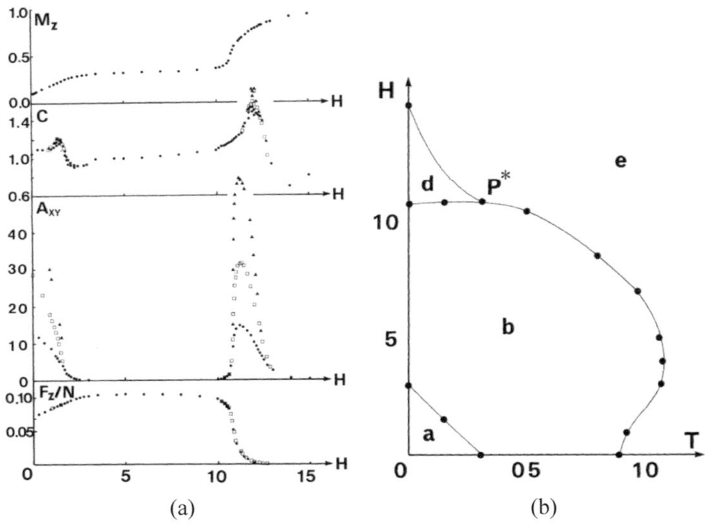


**Fig. 5.16** **a** Magnetic field dependence of quantities for the model with $A = 2$ at $T = 0.15$. $M_z$, the specific heat $C$, $A_{XY} = A_x + A_y$, and $F_z/N$ are shown. Open cicle, closed triangle, open triangle, and open square represent the sizes $L = 12$, 15, 18, and 24, respectively. **b** Phase diagram $(T, H)$ for $A = 2$. It should be noted that, here, 'a' and 'b' denote the configurations of the Y-shape (b) and the uud-shape (c) in Fig. 5.3. 'd' and 'e' are the configurations of V-shape (d) and uuu-shape (e) in Fig. 5.3. (These figures are reproduced from Ref. [30].©1986 The Physical Society of Japan)

### 5.9.2 *Magnetization Process ($M(H)$) of the XY-Like Heisenberg Model ($A < 1$ or $D < 0$)*

In the case of planar anisotropy ($A < 1$), although energetically the umbrella structure is favorable, entropy favors a planar structure in the plane including the $z$-axis as we studied in Fig. 5.13. It was found that non-trivial phase transitions due to entropy difference occur under the competition between them. We review the results of a Monte Carlo study of the $M(H)$ for a layered triangular lattice, i.e., the hexagonal lattice [35].

As explained in Sect. 5.7.2, we use the model with the single-ion type anisotropy (5.232). The systems with $D \geq 0$ exhibit phase diagrams similar to that of $A \geq 1$. The phase diagram for $D = 0$, shown in Fig. 5.17a, is the same model as $A = 1$. For $D < 0$, the ground state in the magnetic field along the $z$-axis is the umbrella state. When the anisotropy is large, e.g., $D = -0.1$, the system remains in the umbrella state at low fields. However, at high fields, before reaching the saturated ferromagnetic state, the V-shaped state appears, as shown in Fig. 5.17b. The transition between the umbrella state and the V-shape state was predicted in the ground state by considering the energy gain due to quantum fluctuations in the spin-wave analysis [36].

When $|D|$ $(D < 0)$ is very small, the energy difference between the umbrella state and the coplanar states becomes small. In this case, the entropy effect becomes

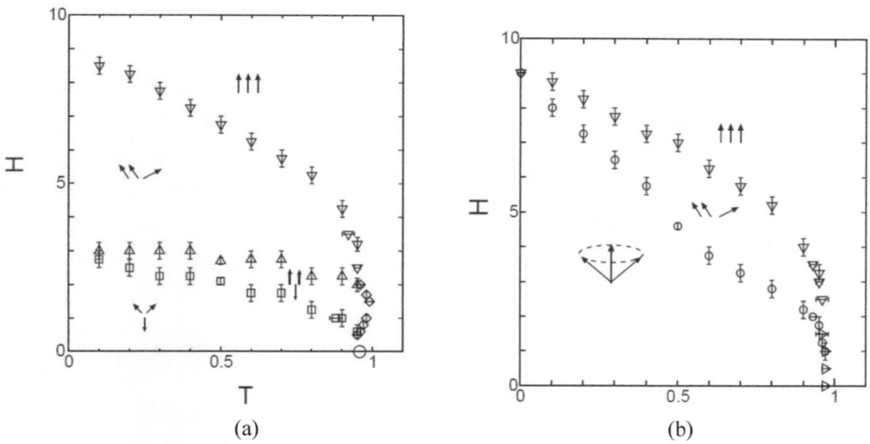


**Fig. 5.17** **a** Phase diagram in the $(T, H)$ plane for the isotropic Heisenberg system ($A = 1$, $D = 0$). **b** Phase diagram in the $(T, H)$ plane for an $xy$-like system ($D = -0.1$). Open circle, closed triangle, open triangle, and open square denote the sizes $L = 12$, 15, 18, and 24, respectively. (These figures are reproduced from Ref. [35]. ©2001 The Physical Society of Japan)

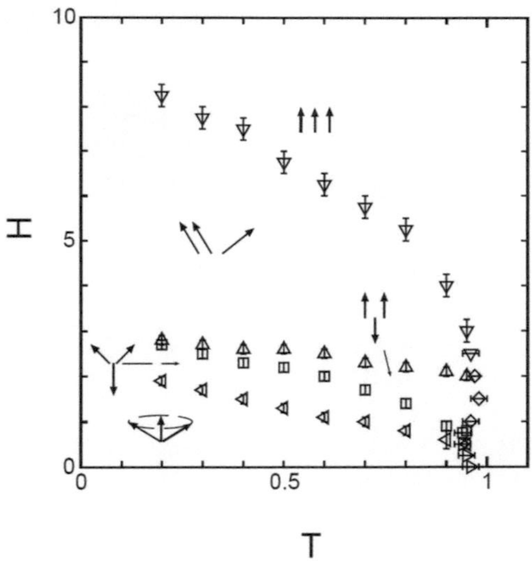


**Fig. 5.18** Phase diagram in the $(T - H)$ plane for a weakly $xy$-like system ($D = -0.01$). Open circle, closed triangle, open triangle, and open square denote the sizes $L = 12$, 15, 18, and 24, respectively. (This figure is reproduced from Ref. [35].©2001 The Physical Society of Japan)

significant. As a result, a more complicated phase diagram emerges. The phase diagram for a very small $D(= -0.01)$ is shown in Fig. 5.18, where the umbrella, Y-shape, and uud states appear. This phase diagram is a remarkable example of the order-by-disorder phenomenon. The transition between the umbrella state and the Y-shape state was observed also in a corresponding quantum system by the DMRG method [37], which will be reviewed in the next chapter.

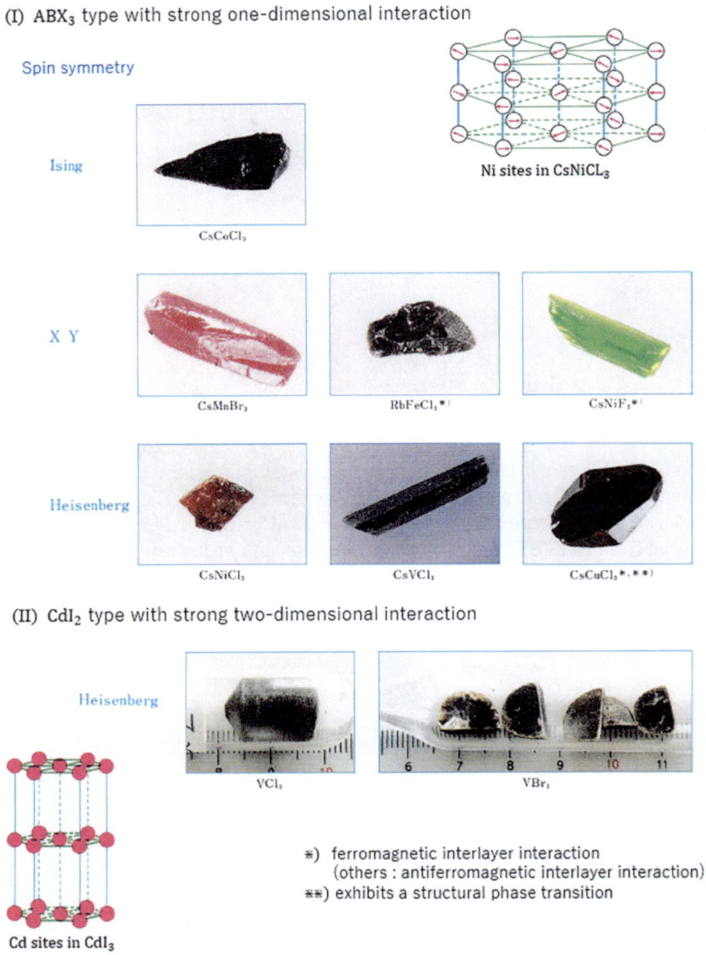


**Fig. 5.19** Examples of $ABX_3$-type materials. (This figure is reproduced from BUTSURI (Monthly magazine published by the Physical Society of Japan) Vol. **41**, 967 (1986)).©1986 The Physical Society of Japan)

## 5.10 Experiments of AFT Materials

As candidates for AFT, various materials have been studied experimentally [38, 39]. First, the $ABX_3$-type materials (Fig. 4.15) were studied; typical examples of materials are depicted in Fig. 5.19. These materials consist of strongly coupled one-dimensional chains that form a triangular lattice.

Depending on the type of B atom in $ABX_3$, the anisotropy in the spin space changes. For example, $CsCoCl_3$ exhibits Ising-like anisotropy. $RbFe(MoO_4)_2$ and $CsFe(SO_4)_2$ [31] and $GdPd_2Al_3$ [32] have Ising-like Heisenberg-type anisotropy.

$CsMnBr_3$, $RbFeCl_3$, and $CsNiF_3$ exhibit XY-like anisotropy, while $CsNiCl_3$, $CsVCl_3$, and $CsCuCl_3$ have nearly isotropic Heisenberg interactions. However, $CsCuCl_3$ exhibits weak anisotropy in the $xy$-plane [40–45], and its spin structure has been studied extensively [46–49]. Moreover, a jump in the magnetization process was observed [50–53], which will be discussed in the next chapter. In the magnetization process, an interesting spin-chirality-driven ferroelectricity has been observed [33], which is a realization of magnetoferroelectricity [54].

As for two-dimensional-like materials, the pioneering work by Hirakawa on the synthesis of $VCl_2$ [55] ($S = 1$) was a significant development. The Hirakawa group also synthesized oxides such as $NaTiO_2$ and $LiNiO_2$ as candidates for materials with $S = 1/2$ to study the quantum properties of AFT models [56]. Since then, various materials have been investigated. In particular, $NiGa_2S_4$ is a quasi-two-dimensional triangular antiferromagnet with $S = 1$, consisting of layered structures with central $NiS_2$ blocks made of edge-sharing $NiS_6$ octahedra and outer sheets of $GaS_4$ tetrahedra [57]. Reflecting the effects of frustrated interactions, unusual spin dynamics have been observed [58]. $NaCrO_2$ is an isotropic triangular Heisenberg antiferromagnet ($S = 3/2$) with very weak interlayer coupling [59]. $\kappa$-$(BEDT\text{-}TTF)_2Cu_2(CN)_3$ is a model system for the $S = 1/2$ antiferromagnetic triangular-lattice Heisenberg model [60]. $EtMe_3Sb[Pd(dmit)_2]_2$ is a Mott insulator that forms a scalene-triangular system with $S = 1/2$ [61].

## 5.11 Ordering in Kagome Antiferromagnets

### *5.11.1 Kagome Lattice*

The kagome lattice is depicted in Fig. 5.20a. The kagome lattice with bond length $a$ consists of triangles forming a triangular lattice with bond length $2a$, as shown by the blue circles. We may take the upward triangle indicated by the blue circles as a unit cell of the kagome lattice. However, the magnetic structure is highly degenerate, and the magnetic unit cell is generally not uniquely defined.

In the kagome lattice, triangles share a vertex, forming a so-called corner-sharing structure, whereas in the triangular lattice, triangles share an edge, referred to as the edge-sharing structure. In the corner-sharing structure, frustration causes macroscopic degeneracy even in systems with continuous spins, unlike the case of lattices with edge-sharing geometry.

As discussed in previous sections, in systems with continuous spins, frustration leads to non-collinear structures, such as the 120° configuration, which resolve the degeneracy seen in the Ising model. In the triangular lattice, once the spin configuration in a single triangle is fixed, all spin configurations are uniquely determined, i.e., all spins on the same sublattice align.

In contrast, in the kagome lattice, even if the spin configuration in one triangle is fixed, there remains freedom in choosing the directions of the spins at two sites

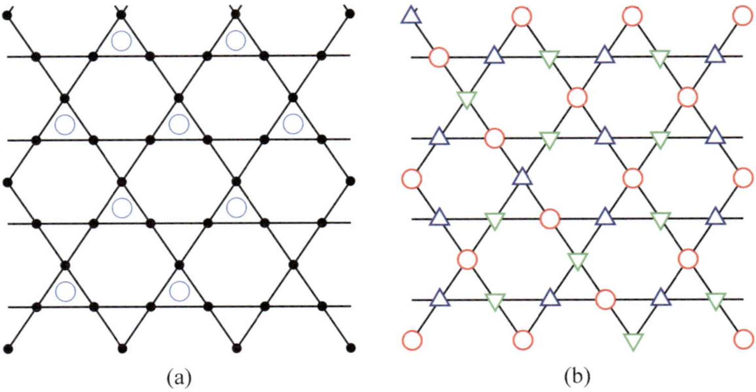


**Fig. 5.20** **a** Kagome lattice. The circle indicates a unit cell of the lattice. **b** One of the ground-state configurations. In (**b**), the circle, upward triangle, and downward triangle each represent a different spin direction forming a 120° structure

of the neighboring triangles. As a result, macroscopic degeneracy persists even in continuous spin systems such as XY and Heisenberg models. An example is shown in Fig. 5.20b, where each triangle exhibits a 120° structure, but their overall arrangement is non-uniform.

The nature of the degeneracy depends on the number of spin components, $n$, where $n = 1, 2, 3$ correspond to the Ising, XY, and Heisenberg models, respectively [62].[7]

Although many different spin arrangements are possible, particular uniform configurations, such as those shown in Fig. 5.21, have attracted attention. One possible configuration is depicted in Fig. 5.21a, where all unit triangles share the same configuration. This is known as the $q = 0$ structure. Another possibility is the $\sqrt{3} \times \sqrt{3}$ structure, depicted in Fig. 5.21b.

In the case of XY spins ($J \cos(\theta_i - \theta_j)$), the spins lie in a plane. The degeneracy thus arises from different ways to orient the three directions. The ground-state degeneracy corresponds to that of the 3-state Potts antiferromagnetic model, which has a zero-temperature entropy $S/N \simeq 0.12k_{\rm B}$ [63]. Huse and Rutenberg found that the spin correlations in the $\sqrt{3} \times \sqrt{3}$ state decay algebraically with exponent $\eta = 4/3$ [62].

For Heisenberg spins ($J\boldsymbol{S}_i \cdot \boldsymbol{S}_j$), the spin plane of the 120° structure is not restricted to the XY plane; it can lie in any direction. Thus, it has been suggested that the degeneracy may be lifted via the order-by-disorder mechanism.

Let us consider the configuration shown in Fig. 5.21b, where the spin direction represented by circles corresponds to the $z$ direction, i.e., $\theta_1 = 0$, and the spins denoted by upward and downward triangles are ($\theta_2 = 2\pi/3, \phi$) and ($\theta_3 = 4\pi/3, \phi$),

[7] In the Ising model ($\sigma = \pm 1$), even within a single triangle, the ground state is degenerate. For example, the three spins can be either $(+, +, -)$ or $(+, -, -)$. If one spin is fixed to $+$, the other two can be either $+$ or $-$. The exact number of ground states has been derived, as discussed in Sect. 4.1.3.

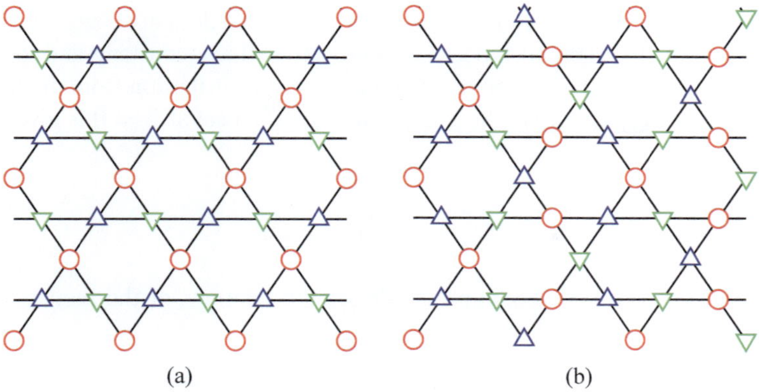

Fig. 5.21 Ground state configurations: **a** $q = 0$ structure, **b** $\sqrt{3} \times \sqrt{3}$ structure

respectively. The spins with angle $\theta_1 = 0$ surround a loop in which the remaining spins alternate between $(\theta_2, \phi)$ and $(\theta_3, \phi)$. Within each loop, the angle $\phi$ must be uniform. Such loops are called "weathervane loops". The angle $\phi$ can be chosen independently for each loop, contributing to the ground-state degeneracy. The number of weathervane loops is maximized in the $\sqrt{3} \times \sqrt{3}$ structure. Thus, this structure is expected to be favored by the order-by-disorder mechanism.

Huse and Rutenberg numerically showed that the $\sqrt{3} \times \sqrt{3}$ structure is preferred [62]. Several theoretical proposals have been made to explain the selection of this ground state via entropy, i.e., the order-by-disorder mechanism [64, 65]. Additionally, the possibility of more complex states, such as the "nine-sublattice coplanar Néel-like ordering," has been discussed [65]. However, no definitive conclusion has been reached thus far.

### 5.11.2 Effect of the Flat Band on the Specific Heat

The kagome antiferromagnet exhibits a peculiar dispersion relation for the harmonic modes of small fluctuations around the ground state. It has been found that a dispersionless mode, known as a flat band, exists.

The Heisenberg kagome model contains three spins per unit cell, each with two degrees of freedom $(\theta_i, \phi_i)$ for $i = 1, 2, 3$, resulting in a total of six degrees of freedom. According to the equipartition theorem, each quadratic mode contributes $(1/2)k_B$ to the specific heat. Therefore, one might naively expect the specific heat per site to be

$$C = \frac{6 \times (1/2)k_B}{3} = k_B.$$

However, in the kagome antiferromagnet, the flat band does not contribute fully to the energy. Only five of the six modes have non-zero stiffness constants and contribute $(5/2)k_B$ per unit cell. It has been pointed out that the contribution from the flat band amounts to only $(1/4)k_B$ [64]. As a result, the specific heat at $T = 0$ is given by

$$C = \frac{5}{6}k_B + \frac{1}{4}k_B = \frac{11}{12}k_B.$$

In the next section, we examine the dispersion relations of the modes in more detail.

#### 5.11.2.1 Eigenenergies of Kagome Antiferromagnetic Heisenberg Model in the Harmonic Analysis

To obtain eigenenergies in the harmonic analysis on kagome antiferromagnetic Heisenberg model ($A = 1$ and $H = 0$), we adopt the $q = 0$ structure as an representative. Here, we consider the coplanar structure including the $z$-axis without losing generality.

The kagome lattice consists of triangles, for which we obtained the ground state minimizing the energy (5.6). The ground-state energy of each triangle is $-3J/2$. Around the ground-state configuration,

$$(\theta_i, \phi_i) = (\theta_i^0, \phi_i^0), \quad (i = 1, 2, 3), \tag{5.236}$$

we consider small deviations:

$$\theta_i = \theta_i^0 + \delta\theta_i, \quad \phi_i = \phi_i^0 + \delta\phi_i, \quad (i = 1, 2, 3). \tag{5.237}$$

In the second order of deviations ($\delta^2$), the energy is given by

$$\delta E_2 = (x_1, x_2, x_3, y_1, y_2, y_3) \begin{pmatrix} A_{11} & A_{12} & A_{13} & 0 & 0 & 0 \\ A_{21} & A_{22} & A_{23} & 0 & 0 & 0 \\ A_{31} & A_{32} & A_{33} & 0 & 0 & 0 \\ 0 & 0 & 0 & C_{11} & C_{12} & C_{13} \\ 0 & 0 & 0 & C_{21} & C_{22} & C_{23} \\ 0 & 0 & 0 & C_{31} & C_{32} & C_{33} \end{pmatrix} \begin{pmatrix} x_1 \\ x_2 \\ x_3 \\ y_1 \\ y_2 \\ y_3 \end{pmatrix} \equiv {}^t\boldsymbol{v} E_2 \boldsymbol{v}, \tag{5.238}$$

with the variables: $x_i = d\theta_i$ and $y_i = \sin\theta_i^0 d\phi_i$ as we did in the case of the triangular lattice.

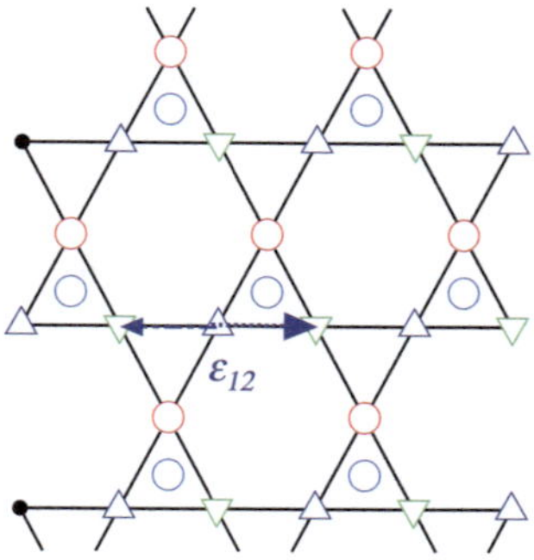


**Fig. 5.22** Vectors connecting unit cells for the $q = 0$ structure

#### 5.11.2.2 Dispersion Relation

To study the specific heat at low temperatures, we obtain the dispersion relation of the harmonic model (5.238) for the $q = 0$ structure. The lattice has translational symmetry, and the modes are classified by the wave vector $\boldsymbol{k}$.

With similar analysis to the case of AFT models, we obtain the Hamiltonian in the form:

$$\mathcal{H}_{\mathrm{HA}} = E_0 + \sum_{\boldsymbol{k}}^{t} \boldsymbol{x}_{\boldsymbol{k}} A_{\boldsymbol{k}} \boldsymbol{x}_{-\boldsymbol{k}} + \sum_{\boldsymbol{k}}^{t} \boldsymbol{y}_{\boldsymbol{k}} C_{\boldsymbol{k}} \boldsymbol{y}_{-\boldsymbol{k}}, \tag{5.239}$$

where

$$A_{\boldsymbol{k}} = \begin{pmatrix} A_{11} & A_{12}\varepsilon_{12}(\boldsymbol{k}) & A_{13}\varepsilon_{13}(\boldsymbol{k}) \\ A_{21}\varepsilon_{21}(\boldsymbol{k}) & A_{22} & A_{23}\varepsilon_{23}(\boldsymbol{k}) \\ A_{31}\varepsilon_{31}(\boldsymbol{k}) & A_{32}\varepsilon_{32}(\boldsymbol{k}) & A_{33} \end{pmatrix}, \quad C_{\boldsymbol{k}} = \begin{pmatrix} C_{11} & C_{12}\varepsilon_{12}(\boldsymbol{k}) & C_{13}\varepsilon_{13}(\boldsymbol{k}) \\ C_{21}\varepsilon_{21}(\boldsymbol{k}) & C_{22} & C_{23}\varepsilon_{23}(\boldsymbol{k}) \\ C_{31}\varepsilon_{31}(\boldsymbol{k}) & C_{32}\varepsilon_{32}(\boldsymbol{k}) & C_{33} \end{pmatrix}. \tag{5.240}$$

Taking into account the locations of unit cells (Fig. 5.22), we have

$$\varepsilon_{12}(\boldsymbol{k}) = e^{2iak_x} + e^{-i2ak_x} = \varepsilon_{21}(\boldsymbol{k}), \tag{5.241}$$

$$\varepsilon_{23}(\boldsymbol{k}) = e^{-iak_x + \sqrt{3}ik_y} + e^{-(iak_x - \sqrt{3}ik_y)} = \varepsilon_{32}(\boldsymbol{k}), \tag{5.242}$$

$$\varepsilon_{31}(\boldsymbol{k}) = e^{-iak_x - \sqrt{3}ik_y} + e^{iak_x + \sqrt{3}ik_y} = \varepsilon_{13}(\boldsymbol{k}). \tag{5.243}$$

The eigenvalues of $A_{\boldsymbol{k}}$ and $C_{\boldsymbol{k}}$, $\lambda_i$ ($i = 1, 2, \cdots, 6$), as functions of $\boldsymbol{k}$, give the stiffness constants for the modes. In Fig. 5.23, the $k_x$ dependencies of $\lambda_i$ are depicted, where we find that one of them remains zero, forming a flat band.

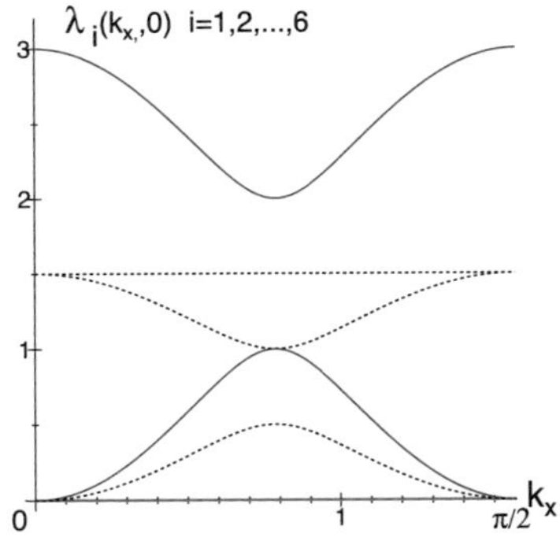


**Fig. 5.23** $k_x$ dependencies of $\lambda_i$. The dotted curves show the eigenvalues of $A_{\boldsymbol{k}}$, and the solid curves show those of $C_{\boldsymbol{k}}$.The red line gives the flat band

#### 5.11.2.3 The Specific Heat

The partition function is given by

$$Z = \int_{-\infty}^{\infty} dx_1 \int_{-\infty}^{\infty} dy_1 \int_{-\infty}^{\infty} dx_2 \int_{-\infty}^{\infty} dy_2 \int_{-\infty}^{\infty} dx_3 \int_{-\infty}^{\infty} dy_3 e^{-\beta E_0 - \beta\left(\sum_{i=1}^{6}\sum_{j=1}^{6}(E_2)_{ij} v_i v_j\right)}. \tag{5.244}$$

Unless the eigenvalues are zero, the integrand is given by the eigenvalues of $E_2$ ($\lambda_i, i = 1, \cdots, 6$) as

$$e^{-\beta E_0 - \beta\left(\sum_{i=1}^{6}\sum_{j=1}^{6}(E_2)_{ij} v_i v_j\right)} = e^{-\beta E_0 - \beta\left(\sum_{i=1}^{6}\lambda_i \xi_i^2\right)}, \tag{5.245}$$

where $-\infty < \xi_i < \infty$. Each mode gives a contribution:

$$\int_{-\infty}^{\infty} d\xi e^{-\beta\lambda\xi^2} \propto (k_B T)^{-1/2}. \tag{5.246}$$

If some of $\lambda_i$ are zero, the integral diverges in the harmonic analysis. However, when we take into account higher order terms, for example, with a positive $c$,

$$\int_{-\infty}^{\infty} d\xi e^{-\beta c\xi^4} \propto (k_B T)^{-1/4}. \tag{5.247}$$

Thus, the partition function is given by

$$Z \propto ((k_B T)^{-1/2})^5 (k_B T)^{-1/4} = (k_B T)^{-11/4})^{N/3} = \beta^{11N/12}, \tag{5.248}$$

and the energy is given by

$$E = -\frac{\partial}{\partial \beta} \ln Z = \frac{11}{12} k_B T. \tag{5.249}$$

Therefore, the specific heat per spin at low temperature is

$$\frac{C}{N} = \frac{11}{12} k_B. \tag{5.250}$$

### 5.11.3 *Phase Transition at a Finite Temperature Exotic Ferromagnetic State*

So far, we have studied the antiferromagnetic kagome model for the Ising model, XY model, and Heisenberg model, where the highly degenerate ground state due to frustration prevents the models from ordering at finite temperatures. However, it has been pointed out that a phase transition occurs at a finite temperature when the system has uniaxial anisotropic interaction [66–68], i.e.,

$$\begin{aligned} \mathcal{H} = & \sin\theta_1 \sin\theta_2 \cos(\phi_1 - \phi_2) + A\cos\theta_1 \cos\theta_2 \\ & + \sin\theta_2 \sin\theta_3 \cos(\phi_2 - \phi_3) + A\cos\theta_2 \cos\theta_3 \\ & + \sin\theta_3 \sin\theta_1 \cos(\phi_3 - \phi_1) + A\cos\theta_3 \cos\theta_1, \end{aligned} \tag{5.251}$$

with $A > 1$.

As we have studied in the section on the triangular lattice, the basic unit is a triangle that has three spins whose configuration is

$$c_1 = 1,\ s_2 = -s_3,\ c_2 = c_3 = -c, \quad \text{or} \quad c_1 = -1,\ s_2 = -s_3,\ c_2 = c_3 = c, \tag{5.252}$$

where $c = A/(A+1)$. There are many arrangements of these three spins on triangles on the kagome lattice, as depicted in Fig. 5.20b. In the case of $A = 1$, the spin configuration $(1, -c, -c)$ can be transformed into $(-1, c, c)$ without encountering an energy barrier. In contrast, when $A > 1$, a domain wall exists between the Y-shape structure and its upside-down counterpart (the $-$Y shape). Therefore, the $+$Y and $-$Y configurations are not compatible, and the ordered states are classified into two distinct sets.

Configurations consisting of $(-1, c, c)$ exhibit a positive magnetization $m = A/(A+1)$, while those with $(1, -c, -c)$ exhibit $m = -A/(A+1)$. This classification causes a two-fold discrete degeneracy of the $Z_2$ type. This $Z_2$ symmetry provides an Ising-like phase transition at a finite temperature, although the low-temperature phase is still macroscopically degenerate. The system exhibits symmetry breaking of the uniform magnetization:

$$M = \sum_{i=1}^{N} \cos\theta_i. \tag{5.253}$$

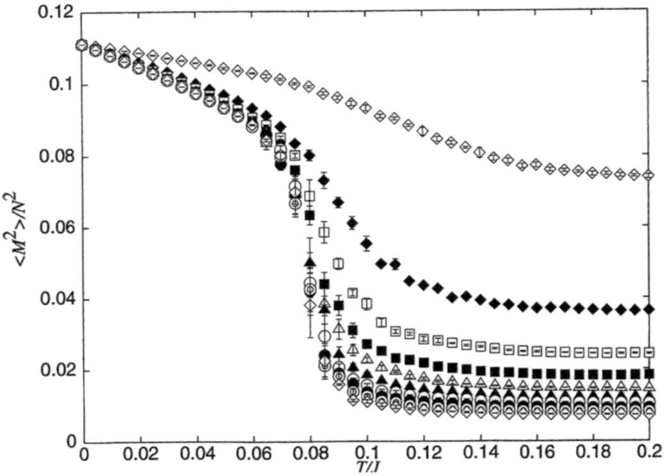


**Fig. 5.24** Size dependence of $\langle M^2\rangle/N^2$ for $A = 2$. Symbols: open diamond, closed diamond, open square, closed square, open triangle, closed triangle, open circle, closed circle, double circle, and open diamond (bottom) represent data for $L = 6, 12, 18, 24, 30, 36, 42, 48, 54$, and 60, respectively. (This figure is reproduced from Ref. [66]. ©1995 The Physical Society of Japan)

The size dependence of $\langle M^2\rangle/N^2$ for $A = 2$ is depicted in Fig. 5.24. This ordered state does not exhibit any special spatial pattern, and thus it is not a type of ferrimagnetic state. Because the average magnetization is uniform, we may regard it as a type of ferromagnetic state despite its local structure of $(\pm 1, \mp c, \mp c)$ and the high degeneracy of each ordered state. In this aspect, it is called "exotic ferromagnetic state".

Magnetic transitions in the kagome material $NH_4Fe_3(OH)_6(SO_4)_2$ [69], potassium jarosite $KFe_3(OH)_6(SO_4)_2$ [70, 71], and $Rb_2M_3S_4$ [72] were observed, which may be caused by the present mechanism.

This macroscopically degenerate ferromagnetic state with non-zero magnetization appears not only in the present Ising-like Heisenberg model but also in the 1/3 plateau in the magnetization process $M(H)$ of Ising, XY, and Heisenberg kagome antiferromagnetic models, where one of the 3-fold degenerate uud configurations appears at each triangle [73].

This exotic ferromagnetic order exhibits a slow relaxation of spin structure due to an entropy-driven ordering process by the reconnection of weathervane loops [68, 74]. Slow relaxation processes were observed in the two-dimensional kagome antiferromagnet, jarosites $(D_3O)Fe_3(OH)_6(SO_4)_2$ [75] by means of ac and dc magnetic measurements. Bekhechi and Southern also found slow dynamics and discussed it in the context of glassy dynamics [67].

Experimental approaches will be reviewed in Sect. 6.4.1.

## References


1. A. Yoshimori, J. Phys. Soc. Jpn. **14**, 807 (1959)
2. J. Villian, R. Bidaux, J.P. Carton, R.J. Conte, J. Phys. (Paris) **41**, 1263 (1980)
3. C.L. Henley: J. Appl. Phys. 61, 3962 (1987). FCC. C. L. Henley: Phys. Rev. Lett. **62**, 2056 (1989)
4. S. Miyashita, H. Shiba, J. Phys. Soc. Jpn. **53**, 1145 (1984)
5. H. Kawamura, J. Phys. Soc. Jpn. **53**, 2452 (1984)
6. S. Miyashita, H. Kawamura, J. Phys. Soc. Jpn. **54**, 3385 (1985)
7. S. Fujiki, K. Shutoh,Y. Abe, S. Katsura, J. Phys. Soc. Jpn. **52**, 1531 (1983). H. Takayama, K. Matsumoto,H. Kawahara, K. Wada, J. Phys. Soc. Jpn. **52**, 2888 (1983). D.P. Landau: Phys. Rev. B **27**, 5604 (1983). S. Miyashita, H. Kitatani, Y. Kanada: J. Phys. Soc. Jpn. **60**, 1523 (1991). M. Nishino, S. Miyashita, Phys. Rev. B **94**, 184434 (2016)
8. J.M. Kosterlitz: J. Phys. C7 (1974) 1046. J.M. Kosterlitz, D.J. Thouless: J. Phys. C6 (1973) 1181. V. L. Berezinskii, Sov. Phys. JETP **32**, 493(1971), ibid. **34**, 610 (1972)
9. J.V. Jose, L.P. Kadanoff, S. Kirkpatrick, D. Nelson, Phys. Rev. B **16**, 1217 (1977)
10. N.D. Mermin, H. Wagner, Phys. Rev. Lett., **17**, 1133 (1966), P.C. Hohenberg, Phys. Rev. **158**, 383 (1967)
11. J. Froehlich, B. Simon, T. Spencer: Phys. Rev. Lett. **36** , 804 (1976). F.J. Dyson, E.H. Lieb, B. Simon, J. Stat. Phys. **18** 804 (1978). K. Kubo, T. Kishi, Phys. Rev. Lett. 61, 2585 (1988)
12. S. Watarai, S. Miyashita, J. Phys. Soc. Jpn. **63**, 1916-1927 (1994). S. Watarai, S. Miyashita, J. Phys. Soc. Jpn. **67**, 816-824 (1998)
13. Y. Ozeki, N. Ito, Phys. Rev. B **68**, 054414
14. Y. Togawa et al., J. Phys. Soc. Jpn. **90**, 014702 (2021)
15. H. Kawamura, S. Miyashita, J. Phys. Soc. Jpn., 53, 4138 (1984). H. Kawamura, S. Miyashita, J. Phys. Soc. Jpn., 54, 4138 (1985)
16. H. Kawamura, J. Phys. Soc. Jpn. **54**, 3220 (1985), **55**, 2095 (1986), **55**, 2175 (1986), **56**, 474 (1987)
17. H. Kawamura, Phys. Rev. B38, 4916 (1988). H. Kawamura, J. Phys. Condens. Matter **10**, 4707 (1998)
18. Y. Ajiro et al., J. Phys. Soc. Jpn. **62**, 2268 (1988)
19. H. Kawamura, A. Yamamoto, T. Okubo, J. Phys. Soc. Jpn. **79**, 023701 (2010)
20. T. Okubo, H. Kawamura, J. Phys. Soc. Jpn. **79**, 084706 (2010)
21. H. Kawamura, A. Yamamoto, J. Phys. Soc. Jpn. **76**, 073704 (2007)
22. H. Yamaguchi, S. Kimura, M. Hagiwara, Y. Nambu, S. Nakatsuji, Y. Maeno, A. Matsuo, K. Kindo, J. Phys. Soc. Jpn. **79**, 054710 (2010)
23. K. Aoyama, H. Kawamura, Phys. Rev. Lett. **124**, 047202 (2020)
24. K. Tomiyasu et al., Phys. Rev. B **106**, 054407 (2022)
25. Y.P. Mizuta, K. Aoyama, K. Tomiyasu, M. Matsuura, H. Kawamura, J. Phys. Soc. Jpn. **91**, 035001 (2022)
26. Y. Ajiro, T. Nakashima, Y. Unno, H. Kadowaki, M. Mekata, N. Achiwa, J. Phys. Soc. Jpn. **57**, 2648 (1988)
27. M. Itakura, J. Phys. Soc. Jpn. **72**, 74 (2003)
28. A. Pelissetto, P. Rossi, E. Vicari, Phys. Rev. B **63**, 140414(R) (2001). P. Calabrese, P. Parruccini, A. Pelissetto, E. Vicari, Phys. Rev. B **70**, 174439 (2004)
29. Y. Nagano, K. Uematsu, H. Kawamura, Phys. Rev. B **100**, 224430 (2019)
30. S. Miyashita, J. Phys. Soc. Jpn. **55**, 3605–3617 (1986)
31. T. Inami, Y. Ajiro, T. Goto, J. Phys. Soc. Jpn. **65**, 2374 (1996)
32. T. Inami, N. Terada, H. Kitazawa, O. Sakai, J. Phys. Soc. Jpn., **78**, 084713 (2009). H. Kitazawa, H. Suzuki, H. Abe, J. Tang, G. Kido, Physica B **259-261** 890 (1999)
33. H. Mitamura, R. Watanuki, K. Kaneko, N. Onozaki, Y. Amou, S. Kittaka, R. Kobayashi, Y. Shimura, I. Yamamoto, K. Suzuki, S. Chi, T. Sakakibara, Phys. Rev. Lett. **113**, 147202 (2014)
34. D. Sellmann, X.-F. Zhang, S. Eggert, Phys. Rev. B **91**, 081104(R) (2015)

35. S. Watarai, S. Miyashita, H. Shiba, J. Phys. Soc. Jpn. **70**, 532–537 (2001)
36. T. Nikuni, H. Shiba, J. Phys. Soc. Jpn. **62**, 3268 (1993), J. Phys. Soc. Jpn. **64**, 3471 (1995), A. E. Jacobs, T. Nikuni, H. Shiba, J. Phys. Soc. Jpn. **62**, 4066 (1993)
37. S. Yoshikawa, K. Okunishi, M. Senda, S. Miyashita, J. Phys. Soc. Jpn. **73**, 1798 (2004)
38. C. Lacroix, P. Mendels, F. Mila, *Introduction to Frustrated Magnetism: Materials, Experiments* (Springer, Theory, 2011)
39. Papers in the SPECIAL TOPICS, Novel States of Matter Induced by Frustration. J. Phys. Soc. Jpn. **79**(1), 011001–011013 (2010)
40. N. Achiwa, J. Phys. Soc. Jpn. **27**, 561 (1969)
41. K. Adachi, N. Achiwa, M. Mekata, J. Phys. Soc. Jpn. **49**, 545 (1980)
42. H. Hyodo, K. Iio, K. Nagata, J. Phys. Soc. Jpn. **50**, 1545 (1981)
43. Y. Tazuke, H. Tanaka, K. Iio, K. Nagata, J. Phys. Soc. Jpn. **50**, 3919 (1981)
44. W. Palme, F. Mertens, O. Born, B. Lüthi, U. Schotte, Solid State Commun. **76**, 1873 (1990)
45. H. Tanaka, U. Schotte, K. Schotte, J. Phys. Soc. Jpn. **61**, 1344 (1992)
46. M. Mino, K. Ubukata, T. Bokui, M. Arai, H. Tanaka, M. Motokawa, Physica B **201**, 213 (1994)
47. H. Ohta, S. Imagawa, M. Motokawa, H. Tanaka, J. Phys. Soc. Jpn. **62**, 3011 (1993)
48. M. Chiba, O. Ohara, Y. Ajiro, T. Morimoto, J. Phys. Soc. Jpn. **62**, 4186 (1993)
49. U. Schotte, N. Stüsser, K.D. Schotte, H. Weinfureter, H.M. Mayer, M. Winkelmann, J. Phys. Conddens Matter. **6**, 10105 (1994)
50. H. Nojiri, Y. Tokunaga, M. Motokawa, J. Phys. (Paris) 49, Suppl. **C8**, 1459 (1988)
51. M. Motokawa, A. Arai, H. Ohta, M. Mono, H. Tanaka, K. Ubukata, Physica B **211**, 199 (1995)
52. H.B. Weber, T. Werner, J. Wosnitza, H. v. Löhneysen: Phys. Rev. B 54 (1996) 15924; T. Werner, H.B. Weber, J. Wosnitza, A. Kelnberger, M. Meschke, U. Schotte, N. Stüsser, Y. Ding, M. Winkelmann: Solid State Commun. **102**, 609. (1997)
53. H. Suematsu, K. Ohmatsu, K. Sugiyama, T. Sakakibara, M. Motokawa, M. Date, Solid State Commun. **40**, 241 (1981)
54. T. Kimura, T. Goto, H. Shintani, K. Ishizaka, T. Arima, Y. Tokura, Nature (London) **426**, 55 (2003)
55. K. Hirakawa, H. Kadowaki, K. Ubukoshi, J. Phys. Soc. Jpn. **52** (1983) 1814. I. Yamada, K. Ubukoshi, K. Hirakawa, J. Phys. Soc. Jpn., **53**, 381 (1984)
56. K. Hirakawa, H. Kadowaki, K. Ubukoshi, J. Phys. Soc. Jpn. **54** (1985) 3526. P. Kalyani, N. Kalaiselvi 2005 Sci. Technol. Adv. Mater. **6** 689 (2005). K. Yamaura, M. Takano, A. Hirano, R. Kanno, J. Solid State Chem. **127** (1997) 109. T. Arimori, S. Miyashita, J. Phys. Soc. Jpn. **69**, 2250 (2000)
57. S. Nakatsuji, Y. Nambu, H. Tonomura, O. Sakai, S. Jonas, C. Broholm, H. Tsunetsugu, Y. Qiu, Y. Maeno, Science **309**, 1697 (2005)
58. H. Takeya, K. Ishida, K. Kitagawa, Y. Ihara, K. Onuma, Y. Maeno, Y. Nambu, S. Nakatsuji, D.E. MacLaughlin, A. Koda, R. Kadono, Phys. Rev. B **77** (2008) 054429. A. Yaouanc, P. de Reotier, Y. Chapuis, C. Marin, G. Lapertot, A. Cervellino, A. Amato, Phys. Rev. B **77** (2008) 092403. D.E. MacLaughlin, Y. Nambu, S. Nakatsuji, R.H. Heffner, L. Shu, O.O. Bernal, K. Ishida: Phys. Rev. B **78**, (2008) 220403. H. Yamaguchi, S. Kimura, M. Hagiwara, Y. Nambu, S. Nakatsuji, Y. Maeno, K. Kindo: Phys. Rev. B **78** (2008) 180404
59. A. Olariu, P. Mendelse, F. Bert, B.G. Ueland, P. Schiffer, R.F. Berger, R.J. Cava: Phys. Rev. Lett. **97**, (2006) 167203. D. Hsieh, D. Qian, R.F. Berger, R.J. Cava, J.W. Lynn, Q. Huang, M.Z. Hasan, Physica B **403**, (2008) 1341
60. Y. Shimizu, K. Miyagawa, K. Kanoda, M. Maesato, G. Saito: Phys. Rev. Lett. **91**, (2003) 107001. S. Yamashita, Y. Nakazawa, M. Oguni, Y. Oshima, H. Nojiri, Y. Shimizu, K. Miyagawa, K. Kanoda: Nat. Phys. **4**, (2008) 459
61. T. Itou, A. Oyamada, S. Maegawa, M. Tamura, R. Kato, Phys. Rev. B **77**, 104413 (2008)
62. A. Huse, A.D. Rutenberg, Phys. Rev. B **45**, 7536 (1992)
63. R.J. Bexter, J. Math. Phys. **11**, 784 (1970)
64. J.T. Chalker, P.C.W. Holdsmorth, E.F. Shender, Phys. Rev. Lett. **68**, 855 (1992)
65. J.N. Reimersand A.J. Berlinski, Phys. Rev. B **48**, 9539 (1993)
66. A. Kuroda, S. Miyashita, J. Phys. Soc. Jpn. **64**, 4509 (1995)

67. S. Bekhechi, B.W. Southern, Phys. Rev. B **67**, 144403 (2003)
68. S. Tanaka, S. Miyashita, J. Phys. Soc. Jpn. **76**, 103001 (2007)
69. S. Maegawa, M. Nishiyama, N. Tanaka, A. Oyamada, M. Takano, J. Phys. Soc. Jpn. **65**, 2776 (1996)
70. T. Inami, M. Nishiyama, S. Maegawa, Y. Oka, Phys. Rev. B **61**, 12181 (2000)
71. M. Nishiyama, S. Maegawa, T. Inami, Y. Oka, Phys. Rev. B **67**, 224435 (2003)
72. W. Bronger, R. Rennau, D. Schmitz, Z. Allg. Chem. **597**, 27 (1991), ibid. **622**, 627 (1996). T. Kato, K. Hondou, K. Iio, J. Magn. Magn. Mater. **177-181**, 591 (1998). T. Fukamachi, Y. Kobayashi, A. Nakamura, H. Harashina, M. Sato, J. Phys. Soc. Jpn. **68**, 3668 (1999). K. Hondou, Y. Fujiwara, T. Kato, K. Iio, A. Saiki, M. Usuda, N. Hamada, J. Alloys Compounds **333**, 274 (2002). T. Hasegawa, M. Inui, K. Hondou, Y. Fujiwara, T. Kato, K. Iio, J. Alloys Compounds **364** 199 (2004)
73. M. Udagawa, M. Ogata, Z. Hiroi, J. Phys. Soc. Jpn. **71**, 2365 (2002)
74. S. Tanaka, S. Miyashita, J. Phys.: Condens. Matter **19**, 145256 (2007)
75. A.S. Wills, V. Dupuis, E. Vincent, J. Hammann, R. Calemczuk, Phys. Rev. B **62**, R9264 (2000). F. Ladieu, F. Bert, V. Dupuis, E. Vincent, J. Hammann, J. Phys.: Condens. Matter **16**, S735 (2004)

# Chapter 6
# Quantum Effects

**Abstract** As we have studied in previous chapters, various configurations are energetically degenerate or nearly degenerate in frustrated systems. This leads to a rich variety of phases. In such systems, thermal fluctuations (entropy effects) play an important role in selecting the thermodynamically stable state, a phenomenon known as "order-by-disorder". At zero temperature, quantum fluctuations play a crucial role in determining the ground state, and quantum phase transitions can occur as a result. Quantum fluctuations in spin systems originate from the non-commutativity of spin operators. These fluctuations reduce the magnitude of the classical order parameter defined by spin orientation, while lowering the ground-state energy. These behaviors are quite counterintuitive but are essential for understanding the ground states of frustrated quantum systems. Degeneracy in classical ground states may be resolved quantum mechanically; that is, some linear combination of the degenerate classical states can become the true ground state. Moreover, even if a state's classical energy is higher, the energy gain due to quantum fluctuations can lower its total energy below that of the classical ground state. Such a reduction in ground-state energy due to quantum fluctuations gives rise to an intriguing phenomenon known as "order-by-quantum fluctuation" (Nikuni and Shiba in J Phys Soc Jpn 62:3268 (1993) [1], Chubukov in Phys Rev Lett 69:832 (1992) [2]) . In this chapter, we briefly overview quantum effects on ordered states and review unusual quantum phase transitions in frustrated quantum systems.

## 6.1 Quantum Fluctuation

First, let us review the fundamental effects of quantum fluctuation. Generally, non-commutativity among the spin components $S^x$, $S^y$, and $S^z$:

$$[S_x, S_y] = \hbar S_z, \quad [S_y, S_z] = \hbar S_x, \quad [S_z, S_x] = \hbar S_y, \tag{6.1}$$

causes a change in the properties of the ground state when the classical order parameter does not commute with the Hamiltonian, such as the Néel order in antiferromagnets.

© The Author(s), under exclusive license to Springer Nature Singapore Pte Ltd. 2025

S. Miyashita, *Physics of Entropy*, Springer Series in Solid-State Sciences 212,
https://doi.org/10.1007/978-981-95-1758-9_6

As a typical example, let us consider a spin pair of $S = 1/2$ coupled by the exchange constant $J$:

$$\mathcal{H} = -J\boldsymbol{S}_1 \cdot \boldsymbol{S}_2. \tag{6.2}$$

In this model, the eigenstates are the singlet state:

$$\mathcal{H}|\text{singlet}\rangle = \frac{3}{4}J|\text{singlet}\rangle, \quad |\text{singlet}\rangle = \frac{|+\rangle_1|-\rangle_2 - |-\rangle_1|+\rangle_2}{\sqrt{2}}, \tag{6.3}$$

and the triplet state:

$$\mathcal{H}|\text{triplet}\rangle = \frac{1}{4}J|\text{triplet}\rangle, \quad |\text{triplet}\rangle = \begin{cases} |\text{triplet}_{+1}\rangle = |+\rangle_1|+\rangle_2 \\ |\text{triplet}_0\rangle = \frac{|+\rangle_1|-\rangle_2+|-\rangle_1|+\rangle_2}{\sqrt{2}} \\ |\text{triplet}_{-1}\rangle = |-\rangle_1|-\rangle_2 \end{cases}. \tag{6.4}$$

For the antiferromagnetic case ($J < 0$), the ground state of the classical system is the Néel state:

$$|\text{Neel}\rangle = |+\rangle_1|-\rangle_2, \text{ or } |-\rangle_1|+\rangle_2, \tag{6.5}$$

which is not an eigenstate of the Hamiltonian:

$$\mathcal{H}|+\rangle_1|-\rangle_2 = \frac{1}{2}J|-\rangle_1|+\rangle_2 - \frac{1}{4}J|+\rangle_1|-\rangle_2, \tag{6.6}$$

and the energy of the Néel state is given by

$$\langle\text{Neel}|\mathcal{H}|\text{Neel}\rangle = \frac{1}{4}J = -\frac{1}{4}|J|. \tag{6.7}$$

In the quantum system, the ground state is the singlet state (6.3). It should be noted that the singlet state has much lower energy ($-3|J|/4$) compared to that of the Néel state ($-|J|/4$), i.e., three times lower in the present case. This is a typical example where the ground-state energy is much lowered by quantum fluctuation.

In the present model of two spins, the Néel order parameter $\langle G|S_1^z S_2^z|G\rangle = 1/4$ in both classical and quantum cases. However, as we will see below, in many cases, the order parameter which does not commute with the Hamiltonian is reduced. This phenomenon is called spin reduction due to quantum fluctuation. In this state, the energy is lowered, although the order parameter is reduced.

### 6.1.1 *Spin-Peierls Transition*

This large energy gain causes the Spin-Peierls transition in a one-dimensional antiferromagnetic Heisenberg chain when the coupling constant can change linearly with lattice deformation [3]. Let us consider the case where a uniform lattice with a lattice constant $a$ distorts by $\pm\delta$, where the lengths of neighboring bonds change as $(a, a) \to (a - \delta, a + \delta)$. Assuming the elastic energy of the bonds increases proportionally to $\delta^2$ and the coupling constant changes as a function of the distance as $J(\delta) = J(0) - \alpha\delta$ (where $\alpha$ is a constant), it is known that the magnetic energy gain in the dimerized state is proportional to $\delta^{4/3}$ in one dimension, which is larger than the elastic energy loss ($\propto \delta^2$). Thus, an arbitrarily small spin-phonon coupling causes spontaneous lattice dimerization in the ground state of the one-dimensional chain, and the ground state consists of sets of singlet pairs. When the lattice motion is treated quantum mechanically, interesting phenomena such as quantum narrowing effect has been found [4].

Experimentally, several compounds showing the Spin-Peierls transition have been found, e.g., TTF-CuBDT [5] and $CuGeO_3$ [6]. It has also been pointed out that doping impurities suppress the Spin-Peierls state and induce antiferromagnetic order [7]. Several phases appear according to the concentration of the impurity, and the coexistence of Spin-Peierls and antiferromagnetic states has been observed [8]. Edge states and doping impurity magnetic structures have been studied extensively [9–11].

### 6.1.2 Reduction of Classical Order Parameters in Regular Lattices

Without lattice deformation, quantum fluctuation still gives significant effects. First, let us study a Heisenberg model on a $2 \times 2$ square lattice (Fig. 6.1a).

$$\mathcal{H} = -J\left(\boldsymbol{S}_1 \cdot \boldsymbol{S}_2 + \boldsymbol{S}_2 \cdot \boldsymbol{S}_3 + \boldsymbol{S}_3 \cdot \boldsymbol{S}_4 + \boldsymbol{S}_4 \cdot \boldsymbol{S}_1\right), \quad |\boldsymbol{S}_i| = \frac{1}{2}. \tag{6.8}$$

The energy levels are depicted in Fig. 6.1b (left for $J > 0$ and right for $J < 0$).

In the case of ferromagnetic interaction ($J > 0$), the ground state is given by completely aligned states with the total spin $S = 2$, which are 5-fold degenerate: $|S = 2, M = 2\rangle, \cdots, |S = 2, M = -2\rangle$. The ground-state energy is $-J$ (Fig. 6.1b-left). In this state, the ferromagnetic order parameter is

$$M = S_1^z + S_2^z + S_3^z + S_4^z, \tag{6.9}$$

which commutes with the Hamiltonian, and the order parameter along the observing direction (i.e., the $z$ axis in the present case) has the same value as those of the classical model:

$$\left(S_1^z + S_2^z + S_3^z + S_4^z\right)^2 = \left(\frac{1}{2} \times 4\right)^2 = 4. \tag{6.10}$$

On the other hand, in the antiferromagnetic case ($J < 0$), the classical order parameter is the Néel order $S_1^z - S_2^z + S_3^z - S_4^z$, which does not commute with the

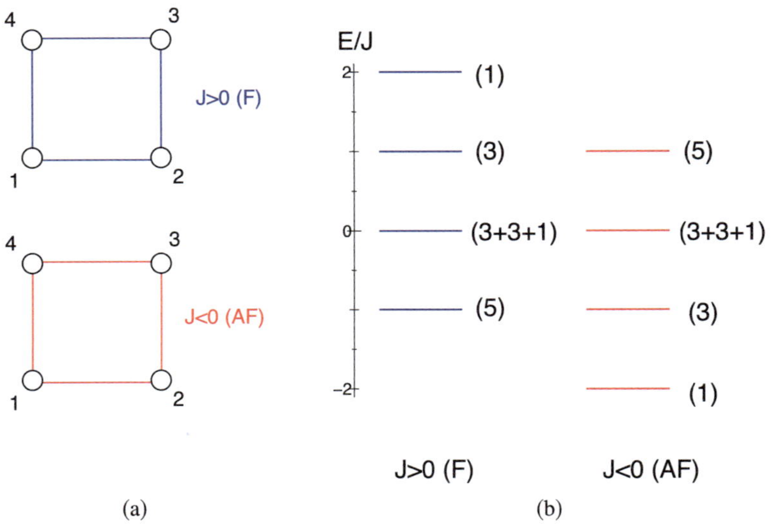


**Fig. 6.1** **a** $2 \times 2$ square lattice, **b** energy levels of the model (6.8). The left part is for $J > 0$ and the right part is for $J < 0$. The numbers beside the levels denote the number of degenerate states

Hamiltonian. In this case, the energy levels are upside down compared to the ferromagnetic case, as depicted in Fig. 6.1b-right, and the ground state is a singlet state with the total spin $S = 0$: $|S = 0, M = 0\rangle$. In this state, the expectation value of the antiferromagnetic order parameter is reduced as

$$\langle (S_1^z - S_2^z + S_3^z - S_4^z)^2 \rangle = 2.6666 \cdots < 4. \tag{6.11}$$

This comparison shows that quantum fluctuation lowers energy (increases the absolute value (energy gain)), while it reduces the classical order parameters (spin reduction).

Above, we studied the properties of a model on a finite lattice. Such properties also hold in infinite lattices. For example, in one-dimensional systems with an easy-plane type XXZ model with $A < 1$, whose Hamiltonian is given by

$$\mathcal{H} = -J \sum_{\langle ij \rangle} \left( S_i^x S_j^x + S_i^y S_j^y + A S_i^z S_j^z \right), \tag{6.12}$$

the classical long-range order parameters in the ground state are suppressed to be zero, and a power-law relaxation of the correlation function appears. Such a state is called Tomonaga-Luttinger liquid, and its properties have been extensively studied [12]. More surprisingly, in the antiferromagnetic Heisenberg model ($A = 1$) with integer spin $S = 1, 2, \cdots$, the correlation function of the classical order parameter decays exponentially (not in a power law). This property has been studied as Haldane problem, and its mechanism has been clarified theoretically, numerically, and also experimentally [13].

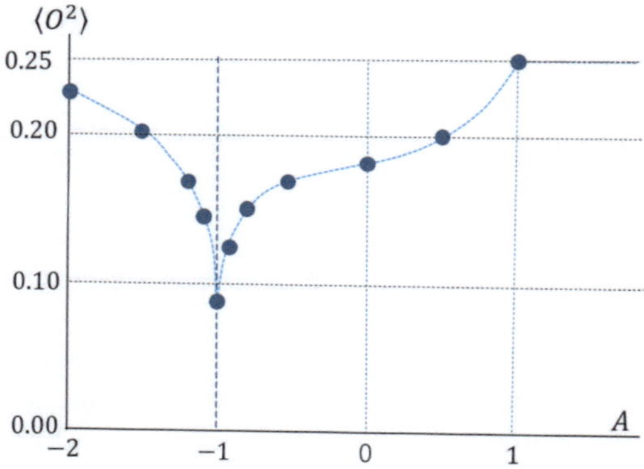


**Fig. 6.2** The long-range order in the ground state of XXZ model on the square lattice as a function of anisotropy $A$. The solid line is drawn for guide of the eye, but it is very close to the results obtained by the spin wave approximation

In two dimensions, there remains a long-range order in the ground state of these models, but the amount is reduced by quantum fluctuations [14], as schematically shown in Fig. 6.2.

For $A > 1$, the long-range order appears as a ferromagnetically ordered $z$ component, and the order parameter is given by $\langle O^2\rangle = \langle(\sum_i S_i^z)^2\rangle/N^2$, where $N$ is the total number of spins.

For $0 < A < 1$, the long-range order appears as a ferromagnetically ordered $xy$ components, and the order parameter is given by $\langle O^2\rangle = \langle(\sum_i S_i^x)^2 + (\sum_i S_i^y)^2\rangle/N^2$. For negative values of $A$, the model (6.12) is transformed into the antiferromagnetic model by replacing the spin operator as

$$\left(S_i^x, S_i^y, S_i^z\right) \rightarrow \left(\varepsilon_i S_i^x, \varepsilon_i S_i^y, S_i^z\right), \tag{6.13}$$

where $\varepsilon_i = \pm 1$ on the different sublattices, denoting the Néel order. This transformation does not change the commutation relation of spins (6.1)[1]. With this transformation, the Hamiltonian (6.12) for $A < 0$ is transformed to

$$\mathcal{H} = J \sum_{\langle ij\rangle} \left(S_i^x S_j^x + S_i^y S_j^y + |A| S_i^z S_j^z\right), \tag{6.14}$$

and the system represents the antiferromagnetic model.

For $-1 < A < 0$, the long-range order appears as an antiferromagnetically ordered $xy$ component, and the order parameter is given by $\langle O^2\rangle = \langle(\sum_i \varepsilon_i S_i^x)^2 + (\sum_i \varepsilon_i S_i^y)^2\rangle/N^2$.

For $A < -1$, the long-range order appears as an antiferromagnetically ordered $z$ component, and the order parameter is given by the Néel order: $\langle O^2\rangle = \langle(\sum_i \varepsilon_i S_i^z)^2\rangle/N^2$.

---

[1] It should be noted that the change of the direction of spin: $\left(S_i^x, S_i^y, S_i^z\right) \rightarrow \left(-S_i^x, -S_i^y, -S_i^z\right)$ is not compatible with the commutation relation of spins (6.1). Therefore, unlike the classical case, the ferromagnetic model and the antiferromagnetic model have different nature of ordering in quantum systems even in bipartite lattices without frustration.

### 6.1.3 Spin Wave Analysis of Antiferromagnetic Heisenberg Model

Low-temperature properties of ferromagnetic systems are governed by small fluctuations from the ordered state, and they are studied by the so-called "spin wave theory" [15]. The mechanism of energy gain and the reduction of the order parameter discussed above is also studied by extending the spin wave analysis to antiferromagnetic cases [16].

In this section, we overview an antiferromagnetic Heisenberg model with spin $S$ in a bipartite lattice where the lattice points are separated into two sublattices. The classical ground state is the Néel state in which the spins on one sublattice are $+S$ and those in the other sublattice are $-S$. As explained in the previous section, the Néel state is not the exact ground state. Here, we introduce a spin wave analysis around the Néel state for the Heisenberg antiferromagnetic model:

$$\mathcal{H} = J\sum_{\langle ij\rangle}^{N} \boldsymbol{S}_i \cdot \boldsymbol{S}_j - D\sum_{i=1}^{N}(S_i^z)^2 - H\sum_{i=1}^{N} S_i^z, \quad \boldsymbol{S}_{N+1} = \boldsymbol{S}_1, \quad J > 0, \tag{6.15}$$

where $\langle ij\rangle$ denotes nearest neighbor sites, and we set the number of nearest neighbor sites to be $z$, e.g., $z = 4$ in the square lattice and $z = 6$ in the cubic lattice. Denoting the odd sites as $(i = l)$ and the even sites as $(i = m)$. the classical ground state is given by $S_l = S$ and $S_m = -S$, the ground-state energy per spin for $S_i^z = -S_j^z = S$ is given by

$$E_0/N = -\frac{1}{2}zJS^2 - DS^2 - HS. \tag{6.16}$$

Now, we consider the quantum fluctuation around this state using the Holstein-Primakoff formula [17], where we introduce quantum fluctuations by boson creation-annihilation operators $(a_l, a_l^\dagger)$ and $(b_m, b_m^\dagger)$ which satisfy commutation relations:

$$[a_l, a_{l'}^\dagger] = \delta_{l,l'}, \quad [b_m, b_{m'}^\dagger] = \delta_{m,m'}, \tag{6.17}$$

and

$$[a_l, b_m] = [a_l, b_m^\dagger] = [a_l^\dagger, b_m] = [a_l^\dagger, b_m^\dagger] = 0. \tag{6.18}$$

With these operators, the spin operators are approximated as

$$\begin{cases} S_l^z &= S - a_l^\dagger a_l \\ S_l^+ &= \sqrt{2S}\, f_l(S)\, a_l, \\ S_l^- &= \sqrt{2S}\, f_l(S)\, a_l^\dagger \end{cases} \qquad \begin{cases} S_m^z &= -S + b_m^\dagger b_m \\ S_m^+ &= \sqrt{2S}\, f_m(S)\, b_m \\ S_m^- &= \sqrt{2S}\, f_m(S)\, b_m^\dagger \end{cases}, \tag{6.19}$$

where

$$f_l(S) = \sqrt{1 - \frac{a_l^\dagger a_l}{2S}}, \quad f_m(S) = \sqrt{1 - \frac{b_m^\dagger b_m}{2S}}, \tag{6.20}$$

up to $a_l^\dagger a_l \leq 2S$ and $b_m^\dagger b_m \leq 2S$.

Using the relations:

$$S_i^x = \frac{S_i^+ + S_i^-}{2}, \quad S_i^y = \frac{S_i^+ - S_i^-}{2i}, \tag{6.21}$$

the spin coupling term in the Hamiltonian (6.15) is rewritten as

$$\boldsymbol{S}_i \cdot \boldsymbol{S}_j = S_i^z S_j^z + \frac{1}{2}\left(S_i^+ S_j^- + S_i^- S_j^+\right). \tag{6.22}$$

Substituting the Holstein-Primakoff formula and taking up to the second order of $(a_l, a_l^\dagger)$ and $(b_m, b_m^\dagger)$, we have

$$\begin{aligned}\mathcal{H} \simeq E_0 + JS \sum_{<lm>} \left(a_l^\dagger a_l + b_m^\dagger b_m + a_l b_m + a_l^\dagger b_m^\dagger\right) \\ + D(2S-1)\left(\sum_l a_l^\dagger a_l + \sum_m b_m^\dagger b_m\right) \\ + H\left(\sum_l a_l^\dagger a_l - \sum_m b_m^\dagger b_m\right).\end{aligned} \tag{6.23}$$

Here, we used the relation $a_l^\dagger a_l a_l^\dagger a_l = a_l^\dagger a_l^\dagger a_l a_l + a_l^\dagger a_l$ and approximated $D(S_l^z)^2 = D(S - a_l^\dagger a_l)^2 \simeq S^2 - (2S-1)a_l^\dagger a_l$.

Because the system is uniform, we introduce Fourier-transformed operators $a_{\boldsymbol{k}}, a_{\boldsymbol{k}}^\dagger$ and $b_{\boldsymbol{k}}, b_{\boldsymbol{k}}^\dagger$ as

$$a_l = \sqrt{\frac{2}{N}} \sum_{\boldsymbol{k}} e^{-i\boldsymbol{k}\cdot\boldsymbol{r}_l} a_{\boldsymbol{k}}, \quad b_l = \sqrt{\frac{2}{N}} \sum_{\boldsymbol{k}} e^{i\boldsymbol{k}\cdot\boldsymbol{r}_l} b_{\boldsymbol{k}}, \quad k = \frac{2\pi}{N} n, \quad n = 1, \cdots, N, \tag{6.24}$$

which satisfy the boson commutation relations.

With these operators $(a_{\boldsymbol{k}}, a^{\dagger}_{\boldsymbol{k}})$ and $(b_{\boldsymbol{k}}, b^{\dagger}_{\boldsymbol{k}})$, the Hamiltonian is expressed as

$$\mathcal{H} \simeq E_0 + JS\sum_{\boldsymbol{k}}\Big((1+h+d)\,a^{\dagger}_{\boldsymbol{k}}a_{\boldsymbol{k}} + (1-h+d)\,b^{\dagger}_{\boldsymbol{k}}b_{\boldsymbol{k}}\Big)$$

$$+ JS\sum_{\boldsymbol{k}}\gamma_{\boldsymbol{k}}\left(a_{\boldsymbol{k}}b_{\boldsymbol{k}} + a^{\dagger}_{\boldsymbol{k}}b^{\dagger}_{\boldsymbol{k}}\right). \tag{6.25}$$

Here,

$$\gamma_{\boldsymbol{k}} = \frac{1}{z}\sum_{\boldsymbol{\rho}} e^{i\boldsymbol{k}\cdot\boldsymbol{\rho}}, \quad d = \frac{D(2S-1)}{JS}, \quad h = \frac{H}{JS}, \tag{6.26}$$

where $\boldsymbol{\rho}$ denotes vectors to the nearest neighbor sites.

Now, we introduce a new set of operators $\alpha_{\boldsymbol{k}}$ and $\beta_{\boldsymbol{k}}$ as

$$\alpha_{\boldsymbol{k}} = c_{\boldsymbol{k}}a_{\boldsymbol{k}} + s_{\boldsymbol{k}}b^{\dagger}_{\boldsymbol{k}}, \quad \alpha^{\dagger}_{\boldsymbol{k}} = c_{\boldsymbol{k}}a^{\dagger}_{\boldsymbol{k}} + s_{\boldsymbol{k}}b_{\boldsymbol{k}},$$

$$\beta_{\boldsymbol{k}} = s_{\boldsymbol{k}}a^{\dagger}_{\boldsymbol{k}} + c_{\boldsymbol{k}}b_{\boldsymbol{k}}, \quad \beta^{\dagger}_{\boldsymbol{k}} = s_{\boldsymbol{k}}a_{\boldsymbol{k}} + c_{\boldsymbol{k}}b^{\dagger}_{\boldsymbol{k}}. \tag{6.27}$$

To make the new operators $\alpha$ and $\beta$ to be boson operators, i.e., to satisfy the commutation relations:

$$[\alpha_{\boldsymbol{k}}, \alpha^{\dagger}_{\boldsymbol{k}}] = 1, \quad [\beta_{\boldsymbol{k}}, \beta^{\dagger}_{\boldsymbol{k}}] = 1, \tag{6.28}$$

the following conditions on the coefficients $c_{\boldsymbol{k}}$ and $s_{\boldsymbol{k}}$ are required:

$$c^2_{\boldsymbol{k}} - s^2_{\boldsymbol{k}} = 1, \tag{6.29}$$

which comes from the following analysis:

$$\alpha_{\boldsymbol{k}}\alpha^{\dagger}_{\boldsymbol{k}} = c^2_{\boldsymbol{k}}a_{\boldsymbol{k}}a^{\dagger}_{\boldsymbol{k}} + c_{\boldsymbol{k}}s_{\boldsymbol{k}}a_{\boldsymbol{k}}b_{\boldsymbol{k}} + s_{\boldsymbol{k}}c_{\boldsymbol{k}}b^{\dagger}_{\boldsymbol{k}}a^{\dagger}_{\boldsymbol{k}} + s^2_{\boldsymbol{k}}b^{\dagger}_{\boldsymbol{k}}b_{\boldsymbol{k}}, \tag{6.30}$$

and

$$\alpha^{\dagger}_{\boldsymbol{k}}\alpha_{\boldsymbol{k}} = c^2_{\boldsymbol{k}}a^{\dagger}_{\boldsymbol{k}}a_{\boldsymbol{k}} + c_{\boldsymbol{k}}s_{\boldsymbol{k}}a_{\boldsymbol{k}}b_{\boldsymbol{k}} + s_{\boldsymbol{k}}c_{\boldsymbol{k}}b^{\dagger}_{\boldsymbol{k}}a^{\dagger}_{\boldsymbol{k}} + s^2_{\boldsymbol{k}}b_{\boldsymbol{k}}b^{\dagger}_{\boldsymbol{k}}, \tag{6.31}$$

and thus

$$[\alpha_{\boldsymbol{k}}, \alpha^{\dagger}_{\boldsymbol{k}}] = c^2_{\boldsymbol{k}}(a_{\boldsymbol{k}}a^{\dagger}_{\boldsymbol{k}} - a^{\dagger}a_{\boldsymbol{k}}) - s^2_{\boldsymbol{k}}(b_{\boldsymbol{k}}b^{\dagger}_{\boldsymbol{k}} - b^{\dagger}_{\boldsymbol{k}}b_{\boldsymbol{k}}) = c^2_{\boldsymbol{k}} - s^2_{\boldsymbol{k}} = 1. \tag{6.32}$$

This condition is automatically satisfied when we parametrize them as[2]

$$c = \cosh\theta, \quad s = \sinh\theta. \tag{6.33}$$

Because

$$\begin{pmatrix} c & s \\ s & c \end{pmatrix}\begin{pmatrix} c & -s \\ -s & c \end{pmatrix} = \begin{pmatrix} c^2 - s^2 & 0 \\ 0 & -s^2 + c^2 \end{pmatrix} = \begin{pmatrix} 1 & 0 \\ 0 & 1 \end{pmatrix}, \tag{6.34}$$

the inverse transformation is given by

$$a_{\boldsymbol{k}} = c_{\boldsymbol{k}}\alpha_{\boldsymbol{k}} - s_{\boldsymbol{k}}\beta^{\dagger}, \quad a_{\boldsymbol{k}}^{\dagger} = c_{\boldsymbol{k}}\alpha_{\boldsymbol{k}}^{\dagger} - s_{\boldsymbol{k}}\beta_{\boldsymbol{k}},$$

$$b_{\boldsymbol{k}} = s_{\boldsymbol{k}}\alpha_{\boldsymbol{k}}^{\dagger} + c_{\boldsymbol{k}}\beta_{\boldsymbol{k}}, \quad b_{\boldsymbol{k}}^{\dagger} = s_{\boldsymbol{k}}\alpha_{\boldsymbol{k}} + c_{\boldsymbol{k}}\beta_{\boldsymbol{k}}^{\dagger}. \tag{6.35}$$

This transformation is called the Bogoliubov transformation, which was introduced for fermion systems to study the BCS theory of superconductivity[3].

Using these operators, the Hamiltonian is rewritten as

$$\mathcal{H} \simeq E_0 + JS\sum_{\boldsymbol{k}}\left[\sqrt{(1+d)^2 - \gamma_{\boldsymbol{k}}^2} - (1+d)\right]$$

$$+ JS\sum_{\boldsymbol{k}}\left[\{\sqrt{(1+d)^2 - \gamma_{\boldsymbol{k}}^2} + h\}\alpha_{\boldsymbol{k}}^{\dagger}\alpha + \{\sqrt{(1+d)^2 - \gamma_{\boldsymbol{k}}^2} - h\}\beta_{\boldsymbol{k}}^{\dagger}\beta\right].$$

Thus, the ground-state energy is lower than the classical one by

$$\Delta E = JS\sum_{\boldsymbol{k}}\left[\sqrt{(1+d)^2 - \gamma_{\boldsymbol{k}}^2} - (1+d)\right], \tag{6.36}$$

and the spin wave dispersion relation is given by

$$\omega_{\pm}(k) = JS\sqrt{(1+d)^2 - \gamma_{\boldsymbol{k}}^2} \pm h. \tag{6.37}$$

Amplitudes of magnetic moments of sublattices in the ground state for the classical system are $\pm S$, while those in the quantum system are

---

[2] The condition can be generalized with a phase factor $e^{i\phi}$, but here we use real coefficients.

[3] For fermion systems, the coefficients need to satisfy $c_{\boldsymbol{k}}^2 + s_{\boldsymbol{k}}^2 = 1$.

$$S - \langle a_l^\dagger a_l \rangle, \quad -S + \langle b_m^\dagger b_m \rangle. \tag{6.38}$$

Because the system is uniform, we calculate the deviations as follows:

$$a_l^\dagger a_l = \frac{2}{N} \sum_l a_l^\dagger a_l = \frac{2}{N} \sum_l \left( \sqrt{\frac{2}{N}} \sum_{\boldsymbol{k}} e^{i\boldsymbol{k}\cdot\boldsymbol{r}_l} a_{\boldsymbol{k}}^\dagger \right) \left( \sqrt{\frac{2}{N}} \sum_{k'} e^{-ik'l} a'_{\boldsymbol{k}} \right)$$

$$= \frac{N}{2} \sum_{\boldsymbol{k}} a_{\boldsymbol{k}}^\dagger a_{\boldsymbol{k}}. \tag{6.39}$$

Using the relation (6.35),

$$\langle a_{\boldsymbol{k}}^\dagger a_{\boldsymbol{k}} \rangle = \langle (c_{\boldsymbol{k}} \alpha_{\boldsymbol{k}}^\dagger - s_{\boldsymbol{k}} \beta_{\boldsymbol{k}})(c_{\boldsymbol{k}} \alpha_{\boldsymbol{k}} - s_{\boldsymbol{k}} \beta_{\boldsymbol{k}}) \rangle$$

$$= \langle (c_{\boldsymbol{k}}^2 \alpha_{\boldsymbol{k}}^\dagger \alpha_{\boldsymbol{k}} + s_{\boldsymbol{k}}^2 \beta_{\boldsymbol{k}}^\dagger \beta_{\boldsymbol{k}} - s_{\boldsymbol{k}} c_{\boldsymbol{k}} (\alpha_{\boldsymbol{k}} \beta_{\boldsymbol{k}} + \alpha_{\boldsymbol{k}}^\dagger \beta_{\boldsymbol{k}}^\dagger)) \rangle. \tag{6.40}$$

In the ground state,

$$\langle \alpha_{\boldsymbol{k}}^\dagger \alpha_{\boldsymbol{k}} \rangle = \langle \beta_{\boldsymbol{k}}^\dagger \beta_{\boldsymbol{k}} \rangle = 0, \quad \langle \alpha_{\boldsymbol{k}} \beta_{\boldsymbol{k}} \rangle = \langle \beta_{\boldsymbol{k}}^\dagger \beta_{\boldsymbol{k}}^\dagger \rangle = 0, \tag{6.41}$$

from which

$$\langle a_{\boldsymbol{k}}^\dagger a_{\boldsymbol{k}} \rangle = s_{\boldsymbol{k}}^2, \quad \langle b_{\boldsymbol{k}}^\dagger b_{\boldsymbol{k}} \rangle = s_{\boldsymbol{k}}^2. \tag{6.42}$$

Therefore, the amplitudes in both sublattices of magnetization are reduced to

$$S - \frac{2}{N} \sum_{\boldsymbol{k}} s_{\boldsymbol{k}}^2. \tag{6.43}$$

These results realize the lowering of the ground-state energy and the reduction of the classical order parameters. The spin wave analyses have been done for various systems, e.g., antiferromagnetic Heisenberg model on fcc lattice [18] and also frustrated square lattice [19].

### *6.1.4 Quantum Fluctuation and Thermal Fluctuation*

In the path-integral representation, the partition function of a $d$-dimensional quantum system is expressed by that of a $(d+1)$-dimensional classical system:

$$Z = \mathrm{Tr} e^{-\beta\mathcal{H}} = \sum_{i}^{M} \langle i | e^{-\beta\mathcal{H}} | i \rangle$$

$$= \sum_{i}^{M} \sum_{k_1}^{M} \cdots \sum_{k_n-1}^{M} \langle i | e^{-\beta\mathcal{H}/n} | k_1 \rangle \langle k_1 | e^{-\beta\mathcal{H}/n} | k_2 \rangle \langle k_2 | \cdots \langle k_{n-1} | e^{-\beta\mathcal{H}/n} | i \rangle, \quad (6.44)$$

where the intermediate states $k_1, k_2, \cdots k_{n-1}$ give a new direction [20]. In the $(d+1)$-dimensional configuration, quantum fluctuation is visualized as demonstrated in Sect. 6.12 of my previous book [21]. This transformation makes Monte Carlo simulation in quantum systems possible. Magnetization reversal in classical Ising systems occurs by domain formation, which costs energy of $L^{d-1}$, while in quantum systems, it occurs by quantum tunneling of the whole system, which costs energy of $L^d$, where $L$ is the linear dimension of the system. Thus, the nature of fluctuation is different, which causes various interesting phenomena, e.g., the so-called quantum Griffiths singularity [22].

## 6.2 Quantum Fluctuation in Frustrated Systems

In classical frustrated systems, incompatibility among interactions causes macroscopic degeneracy and non-collinear ordered structures as we studied in Chaps. 4 and 5. In quantum systems, quantum fluctuation introduces additional aspects of the ground state. As we studied in Sect. 6.1, by quantum fluctuation the ground state gains a large amount of energy.

Here, we study the case of Villain odd model type depicted in Fig. 6.3a. The energy level structures of Ising and Heisenberg models on this lattice and that of non-frustrated antiferromagnetic case are shown in Fig. 6.3b. Here, we find that the energy levels of the frustrated model of Ising spins have only two levels, $E = \pm J/2$, where each level is 8-fold degenerate. This 8-fold degeneracy arises from the various possible spin configurations in the frustration condition, which allows multiple distinct spin arrangements in the same energy. In the Heisenberg model on the Villain lattice, the degeneracy is resolved, and the ground state becomes a singlet with its energy being significantly reduced. As a reference, the energy levels of the antiferromagnetic Heisenberg model are depicted again (shown in Fig. 6.1b, right). The ground-state energies in the non-frustrated model and frustrated one (Villain model) is $E_\mathrm{G} = -1$ and $-1/2$ in the case of Ising spins, while they are $E_\mathrm{G} = -2$ and $-1.74\cdots$ in Heisenberg model. This observation indicates that quantum fluctuation redresses energy loss due to frustration, creating a low-energy state in frustrated systems. Such a low-energy state implies possibility of a new type of ground state in frustrated quantum systems such as spin-liquid states.

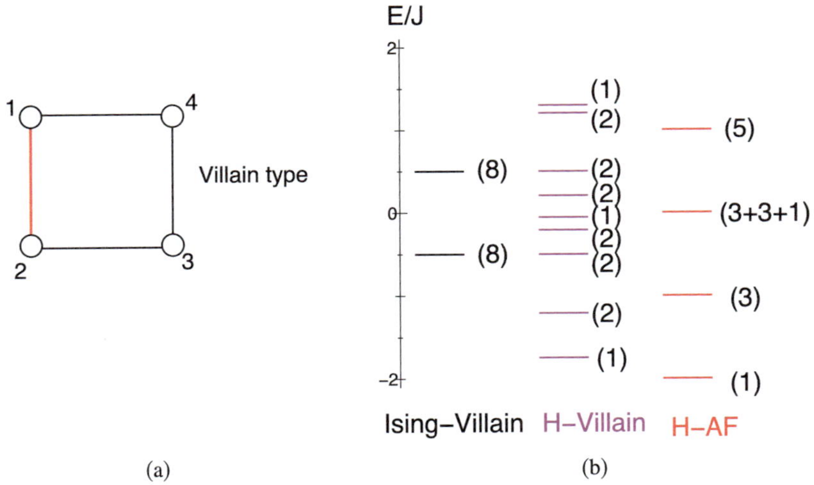


**Fig. 6.3** **a** 2 × 2 Villain-type frustrated lattice, **b** energy levels of the models (Ising model on the lattice (a), Heisenberg model on the lattice (a), and Heisenberg model on the non-frustrated lattice shown in Fig. 6.1a

## 6.2.1 *Ground States of Systems Composed of Triangles*

In Fig. 6.4, we show the energy levels of antiferromagnetic Heisenberg models on a single triangle, a lattice composed of six triangles (pyramid), and an array of triangles (SAW).

For the lattice of a single triangle, the ground state is regarded as a state in which one bond forms a singlet state, and the remaining site is free. The formation of the singlet state, ignoring the other two bonds, gives the ground-state energy $-(3/4 + 0 + 0)/3 = -0.25$. Here, we find that the energy of the ferromagnetic ground state (top of the levels) is equal to the ground state of the antiferromagnetic Heisenberg model. This fact is very different from the case of the Ising model, where the ground state of the antiferromagnetic Ising model is $-(1 + 1 - 1)/4/3 = -0.0833$, which is 1/3 of the ferromagnetic case. This observation indicates that the energy gain due to quantum fluctuations is very large in frustrated systems. With this large energy gain, one may expect that the ground state would be given by an ensemble of singlet pairs, which is called the valence bond state, introduced by Pauling [23]. Such a large reduction of energy is also found for the extended lattices (pyramid and SAW) in Fig. 6.4b. The energy structure of kagome-type antiferromagnets on the sphere, e.g., the cuboctahedron and icosidodecahedron, has been studied in detail [24].

How much the long-range order is reduced by the combination of quantum fluctuations in frustrated systems is an important issue. Anderson proposed the idea of the resonating valence bond (RVB) state [25] for the $S = 1/2$ AFTH model, which is an extension of the pyramid lattice. The RVB state is a new kind of quantum ground state originating from the large energy gain of the singlet state. Indeed, the so-called

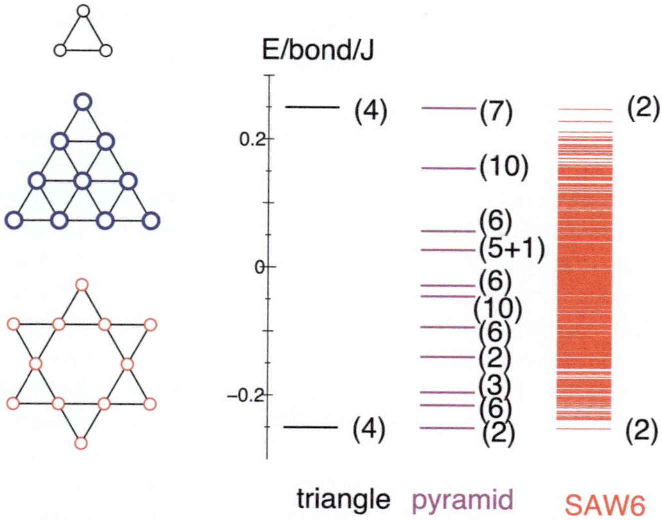


(a) (b)

**Fig. 6.4** **a** Lattices with triangles: a triangle, a pyramid, and a SAW, from top to bottom. **b** Energy levels of antiferromagnetic Heisenberg models on (left) a triangle, (middle) a pyramid lattice, and (right) a SAW lattice

dimer covering ground state, which is given by the product of dimer states, has been known in models such as the Shastry-Sutherland model [26, 27]. The ground state of the SAW lattice depicted in Fig. 6.4a (bottom) is one such state, where two ways of setting singlet pairs exist, making the ground state doubly degenerate.

Experimental attempts to find Anderson's RVB state were conducted by Kinshiro Hirakawa and collaborators by synthesizing materials such as $VCl_2$ ($S = 1$) and oxides like $LiNiO_2$ and $NaTiO_2$ [28]. The RVB state also implies the existence of a ground state of a quantum spin system without any spatial long-range spin order, which is called the quantum spin liquid (QSL) state and has been extensively studied both theoretically and experimentally [29]. Moreover, the concept of a spin nematic state has also been discussed [30].

However, for the AFTH model, it was pointed out that the energy obtained from a variational approximation of the 120° structure gives a comparably low ground-state energy [31–34], and detailed analyses using the exact diagonalization method revealed the existence of long-range order without an energy gap [35]. Thus, for the AFTH Heisenberg model, quantum fluctuations largely reduce the classical order but do not lead to a spin liquid state.

## 6.3 Quantum Fluctuation Induced Phase Transition, Quantum Version of "Order-by-Disorder"

In classical models, entropy plays an essential role in selecting the most thermally stable state (order-by-disorder mechanism). In quantum systems, the focus is on the ground-state properties, and quantum fluctuations play a similar role in selecting the most energetically stable state. By taking into account the effect of non-commutativity among the spin components on the degenerate classical configurations, we may obtain a quantum fluctuation-selected ground state by resolving the degeneracy through quantum effects (quantum order-by-disorder mechanism) [1, 2].

One of the most prominent cases of quantum order-by-disorder was observed in experiments on $CsCuCl_3$ [36]. This material consists of $S = 1/2$ one-dimensional ferromagnetic spin chains along the c-axis. The chains form a triangular lattice in the ab-plane. Because the interaction between chains exhibits weak easy-plane-type (XY-like) anisotropy, the spins classically form a 120° structure in the $xy$ plane at zero field. At very low temperatures, where the system is essentially in the ground state, it was found that the magnetization process $M(H)$ of this material exhibits a sudden jump, in contrast to a plateau often observed in cases with easy-axis anisotropy.

In the corresponding classical models, the ground-state configuration in a field along the c-axis $H$ is the umbrella configuration, as studied in Chap. 5. There, at finite temperatures, the equilibrium configuration shifts to the V-shaped configuration (Fig. 5.3), even though the V-shaped configuration is energetically unfavorable. This change occurs due to entropy effects [37]. The magnetization process $M_z(H_z)$ has also been studied using the large-size cluster mean-field method [38].

However, at $T = 0$, where entropy has no effect, Nikuni and Shiba demonstrated that even if the classical energy of the umbrella state is lower, quantum fluctuations can lower the total energy of the V-shaped configuration, leading to its selection as the ground state (a quantum version of order-by-disorder) by studying the ground-state energy using spin-wave analysis [1].[4]

### 6.3.1 *Spin-Wave Correction of Ground State Energy with Non-collinear Spin Configurations*

In frustrated systems, the ground-state configuration is not necessarily collinear, as in the case of the Néel state studied in Sect. 6.1.3, but can instead be non-collinear. In such cases, it is necessary to extend the spin-wave analysis to non-collinear configurations [1, 39]. Let the direction of the $i$-th spin be given by $(\theta_i, \phi_i)$. Then, we define the local coordinate system $(\xi, \eta, \zeta)$, where the $\zeta$ axis is aligned with the direction

[4] Historically, quantum effects were studied earlier than thermal effects.

$(\theta_i, \phi_i)$ of the corresponding spin in the classical ground state. The spin components $(S_i^x, S_i^y, S_i^z)$ are expressed in terms of the spin components $(S_i^\xi, S_i^\eta, S_i^\zeta)$ in the new coordinate system as [1]:

$$S_i^x = -S_i^\xi \sin\phi_i - S_i^\eta \cos\theta_i \cos\phi_i + S_i^\zeta \sin\theta_i \cos\phi_i, \tag{6.45}$$

$$S_i^y = S_i^\xi \cos\phi_i - S_i^\eta \cos\theta_i \sin\phi_i + S_i^\zeta \sin\theta_i \sin\phi_i, \tag{6.46}$$

$$S_i^z = S_i^\eta \sin\theta_i + S_i^\zeta \cos\theta_i. \tag{6.47}$$

To study small fluctuations around the ground-state configuration, we rewrite the Hamiltonian terms in terms of $\{S^\xi, S^\eta, S^\zeta\}$. For example, the terms in $\mathbf{S}_i \cdot \mathbf{S}_j$ are expressed as

$$\begin{aligned} S_i^x S_j^x &= S_i^\xi S_j^\xi \sin\phi_i \sin\phi_j + S_i^\eta S_j^\eta \cos\theta_i \cos\phi_i \cos\theta_j \cos\phi_j \\ &\quad + S_i^\zeta S_j^\zeta \sin\theta_i \cos\phi_i \sin\theta_j \cos\phi_j \end{aligned}$$

$$+ S_i^\xi S_j^\eta \sin\phi_i \cos\theta_j \cos\phi_j + S_i^\eta S_j^\xi \sin\phi_j \cos\theta_i \cos\phi_i$$

$$- S_i^\xi S_j^\zeta \sin\phi_i \sin\theta_j \cos\phi_j - S_i^\zeta S_j^\xi \sin\phi_j \sin\theta_i \cos\phi_i$$

$$- S_i^\eta S_j^\zeta \cos\theta_i \cos\phi_i \sin\theta_j \cos\phi_j - S_i^\zeta S_j^\eta \cos\theta_j \cos\phi_j \sin\theta_i \cos\phi_i, \tag{6.48}$$

$$\begin{aligned} S_i^y S_j^y &= S_i^\xi S_j^\xi \cos\phi_i \cos\phi_j + S_i^\eta S_j^\eta \cos\theta_i \sin\phi_i \cos\theta_j \sin\phi_j \\ &\quad + S_i^\zeta S_j^\zeta \sin\theta_i \sin\phi_i \sin\theta_j \sin\phi_j \end{aligned}$$

$$- S_i^\xi S_j^\eta \cos\phi_i \cos\theta_j \sin\phi_j - S_i^\eta S_j^\xi \cos\phi_j \cos\theta_i \sin\phi_i$$

$$+ S_i^\xi S_j^\zeta \cos\phi_i \sin\theta_j \sin\phi_j + S_i^\zeta S_j^\xi \cos\phi_j \sin\theta_i \sin\phi_i$$

$$- S_i^\eta S_j^\zeta \cos\theta_i \sin\phi_i \sin\theta_j \sin\phi_j - S_i^\zeta S_j^\eta \cos\theta_j \sin\phi_j \sin\theta_i \sin\phi_i, \tag{6.49}$$

$$S_i^z S_j^z = +S_i^\eta S_j^\eta \sin\theta_i \sin\theta_j + S_i^\zeta S_j^\zeta \cos\theta_i \cos\theta_j + S_i^\eta S_j^\zeta \sin\theta_i \cos\theta_j + S_i^\zeta S_j^\eta \sin\theta_j \cos\theta_i, \tag{6.50}$$

and $\boldsymbol{S}_i \cdot \boldsymbol{S}_j$ is written as

$$\boldsymbol{S}_i \cdot \boldsymbol{S}_j = A_{\xi,\xi} S_i^\xi S_j^\xi + A_{\eta,\eta} S_i^\eta S_j^\eta + A_{\zeta,\zeta} S_i^\zeta S_j^\zeta$$

$$+ A_{\xi,\eta} S_i^\xi S_j^\eta + A_{\eta,\xi} S_i^\eta S_j^\xi + A_{\xi,\zeta} S_i^\xi S_j^\zeta + A_{\zeta,\xi} S_i^\zeta S_j^\xi + A_{\eta,\zeta} S_i^\eta S_j^\zeta + A_{\zeta,\eta} S_i^\zeta S_j^\eta, \tag{6.51}$$

where

$$A_{\xi,\xi} = \sin\phi_i \sin\phi_j + \cos\phi_i \cos\phi_j = \cos(\phi_i - \phi_j), \tag{6.52}$$

$$A_{\eta,\eta} = \cos\theta_i \cos\phi_i \cos\theta_j \cos\phi_j + \cos\theta_i \sin\phi_i \cos\theta_j \sin\phi_j + \sin\theta_i \sin\theta_j, \tag{6.53}$$

$$A_{\zeta,\zeta} = \sin\theta_i \cos\phi_i \sin\theta_j \cos\phi_j + \sin\theta_i \sin\phi_i \sin\theta_j \sin\phi_j + \cos\theta_i \cos\theta_j, \tag{6.54}$$

$$A_{\xi,\eta} = \sin\phi_i \cos\theta_j \cos\phi_j - \cos\phi_i \cos\theta_j \sin\phi_j, \tag{6.55}$$

$$A_{\eta,\xi} = \sin\phi_j \cos\theta_i \cos\phi_i - \cos\phi_j \cos\theta_i \sin\phi_i, \tag{6.56}$$

$$A_{\xi,\zeta} = -\sin\phi_i \sin\theta_j \cos\phi_j + \cos\phi_i \sin\theta_j \sin\phi_j, \tag{6.57}$$

$$A_{\zeta,\xi} = -\sin\phi_j \sin\theta_i \cos\phi_i + \cos\phi_j \sin\theta_i \sin\phi_i, \tag{6.58}$$

$$A_{\eta,\zeta} = -\cos\theta_i \cos\phi_i \sin\theta_j \cos\phi_j - \cos\theta_i \sin\phi_i \sin\theta_j \sin\phi_j + \sin\theta_i \cos\theta_j, \tag{6.59}$$

$$A_{\zeta,\eta} = -\cos\theta_j \cos\phi_j \sin\theta_i \cos\phi_i - \cos\theta_j \sin\phi_j \sin\theta_i \sin\phi_i + \sin\theta_j \cos\theta_i. \tag{6.60}$$

Then, we use the Holstein-Primakoff transformation (6.19) for spins in each sublattice unit, e.g., at the $i$-th sublattice:

$$S_i^\zeta = S - a_i^\dagger a_i, \quad S_i^\xi = \frac{\sqrt{2S}}{2}(a_i^\dagger + a_i), \quad S_i^\eta = i\frac{\sqrt{2S}}{2}(a_i^\dagger - a_i), \tag{6.61}$$

and so on. For sublattices, different Bose operators are introduced, e.g., $b_j$, $b_j^\dagger$ for the $j$-th sublattice. The Hamiltonian is expressed by these boson operators. For example, the products of spin components are given as

$$S_i^\xi S_j^\xi = \frac{S}{2}(a_i^\dagger + a_i)(b_i^\dagger + b_i) = \frac{S}{2}\left(a_i^\dagger b_i^\dagger + a_i^\dagger b_i + a_i b_i^\dagger + a_i b_i\right), \tag{6.62}$$

and so on. Using a generalized Bogoliubov transformation, the Hamiltonian is diagonalized and we obtained the ground-state energy.

$CsCuCl_3$ has three sublattices, and the classical ordered states $(\theta_i, \phi_i)$, $i = 1, 2, 3$ form several non-colinear configuration as we studied in Chap. 5. Thus, the quantum fluctuation in the three sublattices are expressed by thress boson operators for the Holstein-Primakoff transformation (6.19): $a_i^\dagger, a_i$, $b_i^\dagger, b_i$, and $c_i^\dagger, c_i$. The energy is expressed as a $6 \times 6$ matrix. By diagonalizing it by a generalized Bogoliubov transformation, Nikuni and Shiba successfully found that the energy gain due to the spin wave correction makes the energy of the V configuration lower than that of the classically stable umbrella structure [1].

### 6.3.2 Quantum Analysis of the Phase Diagram $(H, T)$ of the AFTXY Model

The selection of the most stable state in quantum systems based on the energy gain due to quantum fluctuations was studied by the spin-wave analysis in the previous section. This problem has been studied with more direct methods. For the jump in the magnetization process of $CsCuCl_3$ [36], the antiferromagnetic XXZ model with easy-plane anisotropy ($A < 1$):

$$\mathcal{H} = J \sum_{\langle ij \rangle} \left( S_i^x S_j^x + S_i^y S_j^y + A S_i^z S_j^z \right) \tag{6.63}$$

was studied by exact diagonalization [40], and also by a density matrix renormalization group (DMRG) method [41]. In the DMRG analysis, a $3 \times \infty$ strip of the triangular lattice with periodic boundary conditions was used as an approximation of the triangular lattice. There, the field dependence of the magnetization, the correlation function of the $z$ component, and also the $x$ component of the chirality were obtained. It was found that the strip lattice well represents the features of quantum phase transitions on the triangular lattice.

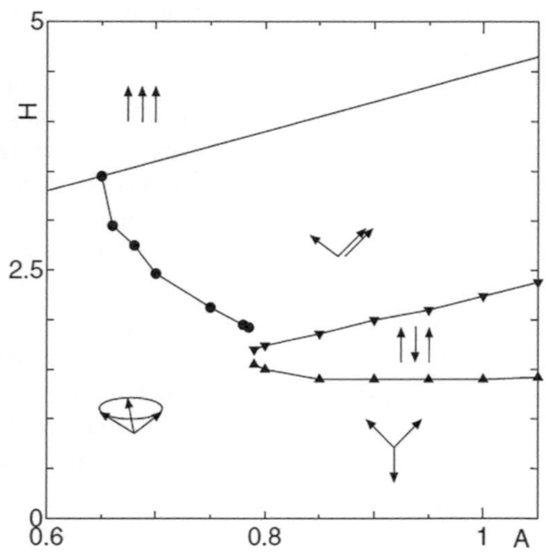


**Fig. 6.5** The ground-state phase diagram in $(A, H)$ coordinates of the $S = 1/2$ antiferromagnet on the triangular lattice. The solid circles represent the boundary of the umbrella phase and the noncollinear phase. The triangles represent the upper and lower edges of the 1/3 plateau. (This figure is reproduced from the reference [41]. ©2004 The Physical Society of Japan)

In Fig. 6.5, the phase diagram of quantum fluctuation-induced phase transitions is shown. For large planar anisotropy ($A < 0.65$), the low-field umbrella phase directly transitions to the field-induced ferromagnetic state, as in the classical case at $T = 0$. When the planar anisotropy is reduced ($0.65 < A < 0.8$), a phase of V-shape appears between the umbrella phase and the field-induced ferromagnetic phase. For small anisotropy ($A > 0.8$), we observe the sequence of phases of the Y-shape, up-down-up phase, and the V-shape phase, as seen in the classical ADTXY model [42], although the identification of phases at low field is ambiguous. For the AFT-XXZ model ($0 < 1/A < 2.3$) with $S = 1/2$, a detailed DMRG study has been done, which produced more details of the phase diagram [43]. Thus, we find that quantum fluctuation in the ground state plays a similar role to thermal fluctuations (entropy effects) at finite temperature.

The $3 \times \infty$ strip of the triangular lattice with periodic boundary conditions in the work [41] is regarded as a twisted three-leg lattice configuring a spin-tube consisting of alternate triangles, which models the material $[(CuCl_2tachH)_3Cl]Cl_2$ [44], although the parameter set is different. For such one-dimensional frustrated spin systems, the existence of an energy gap between the ground state and the excited state is an important issue. It has been shown that a simple three-leg lattice has an energy gap [45]. For the present twisted three-leg lattice, detailed analysis has been done, and it has been clarified that the existence of a gap depends on the lattice parameters [46, 47].

## 6.4 Quantum Kagome Antiferromagnets

Because of the highly degenerate ground states of antiferromagnets in the kagome lattice, which has a corner-sharing structure, the effects of quantum fluctuations are expected to be significant (quantum kagome antiferromagnets), and various approaches have been taken [48–55]. The ground state of a system with $N = 36$ [54] does not possess magnetic order, indicating a dimer-liquid model.

The magnetization process $M(H)$ of the present model has attracted considerable interest and has been studied extensively [56–62]. The existence of both magnetic and non-magnetic gaps has been discussed in detail [53, 55]. A detailed analysis using DMRG has indicated the presence of a finite magnetic gap [59].

For this model, peaks in the specific heat at low temperatures have attracted interest. Because of the synergistic effect of frustration and quantum fluctuations, the energy level structure exhibits peculiar properties. For example, the specific heat of the lattice depicted in Fig. 6.4a (bottom), which is a part of the kagome lattice, shows double peaks as illustrated in Fig. 6.6. We have seen that the double peaks in the specific heat originate from a special structure of the three-level system in Sect. 3.1.5. We expect that such a structure may be effectively realized in the present model. The high-temperature expansion method and the exact diagonalization method show a curious size dependence of the specific heat at low temperatures [63]. This double-

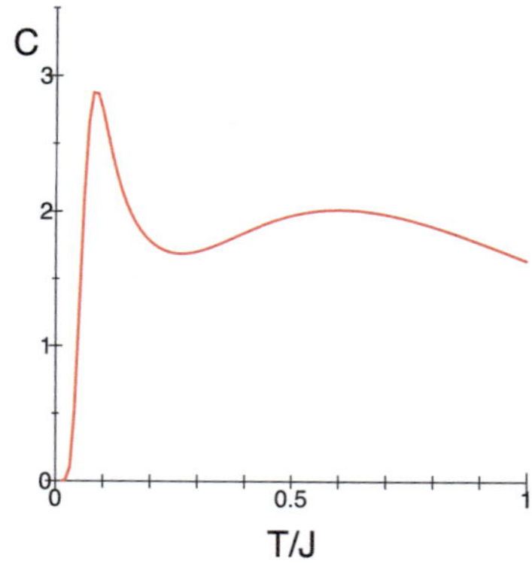


**Fig. 6.6** Double peaks in the specific heat of the lattice in Fig. 6.4a (bottom), which is a part of the Kagome lattice

peak structure has been extensively studied [63–66]. For the SAW lattice, which is an array of triangles, the existence of double peaks has been confirmed [67, 68]. In contrast, for the kagome lattice, the situation highly depends on the lattice size, and a definite conclusion has not yet been reached. Such a nature of the specific heat may arise from peculiar low-temperature properties of spin-liquid-type systems [69].

### 6.4.1 *Experiments on Quantum Kagome Antiferromagnets*

Various experimental works have been conducted on possible kagome antiferromagnets [27]. $^3He$ adsorbed on graphite forms a kagome lattice, whose specific heat was observed [70]. In magnetic materials, $SrCr_8Ga_4O_{19}$ (SCGO) is also a possible material for the kagome lattice with $Cr^{3+}$ ($S = 3/2$) [71]. $Rb_2M_3S_4$ (M is a magnetic ion such as Ni, Co, or Mn) [72] ($Ni^{2+}$ ($S = 1$)) also has a kagome plane. $KV_3Ge_2O_9$ is another $S = 1$ kagome antiferromagnet [73].

Various Jarosites, $AM_3(OH)_6(SO_4)_2$, ($A^+$ is a cation such as $K^+$, $Rb^+$, $NH_4^+$, $Tl^+$, $Ag^+$, or $Na^+$, and $M^{3+}$ is a magnetic ion such as $Fe^{3+}$ or $Cr^{3+}$), also form kagome lattices [27, 74–78].

For materials with $S = 1/2$, various compounds have been synthesized, e.g., $Cs_2Cu_3MF_{12}$ (M=Zr, Hf, and Sr) [79], $[Cu_3(titmb)_2\ (OCOCH_3)_6]\cdot H_2O$ [80], $Zn_xCu_{4-x}(OH)_6Cl_2$ [81–88] and so on.

However, as we studied above, the combination of strong frustration in the corner-sharing lattice structure and large quantum fluctuations creates very subtle conditions, and a small perturbation could cause a significant change in the properties of the system. The effect of lattice distortion has also been reported [89]. As a result, a clear comparison between theoretical and experimental works has not yet been realized.

## References


1. T. Nikuni and H. Shiba, J. Phys. Soc. Jpn. **62**, 3268 (1993), J. Phys. Soc. Jpn. **64**, 3471 (1995), A. E. Jacobs, T. Nikuni and H. Shiba, J. Phys. Soc. Jpn. **62**, 4066 (1993)
2. A. Chubukov, Phys. Rev. Lett. **69**, 832 (1992)
3. M.C. Cross, D.S. Fisher: Phys. Rev. B **19**, 402 (1979). T. Nakano, H. Fukuyama: J. Phys. Soc. Jpn. **49**, 1679 (1980). T. Nakano, H. Fukuyama: J. Phys. Soc. Jpn. **50**, 2489. (1981)
4. H. Onishi, S. Miyashita, J. Phys. Soc. Jpn. **72**, 392 (2003)
5. J.W. Bray, H.R. Hart Jr., L.V. Interrante, I.S. Jacobs, J.S. Kasper, G.D. Watkins, S.H. Wee, J.C. Bonner, Phys. Rev. Lett. **35**, 744 (1975)
6. M. Hase, I. Terasaki, K. Uchinokura, Phys. Rev. Lett. **70**, 3651 (1993)
7. M. Hase, N. Koide, K. Manabe, Y. Sasago, K. Uchinokura, Physica B **215**, 164 (1995)
8. L.P. Regnault, J.P. Renard, G. Dhalenne, A. Revcolevschi, Europhys. Lett. **32** 579 (1995). Y. Sasago, N. Koide, K. Uchinokura, M.C. Martin, M. Hase, K. Hirota, G. Shirane: Phys. Rev. B **54**, R6835 (1996)
9. H. Onishi, S. Miyashita, J. Phys. Soc. Jpn. **69**, 2634 (2000)
10. M. Nishino, H. Onishi, P. Roos, K. Yamaguchi, S. Miyashita: Phys. Rev. B **61** 4033 (2000). M. Nishino, H. Onishi, K. Yamaguchi, S. Miyashita, Phys. Rev. B **62**, 9463–9471 (2000)
11. J. Zeisner, O. Pilone, L. Soriano, G. Gerbaud, H. Vezin, O. Jeannin, M. Fourmigué, B. Büchner, V. Kataev, S. Bertaina, Phys. Rev. B **100**, 224414 (2019)
12. S. Tomonaga, Prog. Theor. Phys. 5, 544 (1950). J. M. Luttinger, J. Math. Phys. 4, 1154 (1963). A. Luther, I. Peschel, Phys. Rev. B 9, 2911, Phys. Rev. **12**, 3908 (1974)
13. F.D.M. Haldane, Phys. Lett. **93A**, 464 (1983). F.D.M. Haldane, Phys. Rev. Lett. **50**, 1153 (1983). I. Affleck, T. Kennedy, E.H. Lieb, H. Tasaki, Phys. Rev. Lett. **59**, 799 (1987). K. Katsumata, H. Hori, T. Takeuchi, M. Date, A. Yamagishi, J.P. Renard, Phys. Rev. Lett. **63**, 86 (1989). Y. Ajiro, T. Goto, H. Kikuchi, T. Sakakibara, T. Inami, Phys. Rev. Lett. **63**, 1424 (1989)
14. S. Miyashita, ed. By J.D. Doll, J. E. Gubernatis, World Scientific, 228-237 (1989). M. Kikuchi, Y. Okabe, S. Miyashita, J. Phys. Soc. Jpn. **59**, 492 (1990)
15. F. Bloch, Z. Phys. **61**, 206 (1930). H.A. Kramers, G. Heller, Proc. Roy. Acad. Sci. Amst. **37** (1934) 371
16. P.W. Anderson, Phys. Rev. **86**, 694 (1952). R. Kubo, Phys. Rev. **87**, 568 (1952)
17. T. Holstein, H. Primakoff, Phys. Rev. **58**, 1098 (1940)
18. T. Oguchi, H. Nishimori, Y. Taguchi, J. Phys. Soc. Jpn. **54**, 4492 (1985)
19. K. Kubo, T. Kishi, J. Phys. Soc. Jpn. **60**, 567 (1991)
20. M. Suzuki, Commun. Math. Phys. **51**, 183 (1976), M. Suzuki, Phys. Rev. B **31**, 2957 (1985)
21. S. Miyashita, *Collapse of Metastability, Fundamental Theories of Physicas 211* (SpringerNature, Siggapole, 2022)
22. T. Senthil, S. Sachdev, Phys. Rev. Lett. **77**, 5292 (1996). H. Rieger, A.P. Young, Phys. Rev. Lett. **72**, 4141 (1994). M. Guo, R.N. Bhatt, D.A. Huse, Phys. Rev. Lett. **72**, 4137 (1994). T. Ikegami, S. Miyashita, H. Rieger, J. Phys. Soc. Jpn. **67**, 2671–2677 (1998)
23. L. Pauling, Proc. N.A.S., **39**, 551 (1953)
24. I. Roudochatzakis, A.M. Mauchhli, F. Mila, Phys. Rev. B **77**, 094420 (2008)
25. P.W. Anderson, Mater. Res. Bull. **8**, 153 (1973). P. Fazekas, P.W. Anderson, Philos. Mag., **30**, 423 (1974)
26. B.S. Shastry, B. Sutherland, Physica B+C **108**, 1069–1070 (1981). S. Miyahara, K. Ueda, Phys. Rev. Lett. **82**, 3701–3704 (1999). H. Kageyama, et al., Phys. Rev. Lett. **82**, 3168–3171 (1999)
27. C. Lacroix, P. Mendels, F. Mila, *Introduction to Frustrated Magnetism: Materials, Experiments* (Springer, Theory, 2011)
28. K. Hirakawa, H. Kadowaki, K. Ubukoshi: J. Phys. Soc. Jpn. **52** (1983) 1824. K. Hirakawa, H. Kadowaki, K. Ubukoshi: J. Phys. Soc. Jpn. **54** (1985) 3526. P. Kalyani, N. Kalaiselvi 2005 Sci. Technol. Adv. Mater. **6** 689 (2005). K. Yamaura, M. Takano, A. Hirano, R. Kanno: J. Solid State Chem. **127** (1997) 109. T. Arimori, S. Miyashita, J. Phys. Soc. Jpn. **69**, 2250 (2000)

29. L. Savary, L. Balents, Rep. Prog. Phys., (2017) **80**(1), 016502 (2016). L. Balents, Nature, **464**(7286) 199-208 (2010), 199–208. J. Knolle, R. Moessner, Annual Review of Condensed Matter Physics. **10**, 451–472 (2019). J. Wen, S.-L. Yu, S. Li, W. J.-X. Li, npj Quantum Materials. **4** (1), 12 (2019). D. Chandler, (2012). Phys .org. Retrieved **24** December 2012. P.A.Lee, N. Nagaosa, X.G. Wen, Rev. Mod. Phys. **78** (1), 17–85 (2006)
30. P. Chandra, P. Coleman, Phys. Rev. Lett. **66**, 100 (1991), P. Chandra, P. Coleman, A.I. Larkin, J. Phys. C **2**, 7933 (1990). P. Chandra, P. Coleman, I. Ritchey, J. Appl. Phys. bf **69**(8), 4974 (1991)
31. S. Miyashita, J. Phys. Soc. Jpn. **53**, 44 (1984)
32. D.A. Huse, V. Elser, Phys. Rev. Lett. **60**, 2531 (1988)
33. R.R.P. Singh, D.A. Huse, Phys. Rev. Lett. **68**, 1766 (1992)
34. G. Baskaran, Phys. Rev. Lett. **63**, 2524 (1989)
35. B. Bernu, C. Lhuillier, L. Pierre, Phuys. Rev. Lett. **69**, 2590 (1992). B. Bernu, P. Lecheminant, C. Lhuillier, L. Pierre: Phys. Rev. B **50** (1994) 10048. R. Deutsscher, H.-U. Everts, Z. Physik B**93**, 77 (1993). L. Capriotti, A.E. Trumper, S. Sorella: Phys. Rev. Lett. **82** (1999) 3899
36. H. Nojiri, Y. Tokunaga, M. Motokawa, J. Phys. (Paris) 49, Suppl. **C8**, 1459 (1988)
37. S. Watarai, S. Miyashita, J. Phys. Soc. Jpn. **63**, 1916–1927 (1994). S. Watarai, S. Miyashita, J. Phys. Soc. Jpn. **67**, 816–824 (1998)
38. D. Yamamoto, G. Mormorini, I. Danshita, Phys. Rev. Lett. **112**,127203 (2014). D. Sellmann, X.-F. Zhang, S. Eggart, Phys. Rev. B **91** 081104(R) (2015)
39. A.V. Chubukov, D.I. Golosov, J. Phys.: Condens. Matter **3**, 69 (1991)
40. A. Honecker, M. Kaulke, K.D. Schotte: Eur. Phys. J. B **15**, 423 (2000). D.C. Cabra, A. Honecker, P. Pujol: Eur. Phys. J. B **13**, 55 (2000). A. Honecker, J. Schulenburg, J. Richter: J. Phys.: Condens. Matter **16**, S749 (2004)
41. S. Yoshikawa, K. Okunishi, M. Senda, S. Miyashita, J. Phys. Soc. Jpn. **73**, 1798 (2004)
42. S. Watarai, S. Miyashita, H. Shiba, J. Phys. Soc. Jpn. **70**, 532–537 (2001)
43. D. Sellmann, X.-F. Zhang, S. Eggert, Phys. Rev. B **91**, 081104(R) (2015)
44. J. Schnack, H. Nojiri, P. Kogerler, G.J.T. Cooper, L. Cronin, Phys. Rev. B **70**, 174420 (2004)
45. K. Kawano, M. Takahashi, J. Phys. Soc. Jpn. **66**, 4001 (1997)
46. K. Okunishi, S. Yoshikawa, T. Sakai, S. Miyashita, Prog. Theor. Phys. Suppl. **159**, 297–301 (2005). K. Okunishi, S. Yoshikawa, T. Sakai, S. Miyashita, Int. J. Modern Phys. C, **20**, 1423 (2009)
47. J.-B. Fouet, A. L*ä*uchli, S. Pilgram, R. M. Noack, F. Mila, Phys. Rev. B **73**, 014409 (2006)
48. V. Elser, Phys. Rev. Lett. **62**, 2405 (1989)
49. D.A. Huse, V. Elser, Phys. Rev. Lett. **60**, 2531 (1988)
50. N. Reimers, Phys. Rev. B **45**, 7287 (1992)
51. R.R.P. Singh, D.A. Huse, Phys. Rev. B **76**, 18047(R) (2007)
52. S. Yan, D.A. Huse, S. White, Steven. Science **332**, 1173 (2011)
53. C. Zeng, V. Elser, Phys. Rev. B **42**, 8436 (1990)
54. P.W. Leung, V. Elser, Phys. Rev. B **47**, 5459 (1993)
55. C. Waldtmann, H.-U. Everts, B. Bernu, C. Lhuillier, P. Sindzingre, P. Lecheminant, L. Pierre, Eur. Phys. J. B2, 501 (1998). F. Milla, Phys. Rev. Lett. **81**, 2356 (1998). P. Sindzingre, G. Misguich, C. Lhuillier, B. Bernu, L. Pierre, Ch. Waldtmann, H.-U. Everts, Phys. Rev. Lett. **84**, 2953 (2000). C. Waldtmann, H. Kreutzmann, U. Schollwœck, K. Maisinger, H.U. Everts, Phys. Rev. B **62**, 9472 (2000). P. Sindzingre, C. Lhuillier, Europhys. Lett. **88**, 27009 (2009). E.S. Sørensen, M.J. Lawler, Y.B. Kim, Phys. Rev. B **79**, 174403 (2009). A.M. Laüuchli, J. Sudan, E.S. Sørensen, Phys. Rev. B **83**, 212401 (2011). H. Nakano, T. Sakai, J. Phys. Soc. Jpn. **80**, 053704 (2011)
56. J. Schulenburg, A. Honecker, J. Schnack, J. Richter, H.-J. Schmidt, Phys. Rev. Lett. **88**, 167207 (2002)
57. M.E. Zhitomirsky, H. Tsunetsugu, Phys. Rev. B **70**, 100403(R) (2004)
58. H. Nakano, T. Sakai, J. Phys. Soc. Jpn. **79**, 053707 (2010)
59. S. Depenbrock, I.P. McCulloch, U. Schollwœck, Phys. Rev. Lett. **109**, 067201 (2012)
60. S. Nishimoto, N. Shibata, C. Hotta, Nat. Commun. **4**, 2287 (2013)

61. H. Nakano, T. Sakai, J. Phys. Soc. Jpn. **87**, 063706 (2018)
62. K. Morita, J. Phys. Soc. Jpn. **93**, 123706 (2024)
63. N. Elstner, A.P. Young, Phys. Rev. B **50**, 6871 (1993)
64. K. Fukamachi, H. Nishimori, Phys. Rev. B **49**, 651 (1994)
65. T. Nakamura, S. Miyashita, Phys. Rev. B **52**, 9174 (1995)
66. P. Sindzingre, G. Misguich, C. Lhuillier, B. Bernu, L. Pierre, Ch. Waldtmann, H.-U. Everts, Phys. Rev. Lett. **84**, 2953 (2000)
67. K. Kubo, Phys. Rev. B**48**, 10552 (1993). T. Nakamura, Y. Saika, J. Phys. Soc. Jpn. **64**, 695 (1995)
68. S. Miyashita, J. Phys. Soc. Jpn. **63**, 2449 (1994)
69. G. Misguich, B. Bernu, C. Lhuillier, C. Waldtmann, Phys. Rev. Lett. **81**, 1098 (1998); G. Misguich, C. Lhuil lier, B. Bernu, C. Waldtmann, Phys. Rev. B **60**, 1064 (1999)
70. D.S. Graywall, P.A. Busch, Phys. Rev. Lett. **62**, 1868 (1989)
71. A.P. Ramirez, G.P. Espinosa, A.S. Cooper, Phys. Rev. Lett. **64**, 2070 (1990). C. Broholm, G. Aeppli, G.P. Espinosa, A.S. Cooper, Phys. Rev. Lett. **65**, 3173 (1990). C. Broholm, G. Aeppli, G.P. Espinosa, A.S. Cooper, J. Appl. Phys. **69**, 4968 (1991)
72. W. Bronger, R. Rennau, D. Schmitz, Z. Allg. Chem. **597**, 27 (1991), ibid. **622**, 627 (1996). T. Kato, K. Hondou, K. Iio, J. Magn. Magn. Mater. 177–181, 591 (1998). T. Fukamachi, Y. Kobayashi, A. Nakamura, H. Harashina, M. Sato, J. Phys. Soc. Jpn. **68**, 3668 (1999)
73. S. Hara, H. Sato, Y. Narumi, J. Phys. Soc. Jpn. **81**, 073707 (2012)
74. A.S. Wills, V. Dupuis, E. Vincent, J. Hammann, R. Calemczuk, Phys. Rev. B **62**, R9264 (2000). F. Ladieu, F. Bert, V. Dupuis, E. Vincent, J. Hammann, J. Phys.: Condens. Matter **16**, S735 (2004)
75. S. Maegawa, M. Nishiyama, N. Tanaka, A. Oyamada, M. Takano, J. Phys. Soc. Jpn. **65**, 2776 (1996)
76. T. Morimoto, M. Nishiyama, S. Maegawa, Y. Oka, J. Phys. Soc. Jpn. **72**, 2085 (2003)
77. A.S. Wills, A. Harrison, C. Ritter, R.I. Smith, Phys. Rev. B **61**, 6156 (2000)
78. N. Oba, C. Michioka, M. Kato, K. Yoshimura, K. Mibu, J. Phys. Chem. Solids **66**, 1438 (2005)
79. T. Ono, K. Morita, M. Yano, H. Tanaka, K. Fujii, H. Uekusa, Y. Narumi, K. Kindo, Phys. Rev. B **79**, 174407 (2009)
80. Z. Honda, K. Katsumata, K. Yamada, J. Phys.: Condens. Matter **14**, L625 (2002)
81. M.P. Shores, E.A. Nytko, B.M. Bartlett, D.G. Nocera, J. Am. Chem. Soc. **127**, 13462 (2005)
82. P. Mendels, F. Bert, M.A. de Vries, A. Olariu, A. Harrison, F. Duc, J.C. Trombe, J.S. Lord, A. Amato, C. Baines, Phys. Rev. Lett. **98**, 077204 (2007)
83. J.S. Helton, K. Matan, M.P. Shores, E.A. Nytko, B.M. Bartlett, Y. Yoshida, Y. Takano, A. Suslov, Y. Qiu, J.-H. Chung, D.G. Nocera, Y.S. Lee, Phys. Rev. Lett. **98**, 107204 (2007)
84. F. Bert, S. Nakamae, F. Ladieu, D. L'Hôte, P. Bonville, F. Duc, J.-C. Trombe, P. Mendels, Phys. Rev. B **76**, 132411 (2007)
85. T. Imai, E.A. Nytko, B.M. Bartlett, M.P. Shores, D.G. Nocera, Phys. Rev. Lett. **100**, 077203 (2008)
86. A. Olariu, P. Mendels, F. Bert, F. Duc, J.C. Trombe, M.A. de Vries, A. Harrison, Phys. Rev. Lett. **100**, 087202 (2008)
87. M.A. de Vries, K.V. Kamenev, W.A. Kockelmann, J. Sanchez-Benitez, A. Harrison, Phys. Rev. Lett. **100**, 157205 (2008)
88. P. Mendels, F. Bert, J. Phys. Soc. Jpn. **79**, 011001 (2010)
89. M. Matsuda, H. Ueda, A. Kikkawa, Y. Tanaka, K. Katsumata, Y. Narumi, T. Inami, Y. Ueda, S.H. Lee, Nat. Phys. **3**, 397 (2007)

# Part III
# Entropy in Non-equilibrium System

The entropy $S$ was introduced as a thermodynamic quantity to explain equilibrium properties, although it has a rather abstract nature compared to other quantities such as pressure $P$. The concept of entropy is expressed as a function of the distribution function of states, known as Boltzmann–Gibbs entropy in statistical mechanics and as Shannon entropy in information theory. This formulation can be applied to general state distributions and provides insight into non-equilibrium states, such as the time evolution of systems. Various extended forms of entropy have also been introduced. In quantum systems, entanglement entropy has been introduced to characterize the nature of wave functions. The entropy of black holes is briefly reviewed, and for non-extensive systems, generalized forms of entropy, e.g., Tsallis entropy, have been proposed.

# Chapter 7
# Entropy in Non-equilibrium System

**Abstract** In statistical mechanics, entropy is related to the number of states with the same energy, which indicates that entropy reflects a property related to the information and randomness of the system. The concept of entropy is extended to non-equilibrium states, where it is defined as a function of the distribution function of states (Boltzmann-Gibbs entropy, Shannon entropy). In quantum mechanics, entropy as a function of state distribution is defined using the density matrix, known as von Neumann entropy. In stationary states, it reproduces the thermodynamic entropy. Moreover, the relative entropy, known as the Kullback-Leibler divergence (KLD), provides a description of how the system evolves toward its equilibrium state. The idea of so-called entanglement entropy of quantum ground states in interacting systems has been introduced, which characterizes quantum mechanical correlations within the system. The entropy of black holes is briefly reviewed, and entropy is also extended to non-extensive systems, where generalized forms of entropy, such as Tsallis entropy, are introduced.

## 7.1 Boltzmann's $H$-Theorem

Boltzmann introduced an equation to describe the time evolution of the probability distribution function $f(\boldsymbol{p}, \boldsymbol{q}, t)$ for the position and velocity of a particle in a gas system while taking into account collision processes [1]. He introduced the so-called $H$ function, defined as a function of the one-body probability distribution $f(\boldsymbol{p}, \boldsymbol{q}, t)$ as

$$H(t) = \int d\boldsymbol{p}d\boldsymbol{q}\, f(\boldsymbol{p}, \boldsymbol{q}, t) \ln f(\boldsymbol{p}, \boldsymbol{q}, t), \tag{7.1}$$

and showed that $H(t)$ irreversibly decreases to a steady value, indicating that the state relaxes to equilibrium.[1] In fact, it has been shown that the steady state follows the canonical distribution. This fact is known as "Boltzmann's $H$-theorem", which describes the irreversible relaxation to the equilibrium state.

[1] In his paper, the letter $E$ was used instead of $H$, and $E$ was given as a function of energy denoted by $x$.

© The Author(s), under exclusive license to Springer Nature Singapore Pte Ltd. 2025

S. Miyashita, *Physics of Entropy*, Springer Series in Solid-State Sciences 212,
https://doi.org/10.1007/978-981-95-1758-9_7

Let us overview the outline of the theorem. Time-derivative of $H(t)$ is given by

$$\frac{dH(t)}{dt} = \int d\boldsymbol{p}d\boldsymbol{q}\, \frac{\partial f(\boldsymbol{p},\boldsymbol{q},t)}{\partial t}(\ln f(\boldsymbol{p},\boldsymbol{q},t)+1). \tag{7.2}$$

First, we consider the case without collision (i.e., no interaction among particles), where the probability distribution satisfies the equation of continuity:

$$\frac{\partial f(\boldsymbol{p},\boldsymbol{q},t)}{\partial t} = -\dot{\boldsymbol{q}}\frac{\partial f(\boldsymbol{p},\boldsymbol{q},t)}{\partial \boldsymbol{q}} = -\frac{\boldsymbol{p}}{m}\cdot\frac{\partial f(\boldsymbol{p},\boldsymbol{q},t)}{\partial \boldsymbol{q}}. \tag{7.3}$$

Substituting this into (7.2), the equation is written as

$$\begin{aligned}\frac{dH(t)}{dt} &= -\int d\boldsymbol{p}d\boldsymbol{q}\frac{\boldsymbol{p}}{m}\cdot\left(\frac{\partial f(\boldsymbol{p},\boldsymbol{q},t)}{\partial \boldsymbol{q}}\right)(\ln f(\boldsymbol{p},\boldsymbol{q},t)+1)\\ &= -\int d\boldsymbol{p}d\boldsymbol{q}\,\frac{\boldsymbol{p}}{m}\frac{\partial}{\partial \boldsymbol{q}}\left[f(\boldsymbol{p},\boldsymbol{q},t)\ln f(\boldsymbol{p},\boldsymbol{q},t)\right].\end{aligned} \tag{7.4}$$

By applying the Gauss' theorem, the integral of $\boldsymbol{q}$ is expressed by a surface integral. Assuming $f(\boldsymbol{p},\boldsymbol{q},t)=0$ at the surface, we obtain

$$\frac{dH(t)}{dt} = 0. \tag{7.5}$$

Thus, $H(t)$ is a constant in time.

Now, let us take into account the effect of a two-body collision among particles. The change in the momentum distribution of two particles due to the collision is assumed to be given by

$$\begin{aligned}\Delta f[\boldsymbol{p}_1,\boldsymbol{p}_2,\boldsymbol{p}_3,\boldsymbol{p}] = &\int d\boldsymbol{p}_1 d\boldsymbol{p}_2\boldsymbol{p}_3\sigma(\boldsymbol{p}_1,\boldsymbol{p}_2,\boldsymbol{p}_3,\boldsymbol{p})\\ &\left[f(\boldsymbol{p}_1,t)f(\boldsymbol{p}_2,t) - f(\boldsymbol{p}_3,t)f(\boldsymbol{p},t).\right],\end{aligned} \tag{7.6}$$

where two particles with momenta, $\boldsymbol{p}_1$ and $\boldsymbol{p}_2$, collide and are scattered into the state with the momenta, $\boldsymbol{p}$ and $\boldsymbol{p}_3$, respectively. Here, $\sigma(\boldsymbol{p}_1,\boldsymbol{p}_2,\boldsymbol{p}_3,\boldsymbol{p})$ is the scattering cross section of this collision.

With this collision term, the equation of motion for the probability distribution is written as

$$\frac{dH(t)}{dt} = \int d\boldsymbol{p}\int d\boldsymbol{q}\,\sigma(\boldsymbol{p}_1,\boldsymbol{p}_2,\boldsymbol{p}_3,\boldsymbol{p})\Delta f[\boldsymbol{p}_1,\boldsymbol{p}_2,\boldsymbol{p}_3,\boldsymbol{p}](\ln f(\boldsymbol{p},\boldsymbol{q},t)+1). \tag{7.7}$$

Here, $\sigma(\boldsymbol{p}_1,\boldsymbol{p}_2,\boldsymbol{p}_3,\boldsymbol{p})$ has the same value for the following cases:

$$[\boldsymbol{p}_1,\boldsymbol{p}_2,\boldsymbol{p}_3,\boldsymbol{p}], [\boldsymbol{p}_1,\boldsymbol{p}_2,\boldsymbol{p},\boldsymbol{p}_3], [\boldsymbol{p},\boldsymbol{p}_3,\boldsymbol{p}_1,\boldsymbol{p}_2], [\boldsymbol{p}_3,\boldsymbol{p},\boldsymbol{p}_1,\boldsymbol{p}_2].$$

Using this fact, the equation can be reformulated as

$$\frac{dH(t)}{dt} = \frac{1}{4}\int d\boldsymbol{p}\int d\boldsymbol{q}\Big\{\int d\boldsymbol{p}_3\int d\boldsymbol{p}\,\sigma(\boldsymbol{p}_3,\boldsymbol{p})(\ln f(\boldsymbol{p},\boldsymbol{q},t)+1)$$
$$\times\big[f(\boldsymbol{p}_1,\boldsymbol{q},t)f(\boldsymbol{p}_2,\boldsymbol{q},t) - f(\boldsymbol{p}_3,\boldsymbol{q},t)f(\boldsymbol{p},\boldsymbol{q},t)\big]$$
$$+\int d\boldsymbol{p}_3\int d\boldsymbol{p}\,\sigma(\boldsymbol{p}_3,\boldsymbol{p})(\ln f(\boldsymbol{p}_3,\boldsymbol{q},t)+1)$$
$$\times\big[f(\boldsymbol{p}_1,\boldsymbol{q},t)f(\boldsymbol{p}_2,\boldsymbol{q},t) - f(\boldsymbol{p},\boldsymbol{q},t)f(\boldsymbol{p}_3,\boldsymbol{q},t)\big]$$
$$+\int d\boldsymbol{p}_3\int d\boldsymbol{p}\,\sigma(\boldsymbol{p}_3,\boldsymbol{p})(\ln f(\boldsymbol{p}_2,\boldsymbol{q},t)+1)$$
$$\times\big[f(\boldsymbol{p}_3,\boldsymbol{q},t)f(\boldsymbol{p},\boldsymbol{q},t) - f(\boldsymbol{p}_1,\boldsymbol{q},t)f(\boldsymbol{p}_2,\boldsymbol{q},t)\big]$$
$$+\int d\boldsymbol{p}_3\int d\boldsymbol{p}\,\sigma(\boldsymbol{p}_3,\boldsymbol{p})(\ln f(\boldsymbol{p}_1,\boldsymbol{q},t)+1)$$
$$\times\big[f(\boldsymbol{p}_3,\boldsymbol{q},t)f(\boldsymbol{p},\boldsymbol{q},t) - f(\boldsymbol{p}_2,\boldsymbol{q},t)f(\boldsymbol{p}_1,\boldsymbol{q},t)\big]\Big\},$$

which can be further rewritten as

$$\frac{dH(t)}{dt} = -\frac{V}{4}\int d\boldsymbol{p}_1\int d\boldsymbol{p}_2\int d\boldsymbol{p}_3\int d\boldsymbol{p}\sigma(\boldsymbol{p}_1,\boldsymbol{p}_2,\boldsymbol{p}_3,\boldsymbol{p})$$
$$[f(\boldsymbol{p}_1,t)f(\boldsymbol{p}_2,t) - f(\boldsymbol{p}_3,t)f(\boldsymbol{p},t)] \tag{7.8}$$

$$\times\,[\ln(f(\boldsymbol{p}_1,t)f(\boldsymbol{p}_2,t)) - \ln(f(\boldsymbol{p}_3,t)f(\boldsymbol{p},t))].$$

Using the general inequality:

$$(x-y)(\ln x - \ln y) \geq 0, \tag{7.9}$$

we conclude that

$$\frac{dH(t)}{dt} \leq 0. \tag{7.10}$$

Thus, the $H$-function monotonically decreases, which is known as Boltzmann's $H$-theorem.[2]

---

[2] The Boltzmann equation describes the relaxation dynamics of the probability distribution function. The scattering cross section gives the transition rate due to collisions. In general, equations governing the dynamics of probability functions with transition probabilities between states are called master equations, which is a linear equation of the probability function of the whole system as will be explained later. However, here $f(\boldsymbol{v},\boldsymbol{q})$ is not the probability distribution of the whole system but a reduced distribution function of one body, and the equation of motion contains non-linear terms. Generally, the equation of motion of the reduced distribution function is called the Boltzmann equation.

### 7.1.1 The Steady State

In the limit $t \to \infty$, $H(t)$ reaches its minimum value and becomes time independent. The steady state corresponds to the equilibrium state. Let us consider the nature of this state.

In the steady state, the change in the distribution (7.6) is zero: $\Delta f[\boldsymbol{p}_1, \boldsymbol{p}_2, \boldsymbol{p}_3, \boldsymbol{p}] = 0$, and $f_{\text{steady}}(\boldsymbol{p}, \boldsymbol{q})$ satisfies

$$f_{\text{steady}}(\boldsymbol{p}_1, \boldsymbol{q}) f_{\text{steady}}(\boldsymbol{p}_2, \boldsymbol{q}) - f_{\text{steady}}(\boldsymbol{p}_3, \boldsymbol{q}) f_{\text{steady}}(\boldsymbol{p}, \boldsymbol{q}) = 0, \tag{7.11}$$

for any pair of $\boldsymbol{p}$ and $\boldsymbol{p}_3$, where $\boldsymbol{p}_1$ and $\boldsymbol{p}_2$ are determined accordingly.

Assuming spatial uniformity, $f_{\text{steady}}(\boldsymbol{p}, \boldsymbol{q})$ does not depend on $\boldsymbol{q}$. In the scattering process, energy is conserved, which ensures

$$\boldsymbol{p}_1^2 + \boldsymbol{p}_2^2 = \boldsymbol{p}_3^2 + \boldsymbol{p}^2.$$

Thus, $f_{\text{steady}}(\boldsymbol{p}, \boldsymbol{q})$ is expressed in the following form with a constant $A$:

$$f_{\text{steady}}(\boldsymbol{p}, \boldsymbol{q}) = \exp(A\boldsymbol{p}^2). \tag{7.12}$$

This corresponds to the Maxwell-Boltzmann distribution:

$$f_{\text{Maxwell-Boltzmann}}(\boldsymbol{v}, \boldsymbol{q}) = \frac{1}{V} \frac{1}{\left(\sqrt{\frac{m}{2\pi k_{\text{B}} T}}\right)^3} \exp\left(-\frac{m}{2k_{\text{B}}T}\boldsymbol{v}^2\right), \quad m\boldsymbol{v} = \boldsymbol{p}, \tag{7.13}$$

for an ideal gas, where $A = -1/2mk_{\text{B}}T$.

This suggests that the equilibrium state is given by the steady-state solution of the process described by the Boltzmann equation, which incorporates the relaxation mechanism, i.e., the collision term.[3]

In Boltzmann's approach, the system relaxes to the equilibrium state, and the equilibrium state is obtained as the steady state of this relaxation process. The probabilistic view of thermodynamics culminated in 1902 with Josiah Willard Gibbs's statistical mechanics for fully general systems (not just gases), and the introduction of generalized statistical ensembles. Nowadays, the Gibbs ensemble method is widely used to calculate thermodynamic properties. However, the dynamical aspects of the relaxation process remain important, even though calculating the relaxation process for concrete systems is often practically impossible.

[3] It should be noted that in the argument of Boltzmann's $H$-theorem, the total energy of the system is conserved. Thus, the dynamical process occurs in an isolated system, and the temperature $T$ in (7.13) is a constant determined by the energy. If the system is in contact with a thermal bath, $H$ does not necessarily decrease but it approaches the form of (7.13) with the temperature of the bath. This issue will be explained in Sect. 7.4.

### 7.1.2 Origin of Irreversibility

The irreversibility of $H(t)$ faced several criticisms. From the perspective of pure mechanical motion, which is time-reversible, irreversibility seems impossible. This argument was raised by Loschmidt. Also, Zermelo noted that, Poincaré's recurrence theorem, which states that a system will return arbitrarily close to its initial state after some finite time, also appears to contradict irreversibility.

Such arguments have significantly contributed to our understanding of the origin of irreversibility. Boltzmann's $H$-theorem first pointed out the important concept that the contraction (or loss) of information causes the irreversibility of description of states.

Since the pure mechanics of motion is inherently reversible, a complete recognition of all processes would allow the entire system to be described purely by mechanics, without the need for "thermo"-dynamics. As we have studied, the concept of entropy inherently requires some "unrecognized parts of the process", such as heat flow.

If we consider the $N$-body distribution function $f_N(\boldsymbol{p}_1, \boldsymbol{q}_1, \ldots, \boldsymbol{p}_N, \boldsymbol{q}_N)$, where $N$ is the total number of particles, and include all scattering processes correctly (i.e., solve the $N$-body system exactly using Newton's equations), then no relaxation occurs, and all properties of mechanics hold.[4] In Boltzmann's $H$-theorem, the one-body distribution function $f(\boldsymbol{p}_1, \boldsymbol{q}_1)$ is used, where all the information about the other particles $(\boldsymbol{p}_2, \boldsymbol{q}_2, \ldots, \boldsymbol{p}_N, \boldsymbol{q}_N)$ is averaged out. This "loss of information" is the fundamental origin of the relaxation process.

To account for unrecognized processes, several methods have been developed. The Boltzmann equation is one such method, where the scattering cross section is used to describe the one-body distribution function. Nowadays, an equation of the distribution function that includes terms causing relaxation processes, in addition to the equation of continuity, is referred to as the Boltzmann equation. Alternatively, processes can also be described using equations with explicit random noise, such as white Gaussian noise, rather than probability distributions. These approaches are commonly used to study relaxation phenomena, like Brownian motion. Such equations are generally referred to as Langevin equations. In these cases, averaging over the distribution of the random noise results in the emergence of irreversibility. Equations involving transition probabilities, which specify the rates of transitions between states per unit time, are called master equation.

In the $H$-theorem, the distribution function $f(\boldsymbol{p}_1, \boldsymbol{q}_1)$ plays a central role. In a more general framework, we consider the probability that the system is in the $i$-th state, denoted as $p(i)$. In such treatments, describing entropy in terms of the probability distribution function becomes essential.

---

[4] Even in pure mechanical motions, so-called chaotic behavior may occur. In such cases, the invariant measure of the chaotic motion can lead to a type of equilibrium distribution. However, we do not discuss this here.

## 7.2 Entropy Expressed by the Probability Function

For the probability distribution, the "Boltzmann-Gibbs entropy" [2] was introduced as a function of the probability distribution $p(i)$, similar to Boltzmann's $H$-function[5]:

$$S = -k_B \sum_{i=1}^{M} p(i) \ln p(i), \quad \sum_{i=1}^{M} p(i) = 1, \quad p(i) \geq 0, \tag{7.14}$$

where $M$ is the total number of the states, $p(i)$ is the probability of the system being in the $i$-th state, and $k_B$ is the Boltzmann constant.

In the field of information theory, Claude Shannon [3] independently discovered the same expression to measure the amount of information. This is referred to as the "Shannon entropy" and is commonly known as the entropy for information. In this context, the prefactor $k_B$ is often omitted. We may regard the above functional as a measure of the missing information about the system under consideration. Eextremalization of the entropy under constraints gives the equilibrium probability distribution.

### *7.2.1 Equilibrium State as an Extremum of Boltzmann-Gibbs (Shannon) Entropy*

Here, we study the relationship between the equilibrium state and the Boltzmann-Gibbs (Shannon) entropy. In the microcanonical ensemble, where $p(i) = 1/W(E)$, the Boltzmann-Gibbs (Shannon) entropy reduces to the Boltzmann entropy:

$$S = -k_B \sum_{i=1}^{W(E)} \frac{1}{W(E)} \ln \frac{1}{W(E)} = k_B \ln W(E). \tag{7.15}$$

In the canonical ensemble, the free energy becomes the quantity to be minimized instead of entropy. The free energy is expressed in terms of the probability distribution as

$$F = \sum_{i=1}^{M} p_{eq}(i) E_i + k_B T \sum_{i=1}^{M} p_{eq}(i) \ln p_{eq}(i). \tag{7.16}$$

To find its extremum, we impose the normalization condition:

$$\sum_{i=1}^{M} p_{eq}(i) = 1. \tag{7.17}$$

---

[5] Here, $p(i)$ is the probability that the $i$-th state of the whole system is realized, but not reduced one.

The equilibrium distribution $p_{\text{eq}}(i)$ is obtained using the method of Lagrange multipliers. We consider the variation:

$$\frac{\partial}{\partial p_{\text{eq}}(i)}\left(F-\alpha\left(\sum_{i=1}^{M} p_{\text{eq}}(i)-1\right)\right)=E_i+k_{\text{B}}T\left(\ln p_{\text{eq}}(i)+1\right)-\alpha=0, \tag{7.18}$$

from which we obtain:

$$-\beta E_i=\ln p_{\text{eq}}(i)+1-\beta\alpha \quad \Rightarrow \quad p_{\text{eq}}(i)=e^{-\beta E_i+\beta\alpha-1}. \tag{7.19}$$

Using the normalization factor $Z$, we note:

$$e^{\beta\alpha-1}=\frac{1}{Z}. \tag{7.20}$$

Then, the probability of the $i$-th state with energy $E_i$ is given by the canonical ensemble distribution:

$$p_{\text{eq}}(i)=\frac{1}{Z}e^{-\beta E_i}. \tag{7.21}$$

Therefore, the canonical distribution corresponds to the extremum of the free energy (7.16).

## 7.3 Entropy and Randomness

Entropy is often associated with the randomness or disorder of a system. Let us consider a system of $N$ coins, where each coin can be modeled as an Ising spin with values $\pm 1$. The state of the $i$-th coin is denoted by $\sigma_i=\pm 1$, representing heads and tails, respectively. For example, for $N=6\times 6=36$, several configurations are illustrated in Fig. 7.1. In this figure, open circles represent heads, while closed circles represent tails. When comparing the states in Fig. 7.1a–c, one may subjectively describe Fig. 7.1a as "neat," Fig. 7.1b as "slightly disturbed," and Fig. 7.1c as "highly disturbed."

Can this difference in perceived randomness be expressed quantitatively using entropy? To address this, consider that each configuration is one of the $2^N$ possible states of the system. If there is no energy difference between head and tail $\sigma_i=\pm 1$, all configurations are equally probable:

$$p(\sigma_1,\sigma_2,\ldots,\sigma_N)=\frac{1}{2^N}. \tag{7.22}$$

The total entropy of the ensemble is given by the Boltzmann-Gibbs entropy formula (7.14):

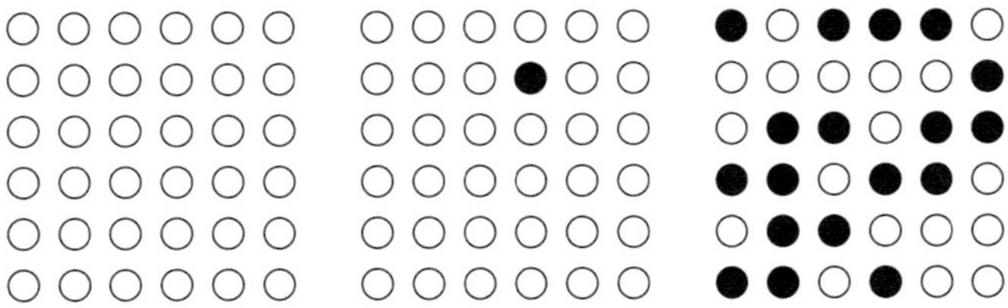

(a) (b) (c)

**Fig. 7.1** Configurations of coins. **a** All coins are heads. **b** One coin is tails. **c** Approximately half of the coins are tails

$$S = -k_B \sum_i p(i) \ln p(i). \tag{7.23}$$

Substituting $p(i) = \frac{1}{2^N}$, we find

$$S = -k_B \sum_{i=1}^{2^N} \frac{1}{2^N} \ln \frac{1}{2^N} = k_B N \ln 2. \tag{7.24}$$

However, assigning entropy to an individual configuration is not meaningful in this framework. If a specific configuration $(\sigma_1, \sigma_2, \ldots, \sigma_N)$ is chosen, its probability becomes

$$p(\sigma_1, \sigma_2, \ldots, \sigma_N) = 1, \tag{7.25}$$

while all other configurations have zero probability. In this case, the entropy evaluates to

$$S = -k_B \sum_i p(i) \ln p(i) = -k_B(\ln 1 + (2^N - 1) \times 0 \ln 0) = 0. \tag{7.26}$$

Thus, from an entropy perspective, the configurations in Fig. 7.1a–c are equivalent, as the entropy for a specific configuration is zero. This demonstrates that entropy, as defined for a probability distribution, cannot distinguish between individual configurations based on their perceived randomness. In this sense, one cannot say which is more random.

### 7.3.1 Pattern Recognition and Perceived Randomness

How does our intuitive sense of randomness arise? It stems from our ability to recognize patterns in configurations. Consider the system of coins as in the previous section. In Fig. 7.2, six examples of configurations with 18 heads and 18 tails are shown. If we do not perceive differences among these configurations and treat them as indistinguishable members of the ensemble with 18 heads and 18 tails, we effectively

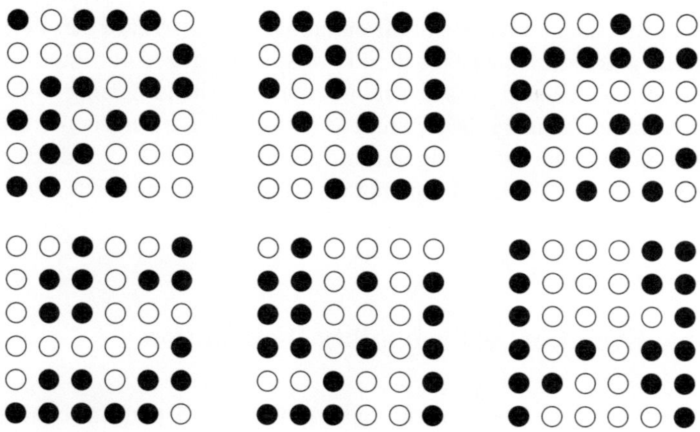

**Fig. 7.2** Examples of configurations with half of the coins showing heads

group them together, ignoring their individual differences. Then, the probability that any state in this group appears is determined by the number of ways to arrange 18 heads and 18 tails:

$$p_{N/2,N/2} = \frac{(N/2)!(N/2)!}{N!}. \tag{7.27}$$

The entropy of this group, representing the "randomness" due to indistinguishability, is given by

$$\begin{aligned} S_{N/2,N/2} &= -k_{\mathrm{B}} \sum_{k=1}^{N!/(N/2)!(N/2)!} \frac{(N/2)!(N/2)!}{N!} \ln \frac{(N/2)!(N/2)!}{N!} \\ &= -k_{\mathrm{B}} \ln \frac{(N/2)!(N/2)!}{N!}, \end{aligned} \tag{7.28}$$

which gives the value obtained by the Boltzmann law, i.e.,

$$S = k_{\mathrm{B}} \ln W = k_{\mathrm{B}} \ln(N!/(N/2)!(N/2)!). \tag{7.29}$$

This value is significantly larger than the entropy for the case where all coins are heads, for which there is only one configuration. and the entropy is

$$S_{N,0} = 0. \tag{7.30}$$

This entropy of groups corresponds to $S(M, \beta = 0)$ in Sect. 3.1.4.

Thus, entropy is defined for an ensemble of indistinguishable states rather than for individual configurations. For example, the six configurations shown in Fig. 7.2

are treated as equivalent members of the ensemble if we disregard the exact arrangement of heads and tails. This abstraction is what gives rise to our intuitive sense of randomness. In this context, it is crucial to note that when referring to the entropy of a "state", we are actually describing the entropy of a macroscopic state that is an ensemble of microscopic configurations grouped together by indistinguishability. A thermodynamic "state" is thus not a single configuration but a macroscopic representation of an ensemble of microscopic states.

## 7.4 Time Evolution of Probability Distribution

To study the dynamics of entropy, we analyze the time-dependent behavior of the Boltzmann-Gibbs entropy or the Shannon entropy. As discussed above, there are several approaches to study the relaxation process towards equilibrium. In this section, we focus on the dynamics of entropy using the framework of the master equation. This method allows us to examine the temporal behavior of entropy during relaxation processes.

The Boltzmann-Gibbs (Shannon) entropy, introduced earlier, corresponds to the entropy which is maximized in the microcanonical ensemble, where each state is equally probable. On the other hand, the relative entropy or Kullback-Leibler divergence (will be discussed in Sect. 7.5) serves as a measure of the free energy which is minimized in the canonical ensemble.

### 7.4.1 *Master Equation*

Let the probability of the system being in state $i$ at time $t$ be $p(i, t)$ $(i = 1, 2, \dots, M)$. For simplicity, we assume that the system has a finite number $M$ of discrete states.[6]

The time evolution of a probability distribution is governed by the so-called master equation, which describes how the probabilities of states $\{p(i)\}$ evolve over according to the "transition probabilities" $\omega_{i\to j}$, which represent the rate of transition from state $i$ to state $j$ per unit time. It should be noted that the transition probability $\omega_{i\to j}$ inherently violates time-reversal symmetry. This violation arises from adopting a probabilistic (stochastic) description of the transition processes, which are derived via a coarse graining of the underlying dynamics.

Over a short time interval $\Delta t$, it is assumed that $\omega_{i\to j}$ remains constant. The probability of transitioning from $i$ to $j$ during this interval is then given by

$$\Delta P_{i\to j} = \omega_{i\to j}\Delta t. \tag{7.31}$$

[6] Extensions to continuous variables, such as the Kramers-Moyal equation or the Fokker-Planck equation, can be made straightforwardly.

At time $t + \Delta t$, the probability of the system being in state $i$, $p(i, t + \Delta t)$, is determined by the total probability flowing out of state $i$ to other states $j \neq i$ and the total probability flowing into state $i$ from other states $j' \neq i$. The former (Loss term) is

$$\Delta P_{\text{out}} = \sum_{j \neq i} p(i, t)\omega_{i \to j}\Delta t, \tag{7.32}$$

and the latter (Gain term) is

$$\Delta P_{\text{in}} = \sum_{j \neq i} p(j, t)\omega_{j \to i}\Delta t. \tag{7.33}$$

Combining these contributions, the probability $p(i, t + \Delta t)$ is updated as

$$p(i, t + \Delta t) = p(i, t) - \Delta P_{\text{out}} + \Delta P_{\text{in}}. \tag{7.34}$$

Thus, we obtain the master equation:

$$p(i, t + \Delta t) = p(i, t) - \sum_{j \neq i} p(i, t)\omega_{i \to j}\Delta t + \sum_{j \neq i} p(j, t)\omega_{j \to i}\Delta t. \tag{7.35}$$

The master equation is a linear equation of $\{p(i, t)\}$. Then, we express the probability distribution by a vector

$$\boldsymbol{P}(t) = \begin{pmatrix} p(1, t) \\ p(2, t) \\ \vdots \\ p(M, t) \end{pmatrix}. \tag{7.36}$$

Using this notation, the master equation (7.35) is expressed by matrix form as

$$\boldsymbol{P}(t + \Delta t) = \mathcal{L}\boldsymbol{P}(t), \tag{7.37}$$

where $\mathcal{L}$ is a $M \times M$ matrix which is called "time-evolution operator" for a time step $\Delta t$. The matrix elements of $\mathcal{L}$ are given by

$$\mathcal{L}_{ij} = \omega_{j \to i}\Delta t \quad i \neq j. \tag{7.38}$$

The diagonal elements $\mathcal{L}_{ii}$ represent the probability of remaining in the original state $i$:

$$\mathcal{L}_{ii} = 1 - \sum_{j \neq i} \omega_{i \to j}\Delta t, \tag{7.39}$$

which should be non-negative. This condition is satisfied by taking small enough $\Delta t$. The matrix $\mathcal{L}$ satisfies the normalization condition:

$$\sum_i \mathcal{L}_{ij} = 1, \tag{7.40}$$

which guarantees the conservation of total probability. This property can be verified explicitly as follows. The total probability at time $t + \Delta t$ is

$$\sum_{i=1}^{M} p(i, \Delta t) = \sum_{i=1}^{M} \left( \sum_{k=1}^{M} \mathcal{L}_{ik} p(k, 0) \right) = \sum_{k=1}^{N} \left( \sum_{i=1}^{M} \mathcal{L}_{ik} \right) p(k, 0). \tag{7.41}$$

Using the normalization condition $\sum_{i=1}^{M} (\mathcal{L})_{ik} = 1$, we obtain

$$\sum_i p(i, 0) = 1 \to \sum_i p(i, t) = 1. \tag{7.42}$$

So far, we studied a stepwise time-evolution ($\Delta t$) of the probability distribution. It is convenient to have the master equation expressed in a form of differential equation taking the limit $\Delta t \to 0$. Expresssing (7.35) as

$$\frac{p(i, t + \Delta t) - p(i, t)}{\Delta t} = -\sum_{j \neq i} p(i, t)\omega_{i \to j} + \sum_{j \neq i} p(j, t)\omega_{j \to i},$$

and taking the limit $\Delta t \to 0$, we have

$$\frac{\partial p(i, t)}{\partial t} = -\sum_{j \neq i} p(i, t)\omega_{i \to j} + \sum_{j \neq i} p(j, t)\omega_{j \to i}, \tag{7.43}$$

which is expressed by the time-evolution operator $L$ as

$$\frac{\partial \boldsymbol{P}(t)}{\partial t} = -L\boldsymbol{P}(t), \quad L \equiv -\lim_{\Delta t \to 0} \frac{\mathcal{L} - 1}{\Delta t}. \tag{7.44}$$

The equation (7.43) is also called "master equation". The discrete-time evolution operator $\mathcal{L}$ and the continuous-time operator $L$ are related as

$$\lim_{\Delta t \to 0} (\mathcal{L}(\Delta t))^{t/\Delta t} = \mathcal{L}(t) = e^{-Lt}. \tag{7.45}$$

### 7.4.2 Eigenstate Decomposition of the Process

The dynamics of the probability distribution is expressed by the eigenstates and eigenvalues of the time-evolution operator $\mathcal{L}$[7]:

[7] It should be noted that, because $\mathcal{L}$ is not a symmetric matrix, the right- and left-eigenvectors are different.

$$\mathcal{L}|\phi_m\rangle = \lambda_m|\phi_m\rangle, \quad |\phi_m\rangle = \begin{pmatrix} \phi_{m1} \\ \phi_{m2} \\ \vdots \\ \phi_{mM} \end{pmatrix}, \quad m = 1, 2, \ldots, M. \tag{7.46}$$

Any initial state $\boldsymbol{P}(0)$ can be expanded in terms of the eigenstates of $\mathcal{L}$:

$$\boldsymbol{P}(0) = \sum_{m=1}^{M} c_m|\phi_m\rangle. \tag{7.47}$$

The probability distribution at $t = n\Delta t$ is obtained by applying $\mathcal{L}$ $n$ times, and it is given by

$$\boldsymbol{P}(t) = \mathcal{L}^n \boldsymbol{P}(0) = \sum_{m=1}^{M} c_m\lambda_m^n|\phi_m\rangle. \tag{7.48}$$

Here, we introduce the idea of ergodicity in the stochastic process. That is, we assume that any state $i$ can be transferred to any other states after some number of iterations of $\mathcal{L}$. In this case, for a large enough $n$, all the matrix elements are positive:

$$\left((\mathcal{L})^n\right)_{ij} > 0. \tag{7.49}$$

If this assumption holds, the matrix is called a 'positive matrix', and the Perron-Frobenius theorem guarantees that the largest eigenvalue $\lambda_1$ of $\mathcal{L}$ is positive and non-degenerate. This means that after many iterations, the probability distribution approaches a unique steady state which is the eigenstate $|\phi_1\rangle$ for $\lambda_1$. Therefore, any initial distribution $\boldsymbol{P}(0)$ converges to the unique stationary distribution $|\phi_1\rangle$ which is regarded as the equilibrium state $\boldsymbol{P}_{\text{eq}}$:

$$\lim_{t\to\infty} \boldsymbol{P}(t) = \boldsymbol{P}_{\text{eq}}, \tag{7.50}$$

which is the principle behind Monte Carlo simulation for the thermal equilibrium state.

To obtain the thermal equilibrium state as the steady state of the master equation $\boldsymbol{P}_{\text{eq}}$, we need to choose adequate form of the transition probabilities. Because the conservation of probability must be satisfied exactly, $\lambda_1 = 1$ and $c_1 = 1$, and contribution to the sum of probabilities from states of $m > 1$, $c_m\lambda_m^N|\phi_m\rangle$, must be zero regardless of $N$. For this condition, the components of the eigenvectors must satisfy the following relation:

$$\sum_k \phi_{mk} = 0, \quad m > 1, \tag{7.51}$$

where $\phi_{mk}$ is the $k$-th component of the vector $\phi_m$. In this case, some components with positive value ($\phi_{mk} > 0$) cause increase of the probabilities of certain states, while those with negative values ($\phi_{mk'} < 0$) cause decrease of the probabilities of other states, and the total sum of the probabilities remains unchanged. Thus, $|\phi_m\rangle$ represents a deformation of the probability distribution while preserving normalization. In this sense, we may regard $|\phi_m\rangle$ as a deforming mode.

As we see above, the largest eigenvalue $\lambda_1$ is 1, and all other eigenvalues are smaller. Then, the relaxation time $\tau_m$ for the $m$-th mode is given by

$$\lambda_m^n = e^{-n|\ln\lambda_m|} = e^{-t/\tau_m} = e^{-n\Delta t/\tau_m} \to \frac{1}{\tau_m} = \frac{|\ln\lambda_m|}{\Delta t}, \quad t = n\Delta t, \tag{7.52}$$

and the evolution of the probability distribution (7.48) is written as

$$\boldsymbol{P}(t) = \mathcal{L}^n \boldsymbol{P}(0) = \sum_{m=1}^{M} c_m e^{-t/\tau_m} |\phi_m\rangle, \quad c_1 = 1. \tag{7.53}$$

Thus, the probability distribution relaxes to the equilibrium state in a multi-exponential manner.

### 7.4.3 Time Evolution of Entropy

Time evolution of the entropy (7.14) is given by

$$\begin{aligned}\frac{dS}{dt} &= -k_B \frac{d}{dt}\left(\sum_{i=1}^{M} p(i,t)\ln p(i,t)\right) = -k_B \sum_{i=1}^{M}\left(\frac{dp(i,t)}{dt}\ln p(i,t) + \frac{dp(i,t)}{dt}\right) \\ &= -k_B \sum_{i=1}^{M} \frac{dp(i,t)}{dt}\ln p(i,t),\end{aligned} \tag{7.54}$$

where we used the conservation of probability, i.e., $d/dt \sum_{i=1}^{M} p(i,t) = 0$. Substituting the master equation (7.43), time evolution of the entropy is given by

$$\frac{dS}{dt} = -k_B \sum_{i=1}^{M}\left(-\sum_{j\neq i} p(i,t)\omega_{i\to j} + \sum_{j\neq i} p(j,t)\omega_{j\to i}\right)\ln p(i,t). \tag{7.55}$$

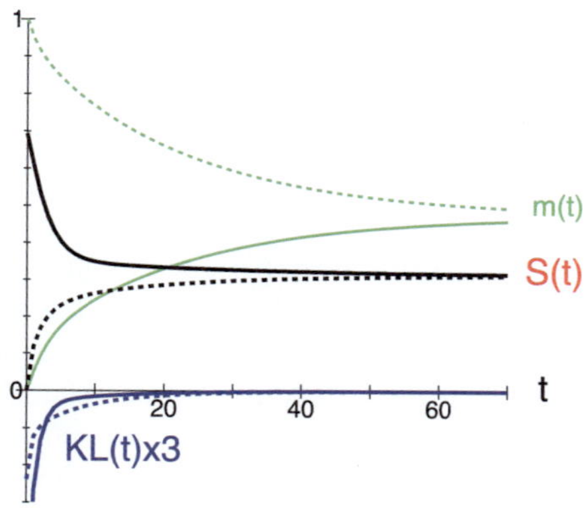


**Fig. 7.3** Time evolution of the entropy (7.14), the relative entropy (7.65) which will be explained in the next section, and the magnetization (7.58) of a kinetic Ising model for a $3 \times 3$ spin system (periodic boundary condition). The data of the relative entropy are magnified by factor 3 to make the dependence more apparent. The solid curves represent data starting from the random state $p(i) = 1/M$, while the dashed curves represent data starting from the perfectly aligned state $p(i) = \delta_{i,1}$. $T = 2.5J/k_B$ and $h = 0.1J$

It should be noted that the entropy does not necessarily increase in the process approaching the equilibrium as depicted in Fig. 7.3. There, we plot the time evolutions of the entropy (7.14) of a kinetic Ising model for a $3 \times 3$ spin system (periodic boundary condition)

$$\mathcal{H} = -J \sum_{\langle ij \rangle} \sigma_i \sigma_j - h \sum_i^N \sigma_i \tag{7.56}$$

with the Glauber transition probability:

$$w_{i \to j} = \frac{e^{-\beta E_j}}{e^{-\beta E_i} + e^{-\beta E_j}}. \tag{7.57}$$

There we also plot the relative entropy (7.65) which will be explained in the next section, and the magnetization

$$m(t) = \sum_{i=1}^{2^N} \left( \frac{1}{N} \sum_{k=1}^{N} \sigma_k \right) P(i,t), \tag{7.58}$$

which provides intuitive understanding of the system's evolution.

When we start from a random state:

$$P(i) = \frac{1}{M} = \frac{1}{2^N}, \quad \text{for all} \quad i, \tag{7.59}$$

$S(t)$ decreases to relax to its equilibrium value as shown by black solid curve. In the figure, we also plot $S(t)$ starting from perfectly aligned state by black dashed curve, where $S(t)$ increases. This observation indicates that in interacting systems, the entropy $S(t)$ does not necessarily increase as it approaches equilibrium.

The concept of entropy increase is valid in the microcanonical ensemble. In the microcanonical ensemble, all states have the same energy, $E_i = E$. This corresponds to the case of $J = 0$ in the kinetic Ising model. In this situation, the transition probabilities satisfy the detailed balance condition:

$$\frac{\omega_{i\to j}}{\omega_{j\to i}} = e^{-\beta(E_j - E_i)} = 1. \tag{7.60}$$

Since $E_i = E_j = E$, all the transition probabilities are identical, and we denote them as

$$\omega_0 = \omega_{i\to j} = \omega_{j\to i}. \tag{7.61}$$

Then, the change of the entropy is given by

$$\begin{aligned}
\frac{dS}{dt} &= -k_B\omega_0 \sum_{i=1}^{M} \left( -\sum_{j\neq i} p(i,t) + \sum_{j\neq i} p(j,t) \right) \ln p(i,t) \\
&= -k_B\omega_0 \sum_{i=1}^{M} \sum_{j\neq i} (-(p(i,t) + p(j,t)) \ln p(i,t) \\
&= -k_B\omega_0 \sum_{i=1}^{M} \sum_{j=1}^{M} (-(p(i,t) + p(j,t)) \ln p(i,t).
\end{aligned} \tag{7.62}$$

By exchanging the summation variable $i$ and $j$, we have

$$\begin{aligned}
\frac{dS}{dt} &= -k_B\omega_0 \sum_{i=1}^{M} \sum_{j=1}^{M} \frac{1}{2} \\
&\quad \{(-(p(i,t) + p(j,t)) \ln p(i,t) + (-(p(j,t) + p(i,t)) \ln p(j,t)\} \\
&= k_B\omega_0 \sum_{i=1}^{M} \sum_{j=1}^{M} \frac{1}{2} ((p(i,t) - p(j,t)) (\ln p(i,t) - \ln p(j,t)) \geq 0.
\end{aligned} \tag{7.63}$$

Thus, the entropy increases monotonically in the microcanonical ensemble, i.e., in the energetically isolated systems where the probabilities of all states become equal in the steady state ($p(1, \infty) = \cdots = p(M, \infty)$), consisting with the principle of equal probability.

## 7.5 Relative Entropy

As we saw in Fig. 7.3, the Boltzmann-Gibbs (Shannon) entropy (7.14),

$$S(t) = -\sum_{i=1}^{M} p(i,t) \ln p(i,t), \tag{7.64}$$

does not necessarily relax to the maximum value. Thus, relaxation to the equilibrium state in the canonical ensemble cannot be fully described by the minimization of the Boltzmann-Gibbs (Shannon) entropy.

To describe relaxation to equilibrium in the canonical ensemble, we introduce the so-called relative entropy, also known as the Kullback-Leibler divergence (KLD), defined as

$$S_{\text{relative}}(t) = -\sum_{i=1}^{M} p(i,t) \ln \frac{p(i,t)}{p_{\text{eq}}(i)}, \tag{7.65}$$

where $p_{\text{eq}}(i)$ is the probability of the steady state (equilibrium state).[8] This relative entropy decreases monotonically over time during the system's evolution governed by the master equation (7.35).

### 7.5.1 *Kullback-Leibler Divergence (KLD)*

In this section, we briefly review the properties of the Kullback-Leibler divergence (KLD) [4]. The KLD is defined for distributions $\{p(k)\}$ and $\{q(k)\}$ as

$$D(\boldsymbol{P}||\boldsymbol{Q}) = \sum_{k=1}^{M} p(k) \ln \frac{p(k)}{q(k)}, \quad \sum_{k} p(k) = 1, \quad \sum_{k} q(k) = 1. \tag{7.66}$$

For any real $x (> 0)$, the inequality $\ln \frac{1}{x} \geq 1 - x$ holds. Using this property, it can be shown that the KLD is always non-negative:

[8] In the microcanonical ensemble, this form reduces to (7.14) plus a constant term, since $p_{\text{eq}}(i)$ is constant.

$$D(\boldsymbol{P}||\boldsymbol{Q}) = \sum_{k=1}^{M} p(k) \ln \frac{p(k)}{q(k)} \geq \sum_{k=1}^{M} p(k) \left(1 - \frac{q(k)}{p(k)}\right) = 0. \tag{7.67}$$

### 7.5.2 Excess Free Energy and KLD

The free energy ($F = E - TS$) for a given distribution can be defined using the Boltzmann-Gibbs (Shannon) entropy as

$$F(t) = \sum_{i=1}^{M} p(i,t)E_i + k_\mathrm{B}T \sum_{i=1}^{M} p(i,t) \ln p(i,t). \tag{7.68}$$

The free energy of the equilibrium state is given by

$$F_\mathrm{eq} = \sum_{i=1}^{M} p_\mathrm{eq}(i)E_i + k_\mathrm{B}T \sum_{i=1}^{M} p_\mathrm{eq}(i) \ln p_\mathrm{eq}(i). \tag{7.69}$$

The difference of $F(t)$ and $F_\mathrm{eq}$, referred to as the excess free energy $\Delta F$ is

$$\Delta F = F(t) - F_\mathrm{eq}. \tag{7.70}$$

For the canonical distribution:

$$p_\mathrm{eq}(i) = \frac{e^{-\beta E_i}}{Z}, \quad F_\mathrm{eq} = -k_\mathrm{B}T \ln Z,$$

the Kullback-Leibler divergence (KLD) between $p(i,t)$ and $p_\mathrm{eq}(i,t)$ is

$$\sum_{i=1}^{M} p(i,t) \ln \left(\frac{p(i,t)}{p_\mathrm{eq}(i)}\right) = \sum_{i=1}^{M} p(i,t) \ln p(i,t) - \sum_{i=1}^{M} p(i,t) \ln p_\mathrm{eq}(i). \tag{7.71}$$

Substituting $p_\mathrm{eq}(i,t) = e^{-\beta E_i}/Z$, we obtain:

$$\sum_{i=1}^{M} p(i,t) \ln \left(\frac{p(i,t)}{p_\mathrm{eq}(i)}\right) = \sum_{i=1}^{M} p(i,t) \ln p(i,t) - \sum_{i=1}^{M} p(i,t) \left(-\beta E_i + \beta F_\mathrm{eq}\right).$$

Using the definitions (7.68) and (7.69), we have the relation:

$$k_\mathrm{B}T \sum_{i=1}^{M} p(i,t) \ln \left(\frac{p(i,t)}{p_\mathrm{eq}(i)}\right) = F(t) - F_\mathrm{eq} = \Delta F. \tag{7.72}$$

Thus, the difference $\Delta F$ is given in the form of KLD.

Because the Kullback-Leibler divergence (KLD) is non-negative, this relation shows that the canonical distribution minimizes the free energy $F_{\text{eq}}$, as discussed in Sect. 7.2.1.

Moreover, during the time evolution governed by the master equation, the free energy $F(t)$ monotonically decreases to $F_{\text{eq}}$ as explained below.

To prove this property, let us examine the time evolution of the KLD [4, 5]:

$$D(\boldsymbol{P}(t)||\boldsymbol{P}_{\text{eq}}) = \sum_{k=1}^{M} p(k,t) \ln \frac{p(k,t)}{p_{\text{eq}}(k)}. \tag{7.73}$$

Assume that the time evolution of $p(i,t)$ is governed by

$$\boldsymbol{P}(i, t+\Delta t) = \mathcal{L}\boldsymbol{P}(i,t), \tag{7.74}$$

where the operator $\mathcal{L}$ is defined as

$$(\mathcal{L})_{ij} = w_{j\to i}\Delta t, \quad i \neq j, \quad (\mathcal{L})_{ii} = 1 - \sum_{i\neq j} w_{j\to i}\Delta t. \tag{7.75}$$

We consider a two-variable probability distribution $P_2(i(t+\Delta t), j(t))$ which represents the probability that the state at the time $t$ is $j$ and the state at the time $t+\Delta t$ is $i$. This distribution is given by

$$P_2(i(t+\Delta t), j(t)) = (\mathcal{L})_{ij}\, p(j,t). \tag{7.76}$$

It should be noted that $(\mathcal{L})_{ij}$ is the transition probability being a positive number (not an operator) and that the above equation represents a simple product of numbers. This two-variable probability distribution satisfies the following properties:

$$\sum_i P_2(i(t+\Delta t), j(t)) = p(j, t+\Delta t), \tag{7.77}$$

$$\sum_j P_2(i(t+\Delta t), j(t)) = p(i,t), \tag{7.78}$$

and the total probability sums to 1, ensuring normalization:

$$\sum_{i,j} P_2(i(t+\Delta t), j(t)) = 1. \tag{7.79}$$

The Kullback-Leibler divergence (KLD) for the two-variable probability distribution $P_2(i(t+\Delta t), j(t))$ is given by

$$D(\boldsymbol{P}_2(t)||\boldsymbol{P}_{2(\mathrm{eq})}) = \sum_{i,j=1}^{M} P_2(i(t+\Delta t), j(t)) \ln \frac{P_2(i(t+\Delta t), j(t))}{P_{2(\mathrm{eq})}(i(t+\Delta t), j(t))} \tag{7.80}$$

$$= \sum_{i,j=1}^{M} (\mathcal{L})_{ij}\, p(j,t) \ln\left(\frac{(\mathcal{L})_{ij}\, p(j,t)}{(\mathcal{L})_{ij}\, p_{\mathrm{eq}}(j)}\right) = \sum_{i,j=1}^{M} (\mathcal{L})_{ij}\, p(j,t) \ln\left(\frac{p(j,t)}{p_{\mathrm{eq}}(j)}\right). \tag{7.81}$$

Because

$$\sum_{i=1}^{M} (\mathcal{L})_{ij} = 1, \tag{7.82}$$

$$D(\boldsymbol{P}_2(t)||\boldsymbol{P}_{2(\mathrm{eq})}) = \sum_{j=1}^{M} p(j,t) \ln\left(\frac{p(j,t)}{p_{\mathrm{eq}}(j)}\right) = D(\boldsymbol{P}(t)||\boldsymbol{P}_{\mathrm{eq}}). \tag{7.83}$$

Next, we introduce $X$ and $X_{\mathrm{eq}}$ which satisfy the following relations:

$$P_2(i(t+\Delta t), j(t)) = X(j(t)|i(t+\Delta t))p(i, t+\Delta t), \tag{7.84}$$

and

$$P_{2(\mathrm{eq})}(i(t+\Delta t), j(t)) = X_{\mathrm{eq}}(j(t)|i(t+\Delta t))p_{\mathrm{eq}}(i, t+\Delta t). \tag{7.85}$$

Note that $X(j(t)|i(t+\Delta t))$ satisfies the relation:

$$\sum_{i=1}^{M} X(j(t)|i(t+\Delta t))p(i, t+\Delta t) = p(j,t). \tag{7.86}$$

So, it may resemble an inverse operation of $\mathcal{L}$. However, it should be noted that $\mathcal{L}$ does not have its inverse operation. Therefore, $X(i(t+\Delta t)|j(t))$ is a quantity defined by (7.84) and (7.85). If $(\mathcal{L})_{ij} = 0$, then $X(i(t+\Delta t)|j(t))$ is arbitrary, and we define $X(i(t+\Delta t)|j(t)) = 0$. It is also important to note that $X(i(t+\Delta t)|j(t))$ depends on both $p(j,t)$ and $p(i, t+\Delta t)$, so that $X_{\mathrm{eq}}(i(t+\Delta t)|j(t))$ and $X(i(t+\Delta t)|j(t))$ are different.

Using $X$, KLD between the two-variable distributions $\boldsymbol{P}_2(t)$ and $\boldsymbol{P}_{2(\mathrm{eq})}$ is written as

$$D(\boldsymbol{P}_2(t)||\boldsymbol{P}_{2(\mathrm{eq})})$$

$$= \sum_{i=1}^{M}\sum_{j=1}^{M} X(j(t)|i(t+\Delta t))p(i, t+\Delta t) \ln\left(\frac{X(j(t)|i(t+\Delta t))p(i, t+\Delta t)}{X_{\mathrm{eq}}(j(t)|i(t+\Delta t))p_{\mathrm{eq}}(i)}\right)$$

$$= \sum_{i=1}^{M} \sum_{j=1}^{M} X(j(t)|i(t + \Delta t)) p(i, t + \Delta t) \ln \left( \frac{p(i, t + \Delta t)}{p_{\mathrm{eq}}(i)} \right)$$

$$+ \sum_{i=1}^{M} p(i, t + \Delta t) \left( \sum_{j=1}^{M} X(j(t)|i(t + \Delta t)) \ln \left( \frac{X(j(t)|i(t + \Delta t))}{X_{\mathrm{eq}}(j(t)|i(t + \Delta t))} \right) \right).$$

Because the sum over $j$ is 1, i.e.,

$$\sum_{j=1}^{M} X(j(t)|i(t + \Delta t)) = 1,$$

the first term is $D(\boldsymbol{P}(t + \Delta t)||\boldsymbol{P}_{\mathrm{eq}})$. And also, the second term is the average of KLD of $X(j(t)|i(t + \Delta t))$ over $p(i, t + \Delta t)$. Thus,

$$D(\boldsymbol{P}_2(t)||\boldsymbol{P}_{2(\mathrm{eq})}) = D(\boldsymbol{P}(t + \Delta t)||\boldsymbol{P}_{\mathrm{eq}}) + \sum_{i=1} p(i, t + \Delta t) D(\boldsymbol{X}(t + \Delta t)||\boldsymbol{X}_{\mathrm{eq}}). \tag{7.87}$$

Since the KLD for any probability distribution is non-negative, we have

$$D(\boldsymbol{P}(t)||\boldsymbol{P}_{\mathrm{eq}}) \geq D(\boldsymbol{P}(t + \Delta t)||\boldsymbol{P}_{\mathrm{eq}}). \tag{7.88}$$

This relation guarantees that the relative free entropy $\Delta F = D(\boldsymbol{P}(t)||\boldsymbol{P}_{\mathrm{eq}})$ monotonically decreases over time.[9]

## 7.6 Entropy and Information

Entropy is a fundamental concept, not only in thermodynamics but also in information theory. In thermodynamics, entropy was originally introduced to handle heat and temperature in a sophisticated manner, resulting from the second law of thermodynamics. The entropy is given by the probability distribution in the form of the expression of Boltzmann-Gibbs entropy [2].

In the study of information theory, Shannon found the same expression, which plays an important role in quantifying the amount of information [3]. Its concept is widely applicable to the study of information. Landauer discussed the irreversibility of the decimation of information and its relation to thermodynamic entropy, e.g., the loss of information of a bit induces heat $k_{\mathrm{B}} \ln 2$ [7, 8]. The relation between

[9] This property is also given in the tex book of van Kampen [6].

information and thermodynamic processes is an important subject and has been extensively studied [9, 10]. This problem has been argued since the problem of "Maxwell's demon" [11]. This type of argument is often made in the context of the possible violation of the second law of thermodynamics [12].

However, it should be remembered that the concept of entropy comes from the behavior of heat, which is defined as macroscopically unrecognized energy in an equilibrium state where only macroscopic quantities are relevant. If we have additional information, i.e., better ability to recognize the state, the amount of unrecognized energy decreases. Thus, if we have a way to obtain additional information about the system, the possibility of extracting additional work arises, which is very natural (even trivial).

Regarding the issue of information-to-energy conversion, we may consider the following examples. Water on the Earth is distributed between the sea and the sky in a stationary manner on a large scale. However, on a smaller scale, we observe the water cycle, i.e., water $\rightarrow$ vapor $\rightarrow$ cloud $\rightarrow$ rain $\rightarrow$ water. As long as we do not recognize such small-scale motion and regard it as microscopic details, the rainwater flows back to the sea without performing any work, and we cannot extract energy from it. If, however, we recognize the occurrence of rain in the mountains, we can build a dam to extract energy from this motion.

A similar scenario can be considered at much smaller scales. Toyabe et al. demonstrated information-to-energy conversion by employing a feedback protocol to manipulate a very small particle [13]. In this experiment, the particle is agitated by thermal fluctuations, which are typically regarded as noise. However, they directly observed the particle's motion and guided it to a location of higher potential energy by controlling the stochastic jumps caused by thermal fluctuations. The crucial point is that they were able to recognize and control the particle's motion, thereby removing it from the conventional scope of thermodynamics.

This remarkable experiment demonstrates that our capacity to recognize and manipulate systems has now extended to such small scales, representing progress toward realizing Maxwell's demon. However, crucially, what happened there does not violate the laws of thermodynamics.

If we could recognize and control the atomic motion of a solvent at the atomic level, the process would amount to a perfect realization of Maxwell's demon. In this case, however, the concept of "thermo" + "dynamics" would no longer apply; only the principles of mechanics would be relevant. In other words, if we were able to observe and account for all degrees of freedom, "thermodynamics" would essentially be reduced to simple "mechanical dynamics."

The manipulation of information in such small systems itself is significant, as it offers potential methods for making more of the energy available and may offer practical benefits. For instance, controlling this ability could lead to more efficient industrial processes, contributing to the achievement of the Sustainable Development Goals (SDGs).

In a broad sense, the process of purifying dirty water using certain bacteria and the elimination of specific atoms using a porous material can be regarded as analogous to the work of Maxwell's demon. In the former case, the bacteria recognize the impurities and break them down, while in the latter, the pores of the porous material recognize the atoms and absorb them. These processes are conceptually equivalent to the separation process of two kinds of solute A and B by a semipermeable membrane depicted in Fig. 3.14 in Sect. 3.5.2.

The topic on the interplay between entropy and our measurement capabilities is sometimes referred to as "microscopic thermodynamics." However, the term "microscopic thermodynamics" is somewhat misleading from the perspective of unrecognized matter. It is essential to be mindful of the definitions of "macro" and "micro" as these terms do not solely refer to size but rather to the level of recognition of the system.

## 7.7 Nonequilibrium Work Relation

Regarding the probability distribution, important rigorous properties have been known as the "nonequilibrium work relation" [14–16]:

$$\exp(-\Delta F) = \langle \exp(-W) \rangle . \tag{7.89}$$

where $\Delta F$ is the difference in the free energies of the initial and final states, $W$ is the work done during the process, and $\langle \cdots \rangle$ denotes average over all the processes. This relation is an equality for any processes, while the second law of thermodynamics states that

$$\Delta F \leq W, \tag{7.90}$$

where equality is satisfied in a quasi-equilibrium process in a large system (thermodynamic limit). Both relations do not conflict with each other.

In small systems, because of the distribution of states, $\Delta F > W$ may happen. As the size of the system increases, the probability of $\Delta F > W$ decreases exponentially with the size. When the parameter of the system changes suddenly, the work $W$ is given by the energy difference

$$W = \Delta \mathcal{H}. \tag{7.91}$$

## 7.8 Entropy in Quantum Systems

Thermodynamics applies to quantum mechanical systems as well. To study the thermodynamic properties of quantum systems, we use statistical mechanics, where the concept of the "state" may not be as intuitive as in classical systems. However, by utilizing the set of eigenstates $\{\phi_i\}$ of the system and the corresponding eigenvalues $\{E_i\}$, the methods of statistical mechanics can be straightforwardly extended to quantum systems.

For instance, the partition function in quantum statistical mechanics is given by

$$Z = \sum_{i:\text{all the states}} e^{-\beta E_i} = \sum_{i:\text{all the eigenstates}} e^{-\beta\langle\phi_i|\mathcal{H}|\phi_i\rangle} = \langle\phi_i|e^{-\beta\mathcal{H}}|\phi_i\rangle = \text{Tr}e^{-\beta\mathcal{H}}, \tag{7.92}$$

which does not depend on the choice of the basis.

### 7.8.1 *Density Matrix*

In quantum systems, the probability distribution of states is described by the following form for a set of eigenstates:

$$\hat{P} = \sum_{i:\text{all the eigenstates}} p(i)|\phi_i\rangle\langle\phi_i|. \tag{7.93}$$

This form is commonly referred to as the density matrix, denoted by $\rho$ (i.e., $\hat{P} = \rho$). The density matrix enables us to describe statistical mixtures of quantum states, where each $p(i)$ represents the probability of the system being in the state $|\phi_i\rangle$.

The natural extension of the Boltzmann-Gibbs-Shannon entropy for a quantum system is expressed in terms of the density matrix as

$$S = -k_\text{B}\text{Tr}\,(\rho \ln \rho)\,. \tag{7.94}$$

which is known as the von Neumann entropy.

In the case that a system is described by a single wave function $|\phi\rangle$, the density matrix $\rho$ is given by

$$\rho = |\phi\rangle\langle\phi|. \tag{7.95}$$

In this case, there is no statistical ambiguity. If we choose a basis set where the state $|\phi\rangle$ is the $m$th state of the basis, the matrix representation of $\rho$ is diagonal with $(\rho)_{mm} = 1$ and all other elements being zero. The entropy in this case is

$$S = 0, \tag{7.96}$$

because $p(m) \ln p(m) = 1 \times 0$ and $p(m' \neq m) \ln p(m' \neq m) = 0 \times \ln 0 = 0$. The density matrix in this case is called the density matrix of a "pure state."

When there is some ambiguity about the state of the system, and the system could be in any state with a certain probability $p(i)$, the density matrix is given by a sum of the density matrices of individual pure states, as expressed in (7.93).

To measure the degree of mixing in a quantum system, "purity" $\gamma$ is commonly used, defined as:

$$\gamma = \mathrm{Tr}\rho^2. \tag{7.97}$$

Purity provides insight into the "closeness" of the state to a pure state. Specifically, when the system is in a pure state, the density matrix satisfies $\rho^2 = \rho$, so the purity is 1 ($\gamma = 1$). For a mixed state, the purity is less than 1. In the case where all the states are equally weighted and each state has the same probability, the purity becomes inversely proportional to the number of states, and the purity decreases as the system size increases.

This property is not exclusive to quantum states. For classical probability distributions, a similar quantity, $\sum_{i=1}^{M} p(i)^2$, can also be used to measure the degree of concentration of the distribution.

### 7.8.2 Entanglement Entropy

In quantum systems, we are often interested in the ground state $|\Psi_{\mathrm{G}}\rangle$. The ground state is a pure quantum state described by a wave function, and only limited information can be extracted from the density matrix of the pure state for the entire system.

To characterize the properties of the ground state, various quantities have been proposed. One such quantity is the entanglement entropy [17]. In quantum systems composed of many interacting local quantities, measurements on one part of the system can affect the states of other local quantities. When this happens, the measured part and the remaining part are said to be entangled. To quantify the degree of this entanglement, the concept of entanglement entropy has been introduced.

To make the explanation more concrete, let us consider a spin system composed of spins $\sigma_i$ ($i = 1, \ldots, N$) with the Hamiltonian $\mathcal{H}$. The ground state is given by

$$\mathcal{H}|\Psi_{\mathrm{G}}\rangle = E_{\mathrm{G}}|\Psi_{\mathrm{G}}\rangle. \tag{7.98}$$

We consider basis vectors constructed as direct products of local states that are eigenstates of the $z$-component of the spins:

$$\sigma_i^z|m_i\rangle_i = m_i|m_i\rangle_i, \quad m_i = \pm 1 \quad i = 1, \ldots, N. \tag{7.99}$$

The ground state is expressed as a linear combination of these basis states:

$$|\Psi_{\mathrm{G}}\rangle = \sum_{k=0}^{2^N-1} c_k |m_1\rangle_1 |m_2\rangle_2 \cdots |m_N\rangle_N, \tag{7.100}$$

where the basis states are labeled by a binary number:

$$k = \frac{m_1 + 1}{2} 2^0 + \frac{m_2 + 1}{2} 2^1 + \cdots + \frac{m_N + 1}{2} 2^{N-1}, \tag{7.101}$$

and we denote the state as

$$|k\rangle = |m_1\rangle_1 |m_2\rangle_2 \cdots |m_N\rangle_N. \tag{7.102}$$

As mentioned above, the entropy of a pure quantum state is always zero, regardless of the coefficients $c_k$:

$$\rho_{\mathrm{G}} = |\Psi_{\mathrm{G}}\rangle\langle\Psi_{\mathrm{G}}|, \quad S_{\mathrm{G}} = k_{\mathrm{B}} \mathrm{Tr}\,(\rho_{\mathrm{G}} \ln \rho_{\mathrm{G}}) = 0, \tag{7.103}$$

irrespective of whether the state is entangled or not.

As the simplest case, let us consider the spin system of $N = 2$ with the Hamiltonian:

$$\mathcal{H} = -J \boldsymbol{\sigma}_1 \cdot \boldsymbol{\sigma}_2. \tag{7.104}$$

Assume that we measure $\sigma_1$ and find $\sigma_1 = +1$. In the ground state of the ferromagnetic case ($J > 0$):

$$|\Psi_{\mathrm{G}}\rangle = |+\rangle_1 |+\rangle_2, \tag{7.105}$$

the measurement does not affect the state of $\sigma_2$, which remains $\sigma_2 = +1$.

In contrast, in the antiferromagnetic case ($J < 0$), where the ground state is given by

$$|\Psi_{\mathrm{G}}\rangle = \frac{|+\rangle_1 |-\rangle_2 - |-\rangle_1 |+\rangle_2}{\sqrt{2}}, \tag{7.106}$$

the measurement of $\sigma_1 = +1$ fixes the other spin to $\sigma_2 = -1$. If the measurement gives $\sigma_1 = -1$, the other spin is fixed to $\sigma_2 = +1$. Thus, in this case, the measurement of $\sigma_1$, i.e., determining the state of $\sigma_1$, affects the state of $\sigma_2$. This indicates that the state (7.106) is entangled.

We now consider more general cases with large $N$. Generally, wave functions of quantum systems composed of many spins are expressed as a linear combination of the basis states $|k\rangle$, and they are likely to be entangled.[10] We now perform a local measurement on a subset of the system A ($\sigma_1, \ldots, \sigma_l$). Let the measured values of $\sigma_1, \ldots, \sigma_l$ be $m_1, \ldots, m_l$, respectively. The resulting wave function of the remaining system (B) ($\sigma_{l+1}, \ldots, \sigma_N$) is a linear combination of the $2^l$ states of system B.

[10] If the state is given as a product of local spin states,

$$|\Psi\rangle = \prod_{i=1}^{N} \left( c_i^+ |+\rangle_i + c_i^- |-\rangle_i \right), \tag{7.107}$$

the wave function is a linear combination of $2^N$ states of $\{|k\rangle\}$, but the state is not entangled.

The coefficient $\Psi_{\text{post measure}}(\sigma_{l+1}, \ldots, \sigma_N)$ of the state $|\sigma_{l+1}\rangle \ldots |\sigma_N\rangle$ depends on $m_1, m_2, \ldots, m_l$:

$$\Psi_{\text{post measure}}(\sigma_{l+1}, \ldots, \sigma_N) = \Psi(m_1, m_2, \ldots, m_l, \sigma_{l+1}, \ldots, \sigma_N), \tag{7.108}$$

and the wave function is given by

$$|\Psi_{\text{post measure}}(m_1, m_2, \ldots, m_l)\rangle$$
$$= \sum_{\sigma_{l+1}} \cdots \sum_{\sigma_N} \Psi(m_1, m_2, \ldots, m_l, \sigma_{l+1}, \ldots, \sigma_N)|\sigma_{l+1}\rangle \cdots |\sigma_N\rangle. \tag{7.109}$$

When we disregard the information about the states of $\sigma_1, \ldots, \sigma_l$, we take the average over $m_i = \pm 1$ for $i = 1, \ldots, l$, leading to the density matrix for this mixed state:

$$\rho_{\text{reduced}}(\sigma_{l+1}, \ldots, \sigma_N)$$
$$= \sum_{m_1=\pm 1} \cdots \sum_{m_l=\pm 1} |\Psi_{\text{post measure}}(m_1, m_2, \ldots, m_l)\rangle\langle\Psi_{\text{post measure}}(m_1, m_2, \ldots, m_l)|, \tag{7.110}$$

where $\rho_{\text{reduced}}$ is an $(N - l) \times (N - l)$ matrix. This is the density matrix of a mixed state of the system B $(\sigma_{l+1}, \ldots, \sigma_N)$.

The entanglement entropy is defined as

$$S_{\text{entanglement}} = -k_B \text{Tr}\,(\rho_{\text{reduced}} \ln \rho_{\text{reduced}}) \geq 0, \tag{7.111}$$

where the equality holds only if system A $(\sigma_1, \ldots, \sigma_l)$ and system B $(\sigma_{l+1}, \ldots, \sigma_N)$ are not entangled.[11]

The amount of entanglement entropy can be used to quantify the degree of entanglement. When $S_{\text{entanglement}}$ is proportional to the area of the boundary between sys-

[11] If the wave function is a direct product of wave functions of the parts A and B:

$$|\Psi(\text{A} + \text{B})\rangle = \left(\sum_j c_j|\Psi(\text{A}_j)\rangle\right)\left(\sum_m c'_m|\Psi(\text{B}_m)\rangle\right), \tag{7.112}$$

then summing over the states of system A gives the density matrix for system B:

$$\rho_{\text{reduced}}(\text{B}) = \sum_{\text{all the states of A}} \rho_{\text{A+B}} = \left(\sum_m c'_m|\Psi(\text{B}_m)\rangle\right)\left(\sum_m (c'_m)^*\langle\Psi(\text{B}_m)|\right). \tag{7.113}$$

This density matrix represents a pure state for part B, and $S_{\text{entanglement}}(\text{B}) = 0$. Thus, there is no entanglement between systems A and B.

tems A and B, the situation is referred to as the "area law" [18]. The area laws are commonly observed in the ground states of gapped many-body systems [19]. In one-dimensional critical systems, a logarithmic correction appears [20]. It also appears in quantum field theory for the black hole [21].

If we diagonalize the reduced density matrix, it is expressed as

$$\rho_{\text{reduced}}(\text{B}) = \sum_m \rho_m |\psi(\text{B}_m)\rangle\langle\psi(\text{B}_m)|, \tag{7.114}$$

where $\rho_m$ and $|\psi(\text{B}_m)\rangle$ are the eigenvalues and eigenstates of the reduced matrix (7.110), respectively. The eigenstates $|\psi(\text{B}_m)\rangle$ with large eigenvalues $\rho_m$ can be regarded as the representative states of part B when it is connected to part A. This idea is utilized in the decimation of irrelevant degrees of freedom in quantum systems when we try to use renormalization group methods, such as the density matrix renormalization group (DMRG) method [22]. This method has been applied in many studies, including the dynamics of Heisenberg antiferromagnets [23]. Entanglement also plays a significant role in quantum computing [24, 25]. In addition to the entanglement entropy, various measures have been introduced to extract information about entanglement, such as concurrence [26] and quantum mutual information [27].

### 7.8.3 Schmidt Entropy

Mathematically, we formulate the entanglement entropy using the Schmidt entropy, defined as follows. Suppose we divide a system into two parts, A and B. Let the basis of system A be $\{|\phi_i\rangle\}$, $(i = 1, \ldots, M_{\text{A}})$, and that of system B be $\{|\xi_j\rangle\}$, $(j = 1, \ldots, M_{\text{B}})$. A state of the whole system A+B, $|\Psi\rangle$, is given by a linear combination of these basis states:

$$|\Psi\rangle = \sum_{i=1}^{M_{\text{A}}} \sum_{j=1}^{M_{\text{B}}} C_{ij} |\phi_i\rangle |\xi_j\rangle. \tag{7.115}$$

Using the idea of Singular Value Decomposition (SVD), this can be rewritten as

$$|\Psi\rangle = \sum_{k=1}^{r} \sqrt{\lambda_k} |\Phi_k\rangle |\Xi_k\rangle, \quad \sum_{k=1}^{r} \lambda_k = 1, \tag{7.116}$$

where $r$ is the Schmidt rank, which is the rank of the matrix $C_{ij}$. The states $|\Phi_k\rangle$ and $|\Xi_k\rangle$ are orthonormal states in subsystems A and B, respectively. The coefficients $\{\lambda_k\}$ are the non-negative singular values (known as Schmidt coefficients).

The Schmidt entropy $S_{\text{Schmidt}}$ is defined as

$$S_{\text{Schmidt}} = -\sum_{k=1}^{r} \lambda_k \ln(\lambda_k), \tag{7.117}$$

which is equal to the von Neumann entropy of subsystem A, and also of subsystem B.

### 7.8.4 *Entropy of the Back Hole*

The concept of entanglement is also employed in the study of black holes. A black hole is a solution to Einstein's equations, and one might initially consider that it has no entropy. However, Bekenstein argued that a thermodynamic treatment of it is necessary and concluded that the entropy of a black hole is proportional to the area of its event horizon [28]:

$$S = k_{\mathrm{B}} \frac{Ac^3}{4\hbar G}, \tag{7.118}$$

which is known as "Bekenstein-Hawking black hole entropy", where $c$ is the speed of light ($= 299, 792, 458$ m/s), and $G$ is the gravitational constant ($= 6.674 \cdots \times 10^{-11} \mathrm{m}^3 \mathrm{kg}^{-1} \mathrm{s}^{-2}$).

This concept has been further developed in the context of the so-called holographic principle [29]. Quantum mechanically, it remains a challenging question about how a mixed state can emerge from a pure quantum state. Hawking proposed the concept of Hawking radiation and its connection to thermodynamics [30]. A microscopic explanation of (7.118) was given by Strominger and Vafa in 1996 within the framework of string theory [31]. Moreover, the relationship between black hole entropy and entanglement entropy was suggested by Bombelli et al. and Srednicki [21], and it has been elucidated through the Anti-de Sitter/Conformal Field Theory (AdS-CFT) correspondence by Ryu and Takayanagi [32].

## 7.9 Variety of Entropy

So far, we have studied the properties of systems for which the thermal equilibrium state provides their stable state. For thermodynamic quantities, extensivity is an important property. For example, if we consider two independent systems A and B, the total entropy of the whole system A+B is given by the sum of the entropies of each system, $S(\mathrm{A})$ and $S(\mathrm{B})$:

$$S(\mathrm{A} + \mathrm{B}) = S(\mathrm{A}) + S(\mathrm{B}). \tag{7.119}$$

First, let us confirm the extensivity of the Boltzmann-Gibbs entropy (7.14):

$$S = -k_{\mathrm{B}} \sum_{i=1}^{M} p(i) \ln p(i), \quad \sum_{i=1}^{M} p(i) = 1, \quad p(i) \geq 0.$$

Let the total number of states of system A be $M_{\rm A}$, and let $p_{\rm A}(i)$ be the probability of being in the $i$-th state of system A. Similarly, let these quantities be defined for system B. Since the systems are independent, $p(i, j)$, representing the probability that system A is in the $i$-th state and system B is in the $j$-th state, is given by

$$p(i, j) = p_{\rm A}(i) p_{\rm B}(j). \tag{7.120}$$

Substituting (7.120) into (7.14), we obtain

$$\begin{aligned}
S_{\rm A+B} &= -k_{\rm B} \sum_{i=1}^{M_{\rm A}} \sum_{j=1}^{M_{\rm B}} p_{\rm A}(i) p_{\rm B}(j) \ln \left( p_{\rm A}(i) p_{\rm B}(j) \right), \\
&= -k_{\rm B} \sum_{i=1}^{M_{\rm A}} \sum_{j=1}^{M_{\rm B}} p_{\rm A}(i) p_{\rm B}(j) \left( \ln p_{\rm A}(i) + \ln p_{\rm B}(j) \right) \\
&= -k_{\rm B} \sum_{i=1}^{M_{\rm A}} p_{\rm A}(i) \ln p_{\rm A}(i) - k_{\rm B} \sum_{j=1}^{M_{\rm B}} p_{\rm B}(j) \ln p_{\rm B}(j) \\
&= S_{\rm A} + S_{\rm B}.
\end{aligned} \tag{7.121}$$

Thus, the Boltzmann-Gibbs entropy (7.14) satisfies the extensive property of entropy (7.119).

Another important property of the form (7.14) is that the state which maximizes the Boltzmann-Gibbs-Shannon entropy (7.14) corresponds to the thermal equilibrium state [33], as we studied in Sect. 7.2.1.

### *7.9.1 Entropy in Nonextensive Systems*

However, there are systems in which extensivity is not realized, e.g., systems with long-range interactions, systems on multifractal structures [34], systems with anomalous diffusion [35], systems at self-organized criticality [36], and so on. If we do not require the entropy to be extensive, we can explore various generalizations of entropy [37].

For instance, the so-called "Tsallis entropy" was introduced by Tsallis [38], where an extension of the logarithmic function was defined as

$$\ln_q(x) \equiv \frac{x^{1-q}-1}{1-q}, \tag{7.122}$$

which reduces to the standard logarithmic function, $\ln(x)$, in the limit $q \to 1$. Using $\ln_q(x)$, the entropy is generalized as

$$S = -k_B \sum_{i=1}^{M} p(i) \ln p(i) = k_B \sum_{i=1}^{M} p(i) \ln \frac{1}{p(i)} \Rightarrow S_q = k_B \sum_{i=1}^{M} p(i) \ln_q \frac{1}{p(i)}. \tag{7.123}$$

Substituting (7.122),

$$S_q = k_B \sum_{i=1}^{M} p(i) \frac{p(i)^{q-1}-1}{1-q} = k_B \sum_{i=1}^{M} \frac{p(i)^q - p(i)}{1-q}. \tag{7.124}$$

Because $\sum_{i=1}^{M} p(i) = 1$, the Tsallis entropy is given in the form:

$$S_q = k_B \frac{1}{q-1} \left( 1 - \sum_{i=1}^{M} p(i)^q \right), \tag{7.125}$$

where $q$ is called the "entropic index." It should be noted that $S_q$ does not satisfy property (7.119) and instead has the following relation:

$$S(\mathrm{A}+\mathrm{B}) = S(\mathrm{A}) + S(\mathrm{B}) + (1-q) S(\mathrm{A}) S(\mathrm{B}). \tag{7.126}$$

In the limit $q \to 1$, using l'Hôpital's rule,

$$S_1 = k_B \lim_{q \to 1} \frac{1}{q-1} \left( 1 - \sum_{i=1}^{M} e^{q \ln p(i)} \right) = -k_B \sum_{i=1}^{M} p(i) \ln p(i), \tag{7.127}$$

which reduces to the standard Boltzmann-Gibbs entropy (7.14), and the relation (7.119) is recovered.

Optimization of the Tsallis entropy (7.125) for the canonical ensemble yields

$$p(i) = \frac{1}{Z_q} \left( 1 + (q-1) \frac{\Delta E_i}{k_B T} \right)^{1/(1-q)}, \quad Z_q = \sum_i \left( 1 + (q-1) \frac{\Delta E_i}{k_B T} \right)^{1/(1-q)}. \tag{7.128}$$

This gives a stationary state of nonextensive thermostatistical systems [39]. This prescription shares many important properties of the standard theory, such as the Legendre thermodynamic structure, the $H$-theorem, Onsager's reciprocity theorem, Kramers-Wannier relations, and thermodynamic stability [40]. This formulation has been applied to various nonextensive systems, e.g., stellar polytropes [41], ferrofluids [42], two-dimensional plasma turbulence [43, 44], anomalous diffusion in Lévy flights [45], cosmic background radiation [46], peculiar velocity functions of galaxy clusters [47], dissipative nonlinear systems [48], and anomalous diffusion in optical lattices [49]. It has also been found that this form of the acceptance probability (energy cost function) works efficiently in simulated annealing [50].

Another generalization is Rényi entropy, defined as

$$S_\alpha = \frac{1}{1-\alpha}\left(\ln \sum_{i=1}^{M} p(i)^\alpha\right), \tag{7.129}$$

where $\alpha$ is a parameter. In particular, in the limit $\alpha \to 1$, we recover the standard entropy:

$$\lim_{\alpha \to 1} S_\alpha = -\sum_{i=1}^{M} p(i) \ln p(i). \tag{7.130}$$

These nonextensive entropies, such as Tsallis and Rényi entropies, describe systems where extensivity is not strictly required and provide a useful framework for understanding dynamical properties in nonequilibrium systems.

## References

1. L. Boltzmann, Wien Ber. **66**, 275 (1872)
2. J.W. Gibbs, *Elementary Principles in Statistical Mechanics*, (New Yprk 1902). J. R.. Waldram, *The Theory of Thermodynamics*, Cambridge University Press, (1985). E.T. Jaynes; American Journal of Physics, **33**, 391 (1965)
3. C.C. Shannon, Bell Syst. Tech. J. **27**(3), 379 (1948)
4. T.M. Cover, J.A. Thomas, *Elements of Information* (Wiley, 2006)
5. S. Yukawa, *Toukeirikigaku* (Statistical Machanics in Japanese, Nihon-hyoronsha, 2021)
6. N.G. Van Kampen, *Stochastic Processes in Physics and Chemistry*, (Northholland Personal Library, 3td Edition, 2007)
7. R. Landauer, IBM J. Res. Dev. **3**, 183 (1961), R. W. Keyes and R. Landauer, **14** (2) 152 (1970)
8. W. Porod, R. Grondin, D. Ferry, G. Porod, Phys. Rev. Lett. **52**, 232 (1984). C.H. Bannett, Phys. Rev. Lett. **53**, 1202 (1984). P. Beni, Phys. Rev. Lett. **48**, 1581 (1982); R. Feynman, Feynman lecture on computation (1983); E. Goto, N. Yoshida, K.F. Loe and W. Hioe, Proc. 3rd Symp. Foundation of Quantum Mechancs, Tokyo, 412–418 (1989)
9. L. Szilard, Z. Phys. **53** 840 (1929). L. Szilard, On the decrease of entropy in a thermodynamic system by the intervention of intelligent beings. Behavioral Science (Wiley Online Library) **9** (4), 301–310
10. T. Sagawa, M. Ueda, Phys. Rev. Lett. **102**, 250602 (2009)
11. H.S.Leff, A.F. Rex, *Maxwell's Demon 2: Entropy, Classical and Quantum Information, Computing*, 2003, IoP Pub.: Bristol. K. Maruyama, F. Nori, and V. Vedral, Rev. Mod. Phys. **81**, 1 (2009)
12. K. Sekimoto, *Stochastic Energetics*, Lecture Notes in Physics, **799**, (Springer, 2010). G. M. Wang, E.M. Sevick, E. Mittag, D.J. Searles, and D.J. Evans, Phys. Rev. Lett. **89**, 050601 (2002)
13. S. Toyabe, T. Sagawa, M. Ueda, E. Muneyuki, M. Sano, Nature Phys. **6**, 988-992 i2010 j. T. Sagawa, and M. Ueda, Phys. Rev. Lett. **100**, 080403 (2008)

14. D.J. Evans, E.G.D. Cohen, G.P. Morriss, Phys. Rev. Lett. **71**, 2401 (1993). D.J. Evans D.J. Searls, Phys. Rev. E **50**, G. Gallavotti, E.G.D. Cohen, Phys. Rev. Lett. **74**, 2694 (1995), G.E. Crooks, Phys. Rev. E **60**, 2721 (1999)
15. C. Jarzynski, Phys. Rev. Lett. **78**, 2690 (1997), Phys. Rev. E **56**, 5018 (1997). J. Liphardt, S, Dumont, S.B. Smith, I. Tinoco, Jr and C. Bustamante, Science **296**, 1832 (2002)
16. S.K. Blau, Physics Today. **55** (2002). (Translated by S. Miyashita, Parity **18**, 33 (2003)
17. L. Amico, R. Fazio, A. Osterloh, V. Vedral, Rev. Mod. Phys.**80**, 517 (2008). R. Horodecki, P. Horodecki, M. Horodecki, K. Horodecki, Rev. Mod. Phys. **81**, 865 (2009)
18. J. Eisert, M. Cramer, M.B. Plenio, Rev. Mod. Phys. **82**, 277 (2010)
19. M.B. Hastings, J. Stat. Mech., P08024 (2007)
20. G. Vidal, J.I. Latorre, E. Rico, A. Kitaev, Phys. Rev. Lett. **90**, 227902 (2003)
21. L. Bombelli, R.K. Koul, J.H. Lee, R.D. Sorkin, Phys. Rev. D **34**, 373 (1986). M. Srednicki, Phys. Rev. Lett. **71**, 666 (1993)
22. S.R. White, Phys. Rev. Lett. **69**, 2863-2866 (1992). U. Schollwöck, U. Rev. Mod. Phys. **77**, 259-315 (2005)
23. V. Subrahmanyam, Phys. Rev. A **69**, 022311 (2004); V. Subrahmanyam, Phys. Rev. A**69**, 034304 (2004)
24. A.K. Ekert, Phys. Rev. Lett., **67**, 661 (1991). A. Ekert and R. Jozsa, Rev. Mod. Phys. **68**, 733 (1996). D. Gottesman, I.L. Chuang, Nature **402**, 390 (1999). C.H. Bennett and D.P. DiVincenzo, Nature **404**, 247 (2000)
25. G. Semeghini, H. Levine, A. Keesling, S. Ebadi, T.-T. Wang, D. Bluvstein, R. Verresen, H. Pichler, M. Kalinowski, R. Samajdar, A. Omran, Science **374**(6572), 1242–1247 (2021)
26. W.K. Wootters, Phys. Rev. Lett. **80**, (1998) 2245. W.K. Wootters, Quant. Inf. Comp. **1**, 27 (2001)
27. B. Groisman, S. Popescu, A. Winter, Phys. Rev. A **72**, 032317, (2005), B. Groisman, S. Strelchuk, Phys. Rev A **92**, 052337 (2015)
28. J.D. Bekenstein, Bekenstein, Phys. Rev. D **7** (1973) 2333, J.D. Bekenstein, Phys. Rev. D **23** (215), 287 (1981)
29. G. 't Hooft, Dimensional Reduction in Quantum Gravity, Utrecht Preprint THU-93/26, gr-qc/9310026, L. Susskind, J. Math. Phys. (N.Y.) **36**, 6377 (1995); R. Bousso, J. High Energy Phys. **07**, 004 (1999)
30. S.W. Hawking, Nature. **248** (5443), 30–31 (1974). S.W. Hawking, Commun. Math. Phys. **25**, 152 (1972). S.W. Hawking, Comm. Math. Phys. **43**(3): 199 (1975). S.W. Hawking, Phys. Rev. D **13**, 191 (1976)
31. A. Strominger, C. Vafa, Phys. Lett. B **379**, 99 (1996)
32. S. Ryu, T. Takayanagi, Phys. Rev. Lett. **96**, 181602 (2006)
33. S. Pressé, K. Ghoshh, J. Lee, K.A. Dill, Rev. Mod. Phys. **85**, 1115 (2013)
34. H.G.E. Hentschel, I. Procaccia, Physica D 8, 435 (1983); T.C. Halsley, M.H. Jensen, L.P. Kadanoff, I. Procaccia, B.I. Shraiman, Phys. Rev. A 33, 1141 (1986) G. Paladin, A. Vulpiani, Phys. Rep. 156, 147 (1987)
35. J.-P. Bouchaud, A. Georges, Phys. Rep. **195**, 127 (1990)
36. P. Bak, C. Tang, K. Wiesenfeld, Phys. Rev. Lett. **59**, 381 (1987)
37. S. Abe, Y. Okamoto (Eds.), *Nonextensive Statistical Mechanics and Its Application*, Lecture notes in Physics Vol 560, Springer (2001). S. Pressé, K. Ghosh, J. Lee, K. Dill, Phys. Rev. Lett. **111** (18), 180604 (2013)
38. C. Tsallis, J. of Stat. Phys., **52**, 479 (1988). C. Tsallis, *Introduction to Nonextensive Statistical Mechanics: Approaching a Complex World* (Springer, New York, 2009)
39. A.K. Rajagopal, XXPhys. A **253**, 271 (1998)
40. E.M.F. Curado, C. Tsallis, J. Phys. A **24**, L69 (1991); **24**, 3187E (1991); **25**, 1019E (1992); A.M. Mariz, Phys. Lett. A **165**, 409 (1992); A.K. Rajagopal, Phys. Rev. Lett. **76**, 3469 (1996); A. Chame, E.V.L. de Mello, Phys. Lett. A **228**, 159 (1997); M.O. Caceres, Phys. Lett. A **218**, 471 (1995); C. Tsallis, Phys. Lett. A **206**, 389 (1995)
41. A.R. Plastino, A. Plastino, Phys. Lett. A **174**, 384 (1993)
42. P. Jund, S.G. Kim, C. Tsallis, Phys. Rev. B **52**, 50 (1995)

43. X.-P. Huang, C.F. Driscoll Phys. Rev. Lett. **72**, 2187 (1994)
44. B.M. Boghosian, Phys. Rev. E **53**, 4754 (1996)
45. E.W. Montroll, M.F. Shlesinger, *Nonequilibrium phenomena* 11. From Stochastics to Hydrodynamics, edited by J. L. Lebowitz and E.W. Montroll (North Holland, Amsterdam, 1984). D. H. Zanette, P.A. Alemany, Phys. Rev. Lett. **75**, 366 (1995); **77**, 2590 (1996); M.O. Caceres, C.E. Budde, Phys. Rev. Lett. **77**, 2589 (1996); C. Tsallis, S. V. F. Lévy, A. M. C. de Souza, R. Maynard, Phys. Rev. Lett. **75**, 3589 (1995), **77**, 5442 (1996)
46. V.H. Hamity, D.E. Barraco, Phys. Rev. Lett. **76**, 4664 (1996)
47. A. Lavagno et al., Astrophys. Lett. Commun. **35**, 449 (1998)
48. M.L. Lyra, C. Tsallis, Phys. Rev. Lett. **80**, 53 (1998). A.R.R. Papa and C. Tsallis, Phys. Rev. E 57, 3923 (1998)
49. E. Lutz, Phys. Rev. A. **67** (5), 051402 (2003). P. Douglas, S. Bergamini, F. Renzoni, Phys. Rev. Lett. **96** (11), 110601(2006)
50. C. Tsallis, D.A. Stariolo, Physica A **233**, 395 (1996). I. Andricioaei, J.E. Straub, Phys. Rev. E **53**, R3055 (1996). H. Nishimori, J. Inoue, J. Phys. A **31**, 5661 (1998)

# Index



© The Editor(s) (if applicable) and The Author(s), under exclusive license to Springer Nature Singapore Pte Ltd. 2025

S. Miyashita, *Physics of Entropy*, Springer Series in Solid-State Sciences 212,
https://doi.org/10.1007/978-981-95-1758-9

MIX
Papier aus verantwortungsvollen Quellen
Paper from responsible sources
FSC® C105338

If you have any concerns about our products,
you can contact us on
**ProductSafety@springernature.com**

In case Publisher is established outside the EU,
the EU authorized representative is:
**Springer Nature Customer Service Center GmbH**
**Europaplatz 3, 69115 Heidelberg, Germany**

Printed by Libri Plureos GmbH
in Hamburg, Germany